Hydrology and Water Management of Deltaic Areas

A.A. BALKEMA / ROTTERDAM / BROOKFIELD / 1993

CUR Report 93-5

Photo on the cover: Delta in the south western part of the Netherlands

Published and distributed for CUR, Gouda by
A.A. Balkema, P.O. Box 1675, 3000 BR Rotterdam, Netherlands
A.A. Balkema Publishers, Old Post Road, Brookfield, VT 05036, USA

ISBN 90 5410 168 7

Printed in the Netherlands

preface

In recent years the habitation of deltaic areas has become a subject of widespread interest and has given rise to many controversial opinions. Many deltaic areas are important sites of intensive economic development and clustering of population. Man seems to be attracted by areas where sea and inland transport facilities meet each other. Agricultural development has been followed by the development of industries, trade and commerce. Deltaic areas present specific technical and environmental problems of their own. They are attacked by floods from all sides: the river, the sea and rainfall. Protection and exploitation of these areas drastically changes the environment, creating a vulnerable situation. In these areas there is often an apparent abundance of water. However, in most cases this water cannot be used for any purpose because of the intrusion of saline water from various sources into the surface water and the groundwater. Deltaic areas are predominantly level without natural slopes, making the drainage of these areas another problem that is not encountered elsewhere.

In their natural state, deltaic areas form valuable ecological assets with unique species of flora and fauna. The ecological significance of these areas extends far beyond their geographical extent. Economic development of deltaic areas cannot be achieved without interfering with the environment. It has been advocated that virgin lands should be left untouched because human occupancy and economic development are incompatible with the preservation of the ecological value. However true that may be, many deltaic areas are occupied and new environments have been created. The question to be addressed is how to optimise the use of these lands, bearing in mind that any action undertaken may be irreversible.

The problems involved are essentially multi-disciplinary and it would be over-ambitious to attempt to cover all the relevant aspects in a single publication. It should however be stressed that the approach to problem solving and decision making for the complex issues involved should be based on a Policy Analysis system. This must take into consideration all relevant aspects and their dimensions, which, as mentioned before, extend far beyond the geographical boundaries. A monofunctional approach will conflict with the concept of sustainable development. The whole system may then backfire in the short or long term and thus nullify the gains obtained by the more short-sighted approach. It is the responsibility of the (political) manager, as well as the scientists and technicians, to understand and embrace the concept of sustainable development as a leading code in the execution of their duties.

The present publication aims at making a modest contribution to the issue by dealing with two specific and interrelated aspects of deltaic areas; namely the hydrology and water management, including the environmental aspects. There exist at present many textbooks, handbooks and manuals on general hydrology and water management. However, little attention has been paid to the relevant problems of deltaic areas, which are very specific and, to some extent, basically different from areas of river basins beyond the effects of the sea. United Nations proceedings of symposia on deltaic areas from the sixties and early seventies constitute a valuable contribution to the multi-disciplinary problems under consideration, however, the need for a systematic, complete and up-to-date outline of the subject is considered necessary.

This publication is intended for use by a broad audience of scientists and technicians, who in one way or another, have to deal with one or more aspects of the conservation, improvement or development of deltaic areas, and who require information on the essential conditions and constraints related to the factor water. It is also hoped that the general public may find something of interest in a field which is not easily accessible. At the same time, the publication aims at making a compilation of what specialists in the field of hydrology and water management of deltaic areas have achieved. It outlines what can be applied in practice and indicates where significant gaps still exist. Although this publication does not deal specifically with subjects such as geology and geomorphology these topics have been included because of their significance for hydrology.

The event which led to the decision to prepare this publication on Hydrology and Water Management of Deltaic Areas was the retirement of Prof.ir. Adriaan Volker from the Delft University of Technology in 1985. Both inside and outside the university it was felt that his considerable knowledge and experience should be passed on to a wider audience. An editorial board was formed, consisting of Professor Volker, ir. R.H. Boekelman, ir. P.C. van Koppen, ir. J.P. van der Linden and ir. M.A.A. Mulder. To cover the different aspects, experts were consulted and found willing to contribute.
The Centre for Civil Engineering Research and Codes (CUR) was later asked to take care of the realisation of this publication. A commitee was installed to execute the task compiling this publication. The commitee F 9 'Hydrology and Water Management of Deltaic Areas' was composed as follows:

The commitee F 9 'Hydrology and Water Management of Deltaic Areas'

Prof.ir. A. Volker, Chairman	Professor Emeritus of Hydrology, T.U. Delft; Senior Engineer in charge of special services Ministry of Publics Works, The Hague (retired)
ir. R.H. Boekelman, Secretary	Delft University of Technology, Faculty of Civil Engineering
ir. A.W. de Haas	Rijkswaterstaat, Directorate Flevoland.
ir. P.C. van Koppen	IWACO, Consultancy for Water & Environment
ir. M.A.A. Mulder	Mitac
Prof.ir. H.J.M. Ogink	Delft Hydraulics
ir. G.J. Schiereck	Delft University of Technology, Faculty of Civil Engineering
A.W. Minns B.Eng, Dip.HE., Editor	International Institute for Infrastructural, Hydraulic and Environmental Engineering (IHE) in Delft
ir. G.J. Verkade, Coordinator	Centre for Civil Engineering Research and Codes (CUR)
Prof.dr.ir. J.F. Agema, Mentor	Centre for Civil Engineering Research and Codes (CUR)

The funding for compilation of this publication was provided by Rijkswaterstaat, Road and Hydraulic Engineering Department (DWW) and Rijkswaterstaat Directorate Flevoland/Centre for Integrated New Land Development.

This publication has been prepared with the support and according to the requirements of the Centre for Civil Engineering Research and Codes (CUR).

September 1993

Centre for Civil Engineering Research and Codes
Ministry of Transport, Public Works and Water Management
Directorate-General for Public Works and Water Management

Authors and contributors

Prof.ir. A. Volker	Professor Emeritus of Hydrology, T.U. Delft; Senior Engineer in charge of special services Ministry of Publics Works, The Hague (retired)
ir. M. Adriaanse	Rijkswaterstaat, Institute for Inland Water Management and Wast Water Treatment (RIZA)
ir. E. Allersma	Delft Hydraulics
ir. W. Andriesse	The Winant Staring Centre for Integral Land, Soil and Water Research (SC-DLO), Wageningen
ir. K.P. Blumenthal	Rijkswaterstaat, Directorate for Water Management and Hydraulic Research (retired)
ir. R.H. Boekelman	Delft University of Technology, Faculty of Civil Engineering
ir. J. van Duivendijk	Haskoning, Royal Dutch Consulting, Engineering and Architects
Prof.ir. W.F.T. van Ellen	International Institute for Infrastructural, Hydraulic and Environmental Engineering (IHE) in Delft
C.D. Jagroop, B.Sc. Dip.HE (assistant editor)	International Institute for Infrastructural, Hydraulic and Environmental Engineering (IHE) in Delft
dr. S. Jelgersma	Geological Survey of the Netherlands (Rijksgeologische Dienst, Haarlem)
ir. P.C. van Koppen	IWACO, Consultancy for Water & Environment
drs. P. Leentvaar	Institute for Forestry and Nature Research (IBN-DLO) (retired)
ir. J.P.A. Luijten	Rijkswaterstaat, Institute for Inland Water Management and Wast Water Treatment (RIZA)
ir. M.A.A. Mulder	Mitac Waddinxveen b.v.
dr. G.W.N.M. van Moorsel	Bureau Waardenburg b.v. Consultants for Environment and Ecology
Prof.dr. J. Oerlemans	Institute for Marine and Atmospheric Research, Utrecht University (IMOU)
ir. G.H.P. Oude Essink	Delft University of Technology, Faculty of Civil Engineering
Prof.dr. N.A. de Ridder †	International Institute for Land Reclamation and Improvement, Wageningen (ILRI)
ir. G.J. Schiereck	Delft University of Technology, Faculty of Civil Engineering
dr. M.M.A. Shahin	International Institute for Infrastructural, Hydraulic and Environmental Engineering (IHE) in Delft
ir. H. van der Tuin	Rijkswaterstaat, Institute for Inland Water Management and Wast Water Treatment (RIZA) (retired)
ir. W.S. de Vries	International Institute for Infrastructural, Hydraulic and Environmental Engineering (IHE) in Delft

summary

The present publication deals with the many-sided problems involved in the utilisation of deltaic and other low-lying coastal areas. At present, these regions constitute the most rapidly developing areas on earth due to their high economic value and their steadily increasing population. At the same time, they are typically wetland landscapes and should be considered as valuable ecological assets because of their great diversity of fresh, brackish and saline environments. Conservation of these assets may clash with the interests of economic development. A well-balanced and harmonious development of such coastal areas is one of the most important issues facing mankind, for which a satisfactory solution still has to be found.
This publication aims at making a contribution to the resolution of these problems by highlighting a number of these aspects. Some of these aspects belong to entirely different disciplines and the contributing experts have attempted to present their know-how in a form that is readily understandable to a wide audience.

In the first chapter, the characteristic features of deltaic areas are discussed as well as their natural framework of climate, land, rivers and sea.
Much attention is paid to geology and geomorphology since deltaic and other low-lying coastal areas are located on the most rapidly changing parts of the crust of the earth and any human intervention in these systems could have significant repercussions.
Geology and geomorphology also explain the soil and subsoil conditions which play an important role in agriculture and other uses of the land.
The second chapter deals with the water in deltas, both surface water and groundwater, and its relation to the sea, which forms the downstream geographical boundary. In many areas the natural hydrological conditions have already been modified by man to some degree.
The ecological features discussed in this publication are mainly related to the aquatic environment. Human intervention and hydraulic development are discussed in the third chapter. Topics in hydraulic engineering that are required for a full utilisation of the high economic potential of deltas have been considered. These topics include flood control, flood protection, repulsion of salt water intrusion, fresh water supply, storage possibilities and land reclamation for agriculture and other purposes. Attention has been given to water management planning and to the relations between the development of the delta and the upstream portions of the river basin.

Finally, as practical illustrations, six examples of deltas in different regions of the world and in different stages of development are given with sharply contrasting cases from the delta of the Mekong river in southern Vietnam (Appendix 2) to the delta of the Nile river in Egypt (Appendix 6). Both areas are very productive and densely populated. In the Mekong, man has adapted himself to the prevailing hydrologic conditions of floods and droughts through the use of suitable cropping patterns and by modifying his way of life. In the Nile delta and Upper Egypt, under extreme conditions of population pressure, food and power requirements, political constraints and soil salinisation problems, the hydrologic environment has been completely modified.
The case of the delta of the river Rhine in The Netherlands (Appendices 1 and 7), with its man-made lowlands and rigidly controlled river and channel systems, demonstrates that environments created with intense human intervention can still be quite acceptable.

1.1 General

1.1.1 Special features of deltaic areas

The development of deltaic areas is a process which is essentially multi-disciplinary in nature. Out of the various sciences and disciplines involved, only two topics have been selected, namely hydrology and water management, being the most fundamental considerations for the habitation of these areas.

Deltas are formed where a river discharges into a body of standing water such as a sea, ocean or lake. They may also be built up when a river enters a river valley or outfalls into another river. There are various definitions for a deltaic area. These all refer to the lowest reach of a river and the land formed by the deposition of sediments from the river, and often from the sea as well. According to one definition, this reach extends as far as the effect of the body of standing water is felt. In the case of a sea, it is the reach where the tide and/or the intrusion of seawater occurs. It can also be defined as the river reach and the alluvial lands downstream of the apex, i.e. the point where the river approaching the low-lying coastal area separates into two or more branches or tributaries. This definition is not applicable to rivers with a single channel or estuary, which may also build up deltaic areas. The latter definition will be adopted here, but the reach and lands upstream of the delta that are hydrologically related to the downstream area will also be included.

Deltas present some special features that are not found, at least not to the same extent, in other parts of the river basin to which they belong. Whereas the flows and water levels in an upstream river station depend on the rainfall in the contributing watershed, the flows and stages in a deltaic river branch depend on the conditions at the two boundaries, i.e. the water levels and discharges at the apex, and the water levels at the sea.

Deltas are virtually flat with land slopes varying generally from a few centimetres to a few decimetres per kilometre. Such areas are exposed to extensive flooding by the river and also by the sea, where storm surges (extremely high sea-levels) may occur. The river branches of a delta are exposed to the intrusion of seawater, which renders the water unsuitable for many uses. The

Transportation of jars for collecting rainwater (Padma River, Bangladesh). Due to the lack of fresh water in these areas rainwater is often the major source of potable water.

control of the saline intrusion is one of the main problems associated with the water management of deltaic areas. Because of the presence of a system of inter-connected river branches and other water courses, human interference with the local hydrological conditions within a delta is bound to have repercussions in other parts of the delta. Delta development should therefore be based upon a master plan which examines possible alternatives of long term development and the integration of short term projects. Since the quantity and quality of water supplied to a deltaic area depend on the upstream conditions, delta development should be viewed within the framework of a development plan for the entire basin.

Although the present publication focuses on deltaic areas, much of the subjects dealt with are also applicable to other environments of deltaic areas such as tidal inlets and embayments, lagoons, coastal marshes and swamps, and other low-lying wetlands. For this reason, the geology and geomorphology of these environments are also important and will be discussed later in this book.

1.1.2 Economic significance of deltas

The intensive economic development which has taken place in many deltas throughout the world supports the view that deltaic areas constitute some of the most valuable areas for mankind. This is due to a series of natural advantages which are not found in other environments. Deltas are, in fact, meeting points between sea and ocean shipping and the transport of goods to and from the hinterland by inland shipping. This is also the reason for the concentration of industry and offices

Weekly floating market in the delta of Bangladesh.

around harbour areas. Most of the large cities of the world are found in deltas. The largest port in the world, Rotterdam, owes its position to its function as a transit and transfer centre for the industrialised areas of the Rhine basin and other nearby Western European countries.

The natural advantages of a delta may be partly offset by inherent drawbacks. Deltas, being situated at the most downstream point, receive all the fresh water from the river basin. The quality and quantity of this water depend to a large extent on the water use practised upstream. The water reaching the delta may be polluted and a considerable volume of water, if not the entire flow, may be withdrawn from the river for the irrigation of upstream areas, especially during dry periods. The relation between the development of the upstream areas of the river basin and the deltaic areas, will be considered later.

Deltas are blessed with alluvial fine-grained soils which are generally very fertile. Because of the presence of a high salt content, the soils exhibit a high water suction capacity, which means that during dry periods the plants are unable to obtain water from the soil in the root zone. In the case of dry food crops which cannot grow in water, the relatively high salt content makes it necessary to control the water table using a system of field drainage, consisting of deep open trenches and ditches, or placing drain intakes at some depth below the surface (subsurface drainage). Sometimes the soils are too heavy, which makes the drainage and tilling very difficult. Deltaic areas have a predominantly horizontal topography without natural slopes or macro relief. This means that ideal conditions exist for good control of the depth of flooding of rice fields, and the depth of the groundwater table below the land surface in the case of dry food-crops. On the other hand, the level characteristic of the land makes it difficult to generate flows in drainage and irrigation canals.The greatest drawback however, is the extensive flooding which occurs when the river is in spate, or in the event of a sea flood. Flood control measures upstream, and flood protection works locally, are generally regarded as the most essential elements of delta reclamation. The groundwater table in deltaic soils is found at a shallow depth of about 0.5 to 2.0 metres below the surface. The capillary ascent from the water table represents a welcome additional supply of water to the root zone in times of low rainfall, provided that the groundwater is fresh. When the groundwater is brackish or saline, which is often the case, deep drainage sand leaching by infiltrating rain or irrigation water, should control the salt concentration in the root zone.

In the network of water courses present within deltas, inter-connected river branches and creeks are found in which the tides and floods propagate. Because of the backwater effect of the sea-level, the water levels in these watercourses never fall, even when the river discharge is very low, thus providing a natural system of communication. Bad foundation properties of the soils and the risk of flooding provide serious difficulties in the construction of heavy structures. This should be mentioned when explaining the different degrees of economic development which have taken place in deltas throughout the world. A first group comprises the deltas where a high degree of water control has been achieved and a high level of agricultural and industrial production have been developed. In spite of a high population density, a general welfare is found, especially when these areas also act as outfalls of a prosperous hinterland. Examples are the many small deltaic areas along the coast of Japan, the delta of the Rhine river in the Netherlands and the delta of the Mississippi river in the U.S.A.

A second group of deltas exists, where water control has contributed to a high agricultural production but in which little industrial development has taken place to date. These deltas are also densely populated but the standard of living is generally lower than in the deltas of the first group. Examples are the deltas of the Nile, Indus, Euphrates, and Tigris rivers, as well as that of the Red River in the northern part of Vietnam.

An apparent paradox is presented by those deltas which are still in the first stage of hydraulic

The temple of AngkorWat in Cambodia, built by the Khmers (12th century AD) near the shores of the TonleSap in the delta of the Mekong River. Deltaic areas and other coastal lowlands have often been the location of great early civilizations (e.g. Egypt, Mesopotamia and Indus Valley).

development, with incomplete water control, but which are nevertheless relatively highly productive and densely populated. Examples can be found in the large deltas of Southeast Asia like those of the Ganges-Brahmaputra, Irrawaddy and Chao Phya rivers. A striking example is the delta of the Mekong river in the southern part of Vietnam, where almost the entire delta is exposed to flooding by the river. In this area, known as the *rice bowl* of Vietnam, cultivation is possible because local rice varieties have been produced which have adapted to the prevailing hydrologic conditions and because the people have learned through experience how to coexist with floods.

1.1.3 Human intervention

Deltas belong to the most rapidly changing part of the crust of the earth. As a result of natural processes, deltas are growing and expanding into the sea, sometimes at rates of a few tens to a few hundreds of metres per year. At the same time, land areas are being built up and river channels are shifting. In other areas or at other times, land disappears due to erosion by the river or the sea. It is in this area that man interferes with the natural processes because such changes in the land and water configuration are contrary to the envisaged development.

Human occupancy starts before the deltaic areas are protected from floods and before any form of water control in the land areas has been achieved. This is true not only in the climatic zones where land rice can be produced but also in the temperate regions where this is not possible. Man settled first along river courses on the natural levees or ridges that were built up during floods by the deposition of sediments from the river. Here the depth of flooding is smaller than in the back-swamps or depressions further away from the river courses. Also, the soil and the depth of the groundwater are more conducive to the cultivation of food-crops like vegetables and fruit trees than elsewhere in the delta. From this base, man penetrated further inland and this marked the beginning of agriculture and fishing, only interfering with the natural system through the removal of the natural vegetation.

With agricultural land-use expanding, the need for easy communication and transportation, for protection from floods, and for adequate water control in the land areas became apparent. The natural course of the river may be changed during regulation of the river channel for flood protection. Regulation involves making the width of the river bed more uniform at low-water levels and smoothing the floodplains to increase the high-water discharge capacity. In moderate and cold climatic regions, regulation may also be necessary to expedite the flow of ice. Normalisation and canalisation of the river channel may also be necessary to improve the situation for shipping, especially during low-water periods. In the delta of the Mekong for instance, where reclamation started in the second half of the 19th century, an elaborate system of canals has been dredged but there is still no flood protection by embankments (see Appendix 2). A similar system of canals is being laid at present in the coastal marshlands of Sumatra, Kalimantan and Irian Jaya in Indonesia. In the delta of the Chao Phya in Thailand, the first canals were dredged as waterways to allow the passage of flood water from the river through the natural levees to the depressions. This also acted as supplementary irrigation water supply during periods of rainfall deficiency. In the Netherlands, dams with sluice gates were built in the mouths of tidal creeks even before a system of small embankments was established. The system of open canals in free communication with the river branches and the sea modified the distribution of the discharge over various water courses.

A much more profound impact on the natural layout of deltas was brought about by the application of embanking or *dyking*, as a means of flood protection. Although, in the absence of such protection, a restricted land-use is possible, maximum yields and diversified cropping require adequate water control measures. This also applies to high yielding varieties of wetland rice. The provision of a drainage and irrigation system after dike construction is a further step towards achieving adequate water control. In many cases such a provision is an inevitable follow-up to the exclusion of the flood waters.

Rehabilitated coastal embankment in the delta of Bangladesh (south-western region).

The interference of man was not restricted to the low-lying areas. Even in the early stages of development, modifications of the system of river branches, creeks and open water courses were attempted. The diversion of water courses, deepening of channels, and the closure of creeks were commonly applied, even with very modest technical means. The construction of river embankments soon made it necessary to focus attention on the rivers. These tend to meander, braid and shift their channels, in turn, threatening the embankments by bank erosion. This has always been, and still is, a problem in deltas where such embankments exist. An adequate solution can be obtained by so-called river training works consisting of a system of groynes aimed at keeping the erosion away from the banks. Underwater bank protection can also be applied. The disadvantage is that great amounts of stone are required to construct these protection works. Natural stone is rare in deltas and often has to be shipped over hundreds of kilometres. Local measures such as dumping clay, stone, etc. at the threatened spot are seldom successful. In some cases (e.g. the Irrawaddy river in Myanmar, and the coastal embankments in the Netherlands), retired or set-back embankments are constructed, where at some distance away from a first embankment, a second one is built, ready to provide protection should the first embankment fail.

Rivers may pose a problem due to the rise of the river bed that causes higher flood levels with respect to the surrounding land elevation. Silting may be caused by the deposition of sediments from erosion in upstream areas or as a result of embanking. Classical examples are the Yellow River in China, the Red River in the northern part of Vietnam, and the Rhine river in the Netherlands. In the latter case the problem was solved by large scale dredging of sand in the river bed. This became economically feasible in the second half of the 19th century when there was an urgent need for considerable amounts of sand for the construction of roads and structures in city expansion projects. The waterlogged deltaic lands are not ideal sites for building cities as the presence of thick layers of clay, loam or peat give rise to poor soil conditions for the foundations of buildings and other structures.

Cities in deltaic areas can be founded in polders or on landfill (Figure 1.1). Polders are level areas which were originally waterlogged and where the surface and groundwater levels can be controlled. This is achieved by isolating the area from the surroundings and establishing an appropriate drainage system. Protection against floods from the surrounding areas is usually ensured by an encircling embankment. In landfill, the land is raised by the dumping of soil, preferably sand, borrowed from the vicinity. From the point view of safety, comfort and sanitation,

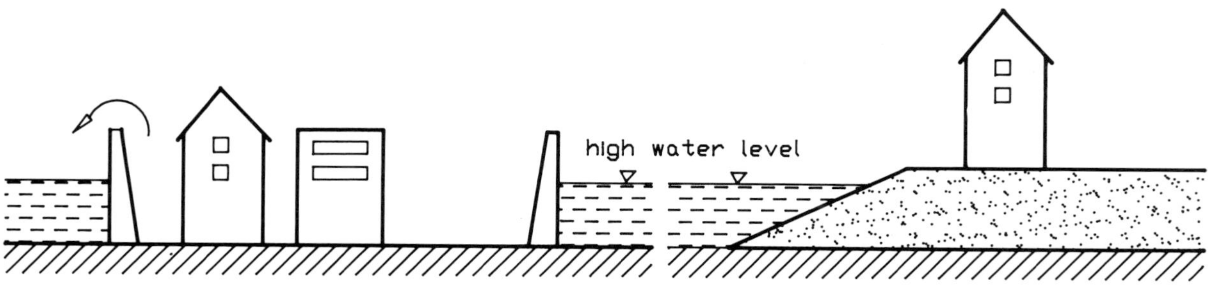

Fig. 1.1
Cities in polders or on landfill.

the landfill system offers the best perspectives. However, for large areas, it is more expensive to implement than the polder system, especially in the presence of a soft subsoil, where a considerable amount of sand is required to replace the soft layers.

When typical delta cities like Amsterdam (13th century), and Bangkok (19th century) were founded in marshy areas a system of canals was excavated and the adjacent land filled with the excavated soil. The extension of the present cities is generally based on the polder concept because of the high cost of fill and the problem of locating suitable borrow areas. Only a thin layer of sand is applied to the water supply and sewerage conduits. Many cities in low-lying coastal areas are subject to subsidence e.g. Osaka, Tokyo, Jakarta, Bangkok, New Orleans, etc. The main reason for this is the large scale extraction of groundwater.

Sea water intrusion into open estuaries is a natural phenomenon which occurs because the density of the sea water is slightly greater than that of the fresh water from the river. The heavier sea water slides along the bottom of the estuary against the river flow, mixing with the fresh water in the process. This penetration is especially pronounced when the river discharge is low, which means that during dry periods, when there is a great need for fresh water, the water in the lowest river reaches is unsuitable for various uses. The intrusion of sea water is increased by hydraulic works in the interest of navigation, such as the deepening of access channels and the dredging of tidal harbour basins.

As in other areas, the mining of minerals is practised in many deltaic areas. In some deltas, like the delta in the Rhine river in the Netherlands, peat was exploited until recently, as an easily accessible and readily available fuel. The removal of the peat created pools which expanded into lakes by wave erosion, threatening the adjacent areas. Also, the mining of clay and sand for use as construction materials resulted in a decrease of land areas, while the extraction of water gave rise to subsidence of these areas.

From the early stages of human occupancy to the present time, the supply of fresh water in the coastal wetlands for domestic, agricultural, and industrial uses has posed many problems, especially in areas near the sea. The water in the estuaries, the lower river courses and in the groundwater is brackish or saline. During dry periods a limited amount of the water can be found in the natural levees where, due to the infiltration of rainwater, a shallow pocket of fresh groundwater is found floating on the saline groundwater. This was one of the reasons for settling on these ridges. Fresh water pockets are also found in beach ridges while fresh water aquifers of a greater extent occur under dunes. Collection of rainwater in small reservoir tanks was also applied

in the early stages of development. Large-scale storage of fresh water can be effected by building reservoirs at suitable sites in the upstream mountainous reaches of the river. In recent times, schemes have been implemented in a number of deltas in which the storage is created within the deltaic area by damming off estuaries, tidal embayments and other water areas. These areas are transformed into estuarine or coastal reservoirs.
Man's desire to exploit the high natural potential of deltaic areas has led to large scale interference of the natural hydrological, ecological and morphological processes. For example, water quality has decreased, flow velocities and river morphology have become more uniform, brackish water areas have been reduced, the temperature of the water has risen, noise and disturbance from shipping activities has increased and land-use around the river banks has intensified. Many interventions generate a chain of repercussions which make further interference necessary. A lot of these effects are inevitable but it is obvious that, in the past, economic impacts have sometimes played a greater role than ecological impacts during the development of the delta. All impacts should be considered in the decision making process. Some deltas, like the delta of the river Rhine in The Netherlands, would not exist today if man had not stopped the natural processes of delta destruction - a so called *drowning delta*.

There is practically no limit to the physical development of deltas and there are consequently no final solutions to the problems of flood protection, drainage, water quality etc. This is partly due to the higher requirements in the course of time, with respect to the degree of safety against flooding, the degree of water control in land areas and the quality of water. It can also be attributed to external factors such as changes in the hydrological conditions and water management of the river basin, climatic changes and a possible accelerated rise in the mean sea-level.

In many deltas, like in other extensively developed areas, the problem of competitive use of the available land and water resources has come to the fore in recent years. Agriculture, industry, urban extension, communications and the preservation of valuable nature reserves all exert a demand for land and water. In deltaic areas this problem presents special challenges because of the dependence of the available water resources on upstream hydraulic developments, and the necessity to drain off a certain portion of the water to the sea, in view of water quality control. This is especially true for deltas in the semi-arid and arid zones.

The fact that in many deltaic areas the present environment is largely man-made does not necessarily imply that this environment is a bad one. It means however, that the environment is a vulnerable one, and in order to implement a development which is sustainable an accurate prediction of the effect of further human interference is required.

1.2 Hydrologic Characteristics

1.2.1 The role of climate in delta formation

Deltas can be found in practically all climatic zones and on the shores of virtually all oceans, seas, lakes and river valleys. The physiographic characteristics of a delta are the product of the climatic factors which are predominant in the river basin, the alluvial valley, the deltaic plain and the receiving basin (Coleman, 1981). Any change in the water-sediment yield within any of the components due to changes in climate or to man-made activities will influence the characteristics of the deltaic plain. During the last decades drastic changes have occurred in the deltas of the Nile and Colorado rivers mainly because of the huge water storage works carried out on them. The disturbance of the equilibrium between the riverine and marine forces is causing a retreat of the shoreline of the Nile delta by about 29 metres per year. Furthermore, there is strong evidence of the development of a new delta where the River Nile enters the huge reservoir formed behind the High Aswan Dam.

The depositional features at or near the receiving basin develop as a result of the influence of the marine dynamic processes which tend to modify and disperse the river-borne sediments. Discharge, wave energy, tidal action and mean water level are the most influential processes in this respect. The balance between the supply of river-borne clastics, and the removal of these sediments by marine processes dictates whether a delta is stable, advancing or retreating. It should be noted however, that the natural equilibrium of deltas is extremely unstable. Equilibrium is only maintained in so far as the dynamic processes and stream input remain unchanged after having attained a stable condition.

Since a climate is defined by a number of parameters, each having a different determining effect, the relation between the type of climate and the corresponding characteristics of deltas is quite complex. A few basic differences between deltas can be related to the type of climatic region as defined in a climatic classification. Many classifications have been proposed, but in the context of this publication, the rather simple systems proposed by Köppen (1931), Thornthwaite (1933), and Way (1973) are considered adequate.

Climatic classification

One of the longest-standing and most useful classifications of the climate is that developed by Köppen (1931). His classification is based upon annual and monthly means of temperature and precipitation. Thornthwaite (1933), proposed some developments to this classification based on two indices, P/E and T/E.

$$\left(\frac{P}{E}\right) = \sum_{i=1}^{n} \frac{P_i}{E_i}$$

$$\left(\frac{T}{E}\right) = \sum_{i=1}^{n} \frac{T_i}{E_i}$$

where:

P_i	is monthly rainfall	(mm)
E_i	is monthly evaporation	(mm)
T_i	is monthly temperature	(°C)
i	is number of months	(1,...,12)

For the purpose of terrain analysis, Way (1973) suggested the use of only four major climatic zones namely, arctic, arid, humid and tropical. This classification is adequate for a number of purposes, of which the categorising of weathering processes is just one. Some of the leading figures of the four major climatic classes are given in Table 1.1.

Climate	Temperature	Precipitation and Evaporation	Geographic Location
Arctic	Dry January: -30°C July : 5°C	Cold Prec. < 250 mm/yr Evap.is very limited	Northern Canada, Siberia
Arid	Cold-warm Jan: - 5°C to 30°C Jul: -10°C to 32°C	Dry Prec. < 500 mm/yr Evap. > Prec.	Widespread in all continents; occupies 25% of the world's
Humid	Temperate - warm Jan: - 5°C to 15°C Jul: -10°C to 28°C	Humid Prec. 1000-2000 mm/yr Prec. > Evap.	Eastern half of the U.S.A., South America, W.Australia, W.Europe
Tropical	Hot January & July 22°C - 32°C	Rainy Rainfall: 1500-2500 mm/yr	Most of South and Central America,South Central part of Africa, South India, South East Asia

Table 1.1 Some leading figures of the arctic, arid, humid and tropical climatic classes.

The four climatic subgroups: sub-arctic, semi-arid, semi-humid and sub-tropical, have their characteristics of temperature and precipitation similar to their respective major types. For example, a sub-humid climate is rather similar to the humid climate, but the precipitation is much less, usually between 500 and 2000 millimetres per year. Figure 1.2 is a map showing the four main climatic groups and the respective sub-groups. Furthermore, the map indicates the locations of the glaciated and mountainous areas.

Run-off in different climates

The hydrologic characteristics of a region are largely determined by its climate and its geological structure. Among the climatic factors which establish the hydrologic features of a region are the amount and distribution of precipitation, the occurrence of snow and ice and the effects upon evaporation and snow-melt of wind, temperature, net radiation, duration of sunshine and humidity. The loss of water by evapo-transpiration is further influenced by the availability of water and the depth to the groundwater table. The many and diverse climate characteristics of any particular drainage basin combined with its geologic and physiographic features lead to a certain rainfall run-off relationship (U.S. Bureau of Reclamation, 1965).

A river system consists of four primary components namely, the drainage basin, alluvial valley, deltaic plain and the receiving basin. It goes without saying that the climatological conditions in large river basins are likely to change from one component to another. Coleman (1981) stated that there are fifty-eight major rivers having drainage areas in excess of 100,000 km^2. It is uncommon to have the same run-off coefficient everywhere in an area of such an order of magnitude.

Weathering

Temperature, humidity, rainfall and frequency of climatic changes, (seasonal, annual, secular, etc.) affect the processes of weathering and erosion. Weathering is the process by which rock is altered through exposure to air, water and organisms. Mechanical or physical disintegration is the reduction of rock into smaller units by physical agents such as temperature, frost, thermal expansion or contraction, pressure release, undermining, and abrasion caused by wind, water and ice. Further fragmentation into much smaller units is caused by chemical processes such as oxidation, carbonation, hydrolysis, hydration and base exchange.

Marine processes and delta characteristics

Waves play a major role in the development of the shorelines of deltaic plains. High, nearshore wave-power leads to sand bodies consisting of clean, well-sorted permeable sands. Examples of deltas subject to high nearshore wave energy are the Magdalena in Columbia and the Sénégal in

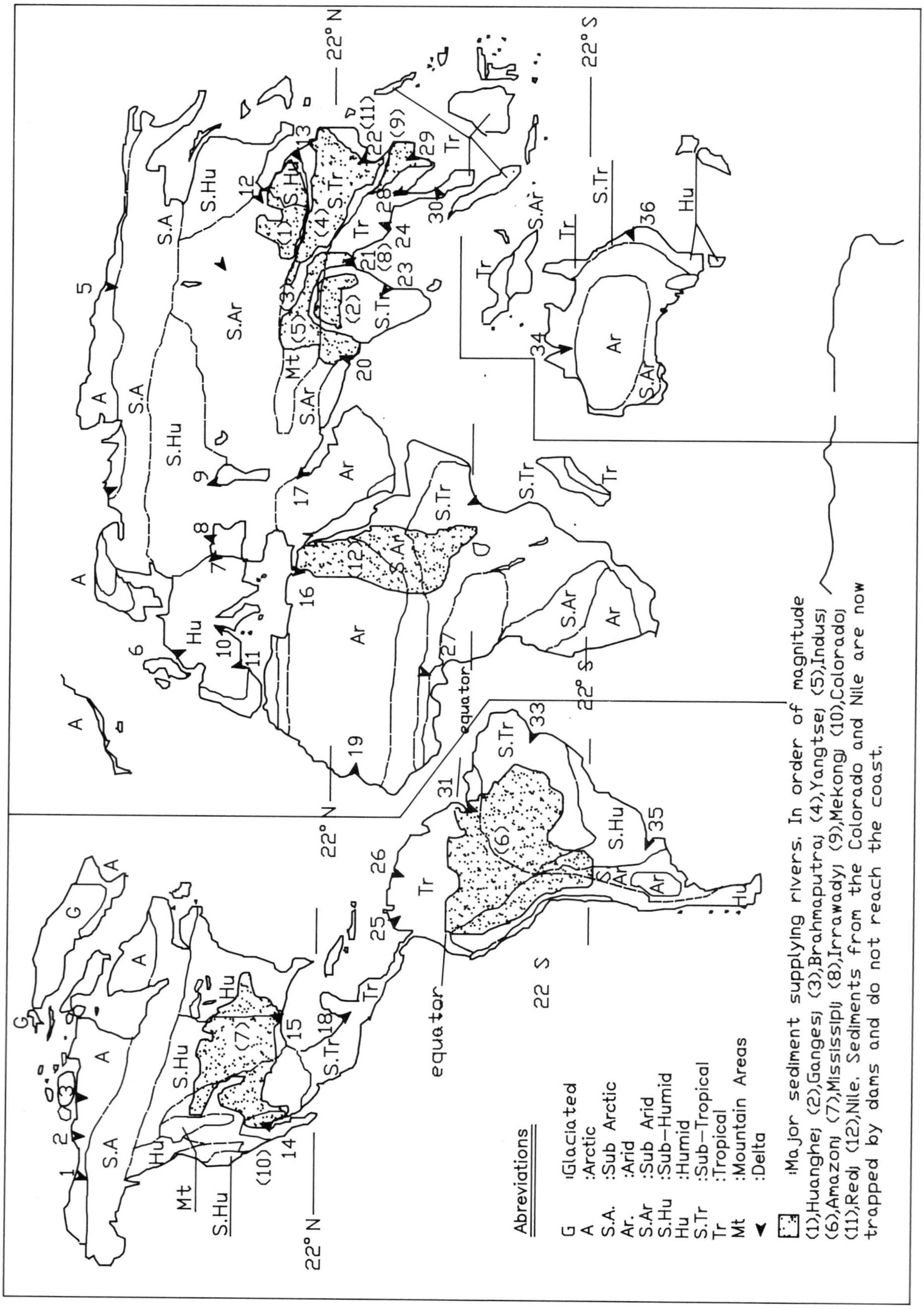

Fig. 1.2
Main climatic groups and subgroups.

Sénégal. On the other hand, low, nearshore wave-power leads to poorly sorted sand bodies as well as clay bound sands. The Yangtze delta in China and the Mississippi delta in the U.S.A. are examples of such deltas. The ratio of the average discharge per unit of channel width to the average nearshore wave power per unit of crest is referred to as the discharge effectiveness index. The higher the index the greater the dominance of the riverine processes on the build-up of the delta. The fact that the index for the Mississippi is about 20,000 times that for the Sénégal, reflects the extremely high wave action in the case of the Sénégal.

Tidal action can be effective in developing deltaic sand bodies. Tides can provide the necessary energy for transporting sediments. Furthermore, they tend to expand the marine riverine interaction zone both horizontally and vertically. Moreover, there is a bi-directional sediment transport process caused by the reversal over each tidal cycle to which the flow, both inland and seaward of the river mouths, is subjected. The tidal processes are practically non-existent in the arctic and sub-arctic deltas. On the other hand, these processes are very active in tropical and sub-tropical deltas, e.g. the Klang in Malaysia, the Yangtze in China and the Ord in Australia. In such cases intense meandering patterns of the tidal channels are found just upstream of the area of maximum tidal influence, and well-defined sandy ridges tend to develop within the deltaic plains.

The wind regime plays an important role in the development of morphological landforms in the coastal plain. This is especially true in arid regions. The most significant influence of the wind system is in the generation of nearshore coastal currents, both at the surface and along the bottom. In those regions where tidal action is extremely weak wind force is responsible for generating currents capable of moving coarse-grained sediments. Strong onshore winds will cause a set-up of water along the coastline, whereas offshore winds cause a lowering or set-down of the water levels. Littoral currents are driven, among other factors, by the wind. Wind driven coastal currents can attain velocities high enough to move and concentrate coarse material in offshore settings. Many of the sand sheets accumulating close to active deltas have been formed as a result of sediment transport by wind-driven currents. Moreover, bottom currents driven by high wind velocities can move sediments far offshore thus concentrating coarse deposits at considerable water depths.

1.2.2 Sea-level change

The Quaternary, the youngest geological era is characterised by the occurrence of an alternating interglacial and glacial climate at the higher latitudes of the Northern Hemisphere. These glacial/interglacial cycles occurred approximately every 100,000 years during the last one million years. The responsible factor is likely to be changes in the earth's radiation budget due to orbital parameter effects. An important phenomenon of the Quaternary is the change in sea-level.

Eustatic and tectonic factors

During the cold periods (glacials), enormous land ice-caps were developed, resulting in an important lowering of the ocean-level of more than 100 metres. Large parts of the Continental Shelf became dry land or coastal swamps. In particular, large parts of the Sunda Shelf in South East Asia became exposed and allowed fauna and flora to migrate from Continental Asia to the present islands of the archipelago (Indonesia and Malaysia). The melting of the ice-caps during the warmer periods (interglacials), resulted in the restoration of the ocean level. These changes in sea-level are called *eustatic*. There are local and regional factors which influence these eustatic sea-level changes such as tectonic movements, upwards or downwards, and the compaction of sediments. As these factors are hard to separate from each other we talk about *relative* sea-level changes when referring to the result of the combination of eustatic, tectonic and compaction factors.

Until 1971, relative sea-level changes were thought to be a combination of eustatic and tectonic movements only. As a result, areas where no tectonic processes occurred were designated *stable* areas. The observed sea-level changes were thought to be due to eustatic effects only. The reported eustatic sea-level changes during the last 6000 years from these stable areas are rather divergent. The reported results indicate that sea-level changes can be classified into three main groups.

Jelgersma (1966) gives these classifications as (see also Pirazolli, 1991):
- an oscillating sea-level;
- a constant sea-level;
- a continuously rising sea-level.

These controversial findings could not be explained until a new concept of sea-level changes came into being as a result of a new geophysical concept. The definition of eustasy in terms of worldwide simultaneous changes in sea-level is not correct. The sea-level is an equipotential surface of the earth's gravity field. Sea-level can only be measured over the oceans. It can however also be defined at the land area because the corresponding equipotential surface is a completely closed surface. This is known as the geoid. Changes in the earth's gravity field bring about a deformation of the geoid. The surface of the geoid is not a perfect sphere. This means that changes in sea-level are not identical over the entire globe and that gravitational changes are compensated over the globe (Figure 1.3).

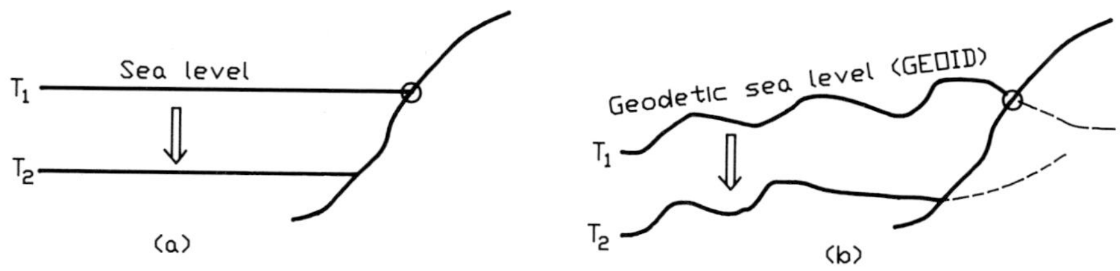

Fig. 1.3
Sea-level movement related to gravity change. The old concept (a) assumes total parallelism while in the new concept (b) sea-levels are never fully parallel.

Removal of glacier ice deforms the geoid. This idea has been emphasised by geophysicists working on the loading effect of ice sheets. The mass of ice deforms the ocean floor and attracts the water in the ocean (perturbs the geoid). The accompanying redistribution of matter within the earth also affects the geoid. Similarly, meltwater loading deforms the ocean floor and perturbs the geoid. Figure 1.4 shows the sea-level movement related to this effect. The deformed geoid is the sea-level position prior to the melting. The un-deformed geoid i.e. the spherical surface, is the sea-level position after the melting of the ice.

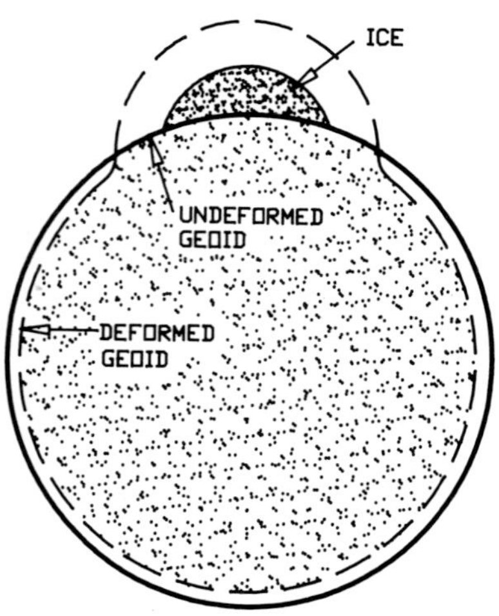

Fig. 1.4
Deformation of the geoid due to ice.

Unlike ice, water flows rapidly until the equilibrium ocean surface is on a gravitational equipotential surface. The redistribution of water causes differential loading of the ocean basins and additional deformation of the ocean floor. This feedback process continues until equilibrium is attained. These factors controlling sea-level are interrelated because both the ice sheets and the water load deform the earth's surface and perturb the geoid. Geophysicists have made a model of the deformation of the geoid and the earth's surface over the last 6000 years as a result of visco-elastic isostatic responses to water and ice loads (Clark et. al., 1978). The Clark model results in six zones in which the relative sea-level rise occurs according to the curves as indicated in Figure 1.5.

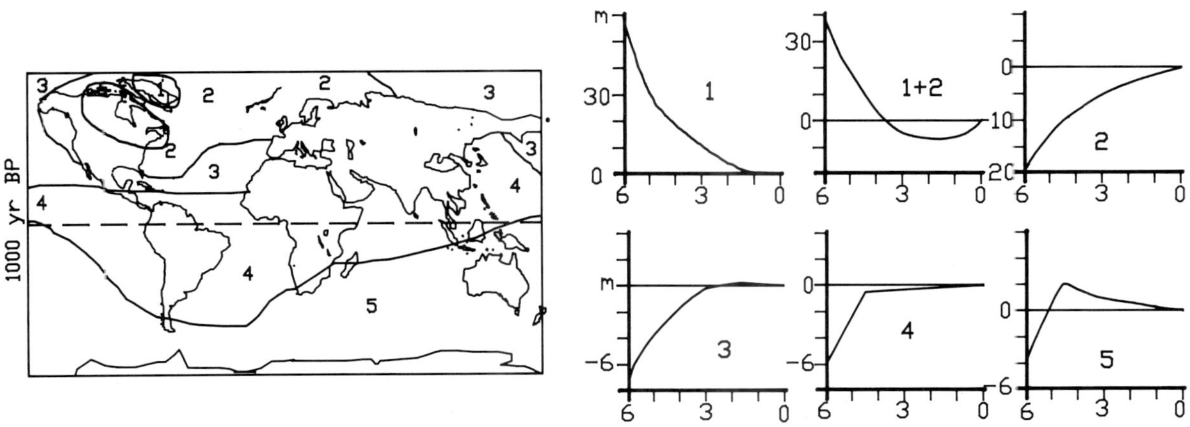

Fig. 1.5
Distribution of the ocean zones as distinguished by sea-level variation of different types.

However, the local tectonic environment of individual coastal areas and islands must also be understood. Figure 1.6 shows some of the processes giving rise to the three different types of eustasy. These, together with crustal movements, determine relative sea-level changes.

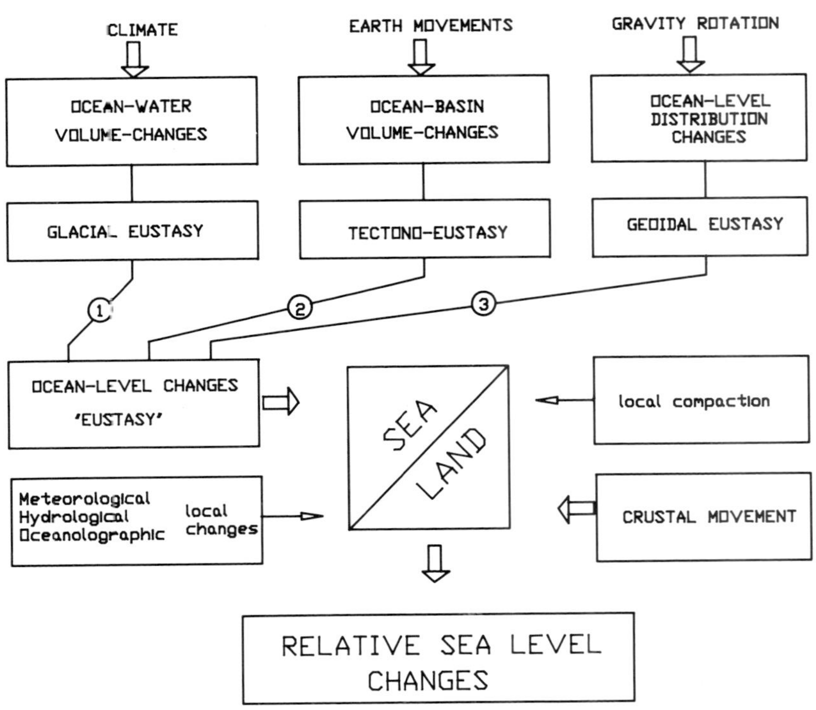

Fig. 1.6
Processes giving rise to the three different types of eustasy.

For example, there are areas with important tectonic movements like the earthquake zones in the Western and South Western Pacific and the Mediterranean. In these areas, shorelines can be uplifted, submerged and/or tilted by earth movements. During the 1960's, earthquakes in Chile resulted in a 1 to 1.5 metre subsidence of a costal belt some 600 kilometres long and 30 kilometres wide. Most deltas are situated in slowly subsiding sedimentary basins with a tectonic subsidence of between 2 and 10 centimetres per century.

It can be concluded that during the last 6000 years, when the land ice-caps had already mostly vanished, sea-level changes were due to the above mentioned local and regional factors. Accordingly, the curves of relative sea-level rise versus time, will show great local and regional variations. It must be emphasised that the best theoretical model cannot replace accurate field observations combined with carbon dating of observed fossil sea-levels.

Careful analysis of recent tide-gauge records from many locations has brought to light the fact that global mean sea-level is rising at a rate of seven to nineteen centimetres per century. Man, by polluting the atmosphere, performs a world-wide climatic experiment with possible consequences for the sea-level. Most important in this context is the production of carbon dioxide and other *greenhouse* gases. Such gases play an important role in the radiation budget of the climate system, the driving force of the hydrological cycle. The carbon dioxide concentration in the atmosphere, for example, has increased from about 270 ppm (pre-industrial value), to about 350 ppm at the present time, (ppm = part per million by volume). Most climatologists agree that a climate warming will result. The global mean value for the temperature increase associated with a doubling of the carbon dioxide concentration, expected at the end of the next century, is estimated to be between 2 to 5 degrees Celsius. Although a greenhouse warming has not yet been detected

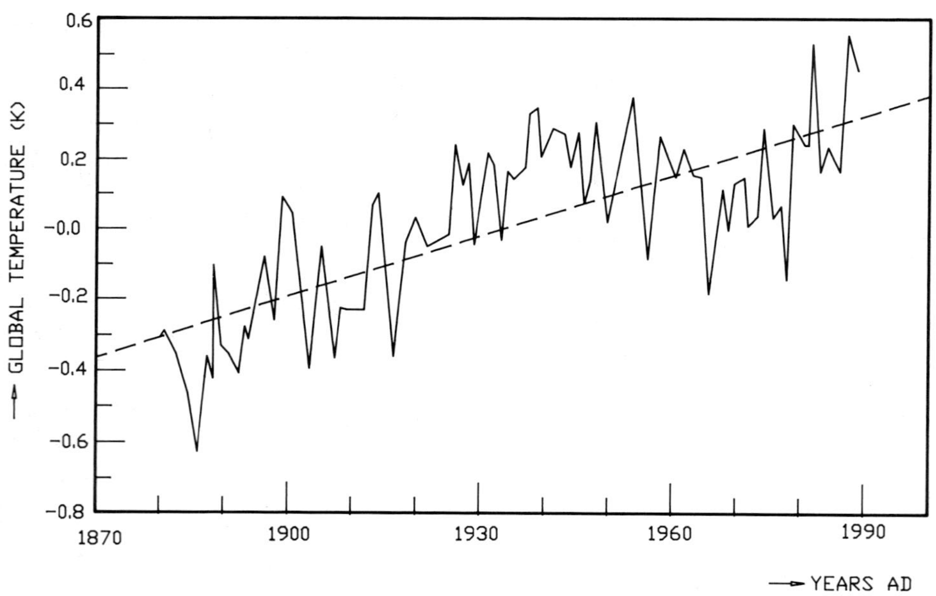

Fig. 1.7. Changes in global temperature.

with definite statistical significance, (the noise in the climate system is large), the observed trend is certainly in line with predictions made so far (Figure 1.7)
The problem has been recognised for some time now and several bodies and national committees have made assessment studies of the greenhouse effect and its associated impacts. This warming-up is expected to have consequences for the magnitude as well as the spatial and temporal variation of hydro-meteorological parameters like rainfall, evaporation, riverflow and storm surges. For example, a 10 % increase in the global rainfall is expected.

Another effect may be the rise in sea-level. As a *most probable* or average scenario, a rise in sea-level of about 0.55 metres in the year 2090 is expected (UN-ESCAP, 1977). However, there is a lot of uncertainty about the magnitude of these changes. The predictions are all based on the so-called General Circulation Models (G.C.M.), which simulate the hydrological cycle. There are several models, all based on sound physical principles but producing different results, because of the many assumptions that have been made in analysing the complex processes of the many systems and sub-systems.

Indeed there are also effects like the increase of precipitation in the form of snow on the Antarctic, which would produce a drop in sea-level. Although a number of fundamental questions concerning the response of the cryosphere (all ice masses) remain unanswered, a projection of future sea-level will be presented here.

To begin with, the physical mechanisms considered to be the most important in linking climatic change to sea-level, on a time scale of decades to a century, are mentioned. These are:
- thermal expansion of the oceans;
- retreat of mountain glaciers and small ice caps;
- changes in the mass budgets on the ice sheets of Greenland and Antarctica.

Other processes can be added, such as the changes in air-pressure patterns, the dynamic height of the ocean surface etc., but these are of minor importance, at least on the global scale. It should be noted that a redistribution of water and ice load on the earth may lead to regional differences in sea-level changes. This involves an elastic response which is instantaneous and a viscous response which is so slow that it can be neglected when changes over a few centuries are considered.

Thermal expansion of the oceans

The thermal expansion coefficient of sea-water is highly dependent on temperature and salinity. However, the density variations associated with the non-uniform salinity distribution in the oceans throughout the world, are less significant. A layer of sea-water, 100 metres thick, at a temperature of 25°C will expand approximately 3 centimetres per degree rise in temperature. The corresponding expansion for a layer at a temperature of 0°C is only about 0.5 centimetres. If climatic warming occurs, it will be some time before the temperature signal penetrates into the deeper layers of the ocean. It is expected that the upper few hundreds metres (the mixed layer, in which the seasonal cycle is evident) will warm quickly (within a few years), while the deeper ocean will react slowly (up to a thousand years). Given the climatic warming mentioned above, ocean models predict a sea-level rise of 20 centimetres for the coming one hundred years as a result of thermal expansion alone.

Retreat of mountain glaciers and small ice caps

Except for the last decades, very few direct observations of changes in the extent of glaciers are available. However, for some glaciers in the Alps and Scandinavia it has been possible to construct records of glacier length dating back to the 18th or even the 17th century. Drawings, etches, paintings and written documents, as well as terminal moraines to infer the exact positions of a glacier snout, have been used to achieve this. In Figure 1.8 some well-known records are shown for the European and North Atlantic sector.

Over a longer time scale, the behaviour of the glaciers is remarkably uniform. In particular, the steady retreat over the last hundred years is evident in all records. Most glaciers reached their neo-glacial maximum between 1600 and 1900 AD and at many places the maximum stand is seen from trimlines. Since about 1850, a large number of mountain glaciers have shown a pronounced retreat. Although many areas exist (e.g. Andes, Himalayas, Alaska) from which data are scarce, it is nevertheless clear that the retreat has been of a world-wide nature. The estimated contribution to the rise in sea-level, associated with this retreat, is between 2 and 5 centimetres over the last one hundred years.

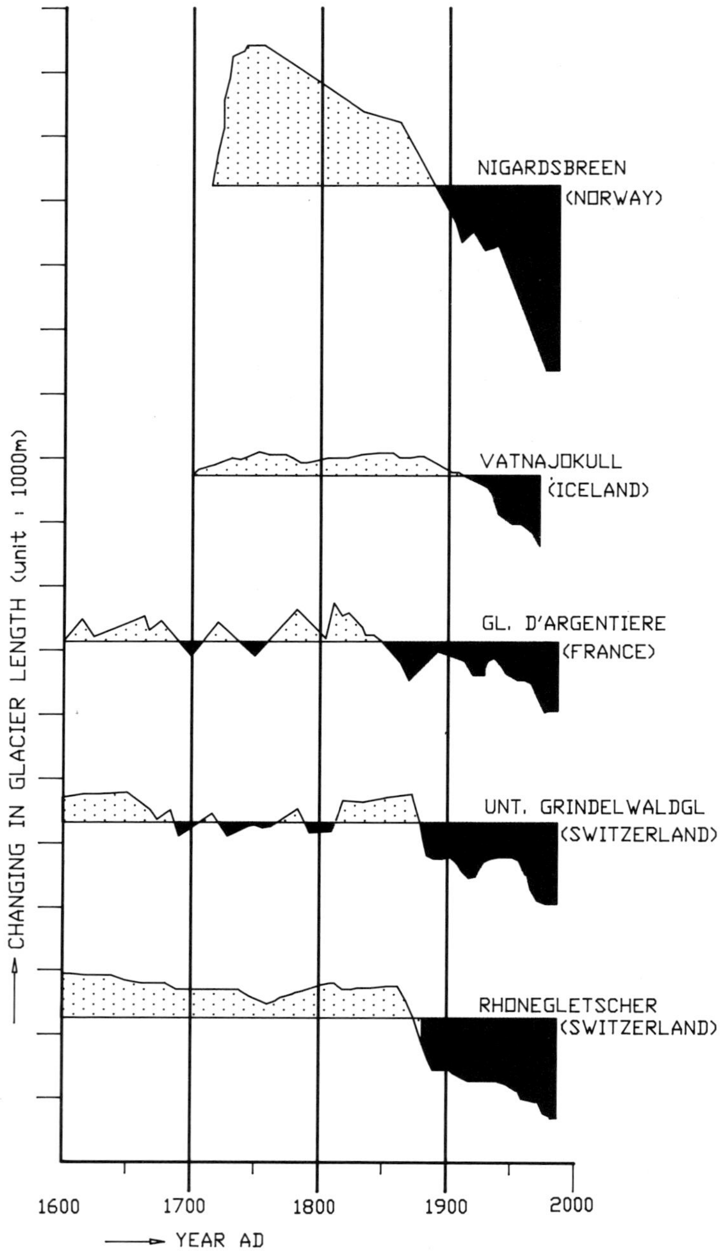

Fig. 1.8
Glacial Records for the European and North Atlantic sector.

It is likely that the retreat of glaciers is due, to a large extent, to the rising air temperature. There are a number of reasons why the mass balance of valley glaciers is so sensitive to small changes in the radiation balance. In the ablation seasons, a surplus of radiation leads to extra melting, but the enhanced temperature difference between the glacier and its surroundings leads to an increasing advective heat flux and thus to an additional increase in melting rate. Due to the fact that the ice surface remains at the melting point, glaciers can make use of surplus radiation in a very efficient way. Therefore, the response of glaciers to the expected carbon-dioxide warming should not be underestimated, even though the total ice mass involved is not very large (estimated to correspond to 30-50 centimetre sea-level equivalent, which includes all ice with the exception of Greenland and Antarctica). The most likely contribution to sea-level rise in the next 100 years is estimated to be 13 centimetres.

	GREENLAND	ANTARCTICA
area (10^6 km^2)	1.7	14
mean ice thickness (m)	1530	2160
ice volume (10^6 km^3)	2.6	30
maximum surface elevation (m)	3300	4000
loss of ice by calving	50%	99%
loss of ice by melting	50%	1%

Table 1.2
Some characteristics of the polar ice sheets.

Changes in the mass budgets of the polar ice sheets

The mass balance of the two major polar ice sheets of Greenland and Antarctica (Table 1.2) will of course also react to a climate warming. First of all, it is important to note that Greenland experiences a much warmer climate than Antarctica. As a consequence, the Greenland ice sheet loses half of the accumulated ice by melting and run-off, and half by calving. In the case of Antarctica, the loss is almost entirely by calving. For most geographical and climatological conditions increasing air temperature will imply decreasing mass balance (Figure 1.9b). This however, does not apply to the coldest polar regions. Over the Antarctic ice sheet the air temperature is very low which limits snow (or rime) accumulation. This leads to a situation of increasing mass balance with air temperature (Figure 1.9a).

In view of this, it is understandable that a climatic warming will cause a decreasing mass balance for Greenland, and an increasing one for Antarctica. Computations suggest that the net effect will be small, i.e. that the accumulation of snow on the Antarctic continent will be offset by increasing ablation in Greenland. However, this is probably a temporary situation. After many centuries, if the warming continues, the mass loss on Greenland will dominate.

Three factors should be taken into account when trying to estimate future sea-level:

- changes associated with the greenhouse warming;
- internal variability of the climate system;
- any systematic changes over long time scales.

The last of these factors immediately poses a problem since it is not known how close the global sea-level is to a steady state. In view of the very long time scales involved, it seems unlikely that the Antarctic ice sheet in its present form is close to equilibrium. It is still reacting to the transition to the present interglacial climate. The problem can be handled by taking the unexplained part of the basic trend as being representative of the long-term changes.

The calculation shown in Figure 1.10 has been carried out for the so-called Bellagio temperature scenario. The uncertainty is represented as a standard deviation, shown as vertical bars. Assuming that a basic trend of 7 centimetres per century will continue in the future, calculations produce the projection of future sea-level. The error bars include the uncertainty in the climate scenario as well as that associated with the scarce knowledge of the response of the large ice masses to climatic changes. In particular, the Antarctic ice sheet contributes much to the uncertainty.

It must be emphasised that the projection given here applies to a global mean situation. Trends of tectonic and isostatic origin, in particular, can differ widely from place to place, and these should be properly taken into account.

(a)

(b)

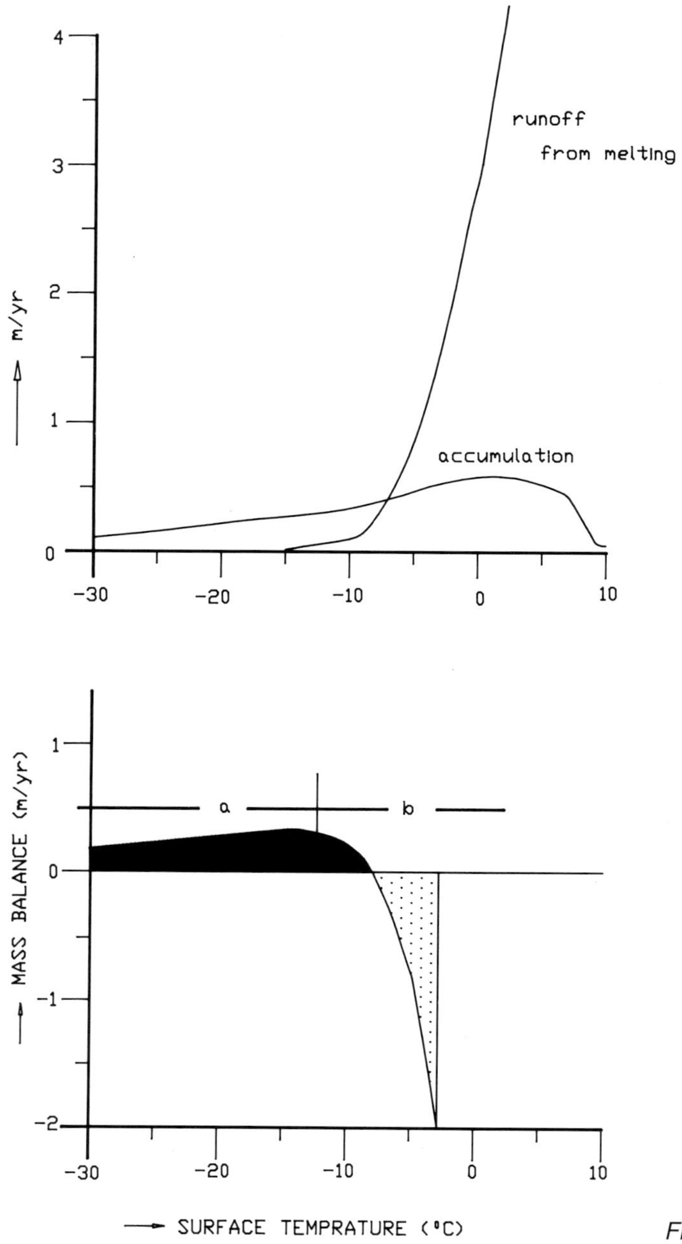

Fig. 1.9
Mass balance of the polar ice sheets.

1.2.3 Ecological aspects

Where a river debouches into a sea or ocean, the sea water penetrates into the river. This process is influenced by the tides. If the ratio of river flow to tidal flow is relatively large, a salt wedge is formed in which the fresh river water flows out on top of the more dense salt water. Saline water is mixed into the fresh water layer. This is compensated by a net flow of sea water moving upstream. Along the bottom, water is either saline or fresh. In areas where a dry season occurs, the salt water wedge is able to penetrate deep inland during this period. The tip of the salt wedge of the Mississippi moves 220 km depending on river flow. In the dry season, the Casamanca River (Senegal) and the Gambia, saline tidal water intrudes more than 200 to 250 kilometres upstream. If tidal flow is relatively strong, stratification is broken up by turbulence and a vertically homogeneous situation results. In these cases a gradient of brackish water is found along the bottom. Changes in river flow result in a shift of the gradient.

In the inter-distributaries of the upper deltaic plain accumulation of toxic salts may occur during the dry season due to high evaporation. An example is the Plain of Reed in the upper deltaic plain

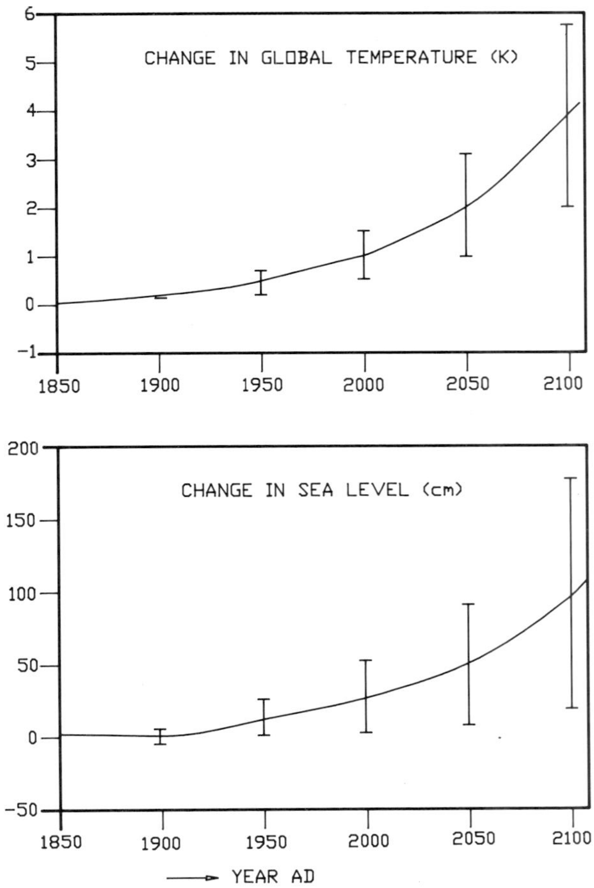

Fig. 1.10
Predicted changes in global temperature and sea-level.

of the Mekong delta. During the dry season the soil becomes strongly acidic and contains high concentrations of aluminium, iron sulphate, sodium and magnesium. These soils cannot be used for agriculture if not leached.

Extensive low plains between the upper and lower deltaic plains are characteristic of most large high tidal deltas. In some cases these inter-distributaries no longer get sediment from the river due to the high elevation of the levees. These saucer-like depressions are flooded by rainwater. In some cases they are drained at low tide by tidal channels in the low deltaic plain. In other cases, drainage is very poor and the water becomes stagnant and is only lowered by evaporation, which may give rise to toxic soils.

In some deltas tidal flats have developed. They may be covered by salt marsh vegetations and in tropical circumstances mangroves may appear. As tidal flats enlarge their effectiveness as a rainfall catchment basin increases. In mangrove clays, salt may leach eventually allowing the establishment of fresh water vegetation. Such fresh water swamps occur in deltas as well as along the coastal plains, e.g. in Malaysia and Indonesia.

1.3 Geology and Geomorphology

1.3.1 Introduction

The coastal zone includes many environments such as deltas, bays, estuaries, marshes, lagoons, beaches and dunes. These environments are controlled by a broad range of factors of which lithology and salinity may be regarded as the most distinct, but highly variable, features.

Most deltaic regions overlie sedimentary basins. Due to slow tectonic subsidence these basins have accumulated huge amounts of sediment. The sediment consists of an alternation of semi-consolidated clay and sand layers. The sand layers in many deltaic plains are important sources of groundwater (aquifers). Offshore, these basins may contain important gas and oil resources.

Most beaches and barrier islands consist of medium to fine grained sands. Lagoons and marshes, on the other hand, are composed of clays, sandy clays, silts and peat. These sediments are generally found in the intertidal and sub-tidal zone. Coastal wetlands are located inland of these sediment zones and show a zonation from salt and brackish water environments to marshes, lagoons and fresh water peat. In some areas, the coastal zone is bordered by dunes. Due to climatic circumstances and the presence of calcite they can be cemented (aeolonites).

Salinity may vary from fresh water in alluvial plains, to brackish and salt water in coastal plains and estuaries in wet climatic zones, and to hypersaline lagoons (Sebkhas) in arid climatic zones. The salt water environment in the supra-tidal zone is strongly influenced by the climate, particularly rainfall. Salt marshes and lagoons occur in areas with sufficient precipitation. In arid and semi-arid areas, high evaporation causes Sebkhas, desiccated mud-flats and algal mats to form.

Variation in morphology and lithology is the result of the combination of the dominant processes of waves, tides, currents and fluvial sediment and water input. This can be seen in important river deltas like the Ganges-Brahmaputra, the Nile and the Niger.
The complex of sub-environments of the shore zone, including intertidal flats, marshes, deltas, channels, lagoons, fresh swamps and Sebkhas, have shifted land inwards due to the Holocene rise in sea-level. The latter was caused by the melting of the continental ice-caps of the last glaciation, which caused a sea-level rise of more than 100 metres.

Sea-level changes can also be caused by tectonic movements. Most deltaic areas are situated in basins that are slowly subsiding at a rate of several centimetres per century. In most sedimentary basins the sedimentation has kept up with the tectonic subsidence. Some coastal areas like the Circum-pacific and the Mediterranean are, however, situated in active tectonic zones, characterised by the occurrence of earthquakes and volcanic eruptions. In these zones important changes in level may occur in a very short time.

1.3.2 Factors influencing delta development

As well as tectonic movements and sea-level changes, the three main factors influencing coastal and deltaic developments are related to:
- land;
- sea;
- biological processes along the shoreline.

Land

Sediment supply from the land deposited in the alluvial plain and the coastal plain depends on the rainfall and the shape, climate and surface geology of the river catchment area and drainage basin (Figure 1.11).

The drainage basin is the area from which water and sediments are derived. In the humid topics there is an intense chemical weathering of rocks under the prevailing wet and warm climate. Therefore, the sediment available to fluvial transportation is great. The rate of sediment yield is a function of runoff, slope gradient and the density of the vegetation cover within the catchment

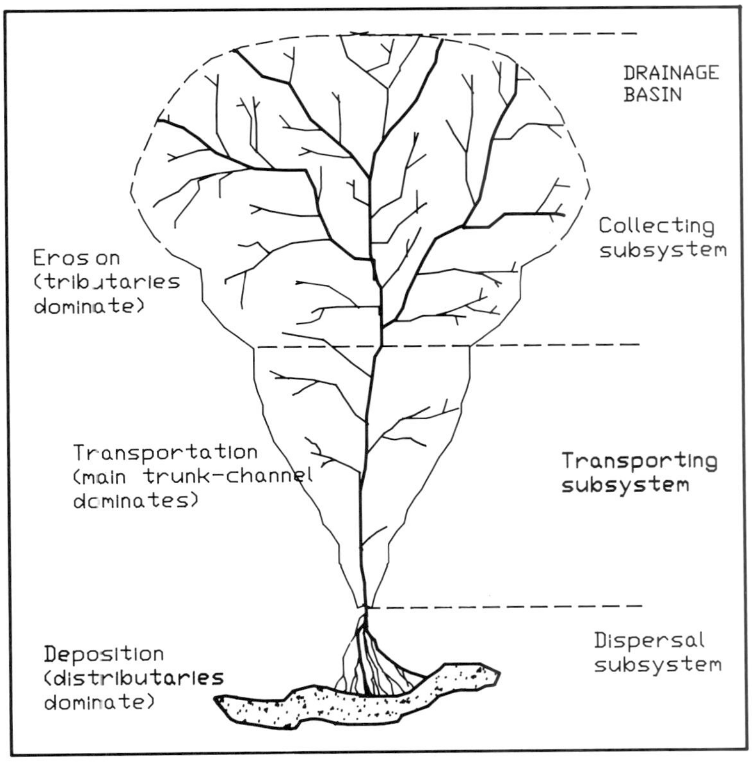

Fig. 1.11
The major parts of a river system.

area. The tributaries in the headwaters collect and funnel water and sediment into a main trunk stream. Erosion is dominant in this area. The main river trunk functions as a transporting subsystem. Both erosion and deposition occur in this area. The lower end of the river is a dispersal subsystem where most of the sediment is deposited in a delta or an alluvial fan and water is dispersed into the sea. Deposition is the dominant process in this part of a river.

The sediment supply can be increased by tectonic movement (uplift), volcanic activity or by reduction of the vegetation cover. The latter can have natural causes (landslide, bushfires, desertification) or be the result of human activities (deforestation, grazing, cultivation). The sediment supply can be reduced by the construction of dams that impound river water and trap sediment, or by the diversion of rivers into canal systems, which disperse the sediment load. In general an increase in sediment supply by the river results in coastal accretion and a decrease in sediment supply in shoreline erosion. Great inputs of sediment and water, as is the case of the big rivers like Ganges-Brahmaputra and Yangtze Kiang, give rise to extensive deltaic plains. In many humid tropical regions the river flow regime shows annual cycles due to changes in precipitation during wet and dry seasons. These changes are governed by the changing monsoon paths.

Sea

The development of the coastline is highly dependent on the energy of the wind, which generates waves and swell, and the tidal range. Onshore directed waves provide most of the energy for shoreline processes. Waves advancing obliquely toward the shore are one of the most important transporting systems along a shoreline. The material delivered by the river into the sea is sorted and transported by the activity of waves and currents. The sand fraction is sorted by these phenomena and distributed along the shore as beaches and spits. The finer sediments like silt and clay are transported until they reach calm water where they are deposited on shoals or in sheltered inlets and embayments. The off-shore characteristics of shelf width and the gradient of the coastal slope are of importance in relation to wave refraction. In the case of steep slopes the shoreline is fully exposed to wave action. A delta will also be more likely to grow further seawards over a broad, gently sloping shelf than over a steep, narrow shelf. Another offshore characteristic is the presence or absence of submarine canyons. These canyons drain sediment to deeper water.

Subsequently, this sediment is lost and no longer available for shoreline building. Good examples are the submarine canyons which are present offshore of the Niger and the Ganges-Brahmaputra deltas.

The frequency of occurrence of storms is of great importance for shoreline stability and flooding. The major damage on a shoreline occurs during storm surges at the time of a high tide. Coastlines situated in a typhoon or hurricane path are very vulnerable. Typhoons are tropical storms originating in the Pacific or Indian Ocean. Hurricanes originate in the Atlantic Ocean.

The tidal range is the height range between high and low water. A macrotidal range is greater than four metres and a microtidal range is less than two metres. A mesotidal range lies between these two extremes. Figure 1.12 presents the world distribution of tidal ranges and the relationship between tidal range and coastal morphology.

Where the tidal range is high, large intertidal flats are exposed at low tide. The associated strong tidal currents shape a complex of shoals and channels in the estuaries and across the near-shore zone. A subdivision can be made between tide-dominated shorelines and wave-dominated shorelines. In regions where tides and waves are of no importance coastlines are classified as zero energy shorelines. In these cases mudflats and sand barriers border the sea.

Biological processes along the shoreline

Biological processes play a direct and active part in coastal geomorphology. Marsh and dune plants and coral and calcareous algae have a constructive effect. This effect and the effects of rock-destroying organisms are measurable over relatively short periods of time. The effects of these biological processes are dependent on the climate. Coral reefs are not only dependent on high sea-water temperature (>19°C) but also on clear water in order to have algal photosynthesis for reef building. The growth of vegetation is very dependent upon temperature and precipitation. In the shore-zone, vegetation plays an important role in sediment trapping, contribution of organic material and reduction of aeolian processes. These facts are clearly demonstrated in humid tropical deltas with extensive mangrove vegetation and arid deltas with nearly no vegetation at all.

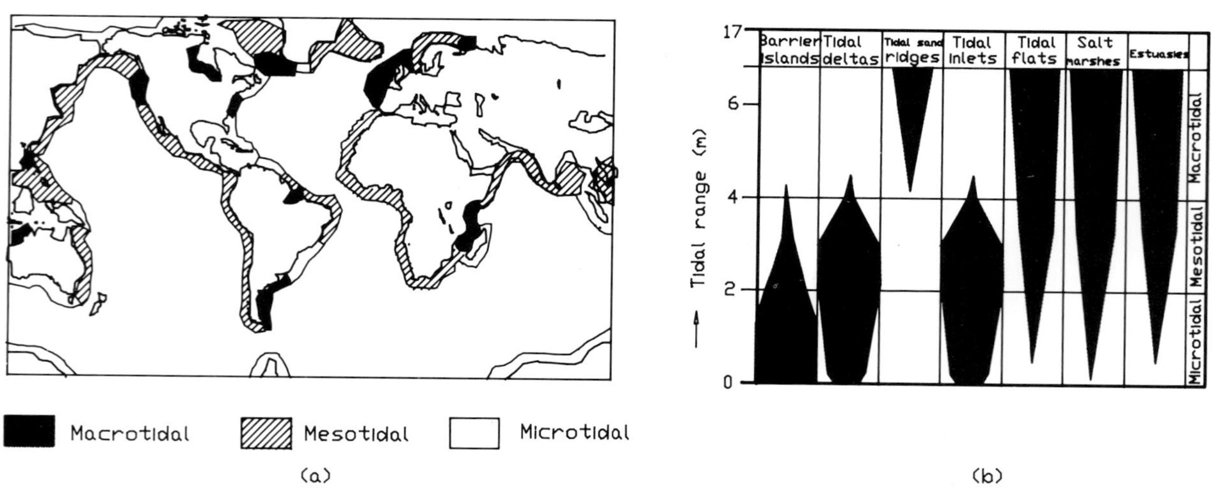

Fig. 1.12
a) World distribution of tidal ranges (Davies, 1964) and
b) the relationship of coastal morphology (Hayes, 1975).

Saltmarshes, or saline wetlands, can be found along the coasts of most continents. They occupy zones that border either fresh water environments or marine/brackish bays and estuaries. Where low-energy conditions occur marshes may be exposed to the open sea. In other areas, the saltmarsh is protected from the open sea by barrier islands and spits. The presence of ancient marsh deposits beneath younger deposits in coastal plains gives not only evidence of fossil rich environments but also of transgressions and regressions and sea-level changes in the past. The coastal saltmarshes are very important to man for two reasons. They help to buffer the force of storm surges before they strike the populous mainland and they provide nurseries for important fish species. Saltmarshes are formed when mudflats are silted up above the mean sea-level. They are colonised by halophytic (salt-tolerant) plants. These plants are grasses, sedges and trees adapted to tolerate salt and brackish water. The vegetation of the mudflats filters the suspended mud out of the tidal waters and build the mudflats up to the high tide level. Important genera in temperate zones are Spartina and Salicornia and the mangroves in tropical zones.

The occurrence of mangrove swamps is controlled by temperature. They can only develop in tropical, humid regions. Mangroves like to settle on soft organic mud in estuarine and deltaic environments. Mangrove trees have a special root structure to enable survival under intertidal salt water immersion and to support the trees in the soft mud. Pneumatophores allow oxygen to reach the roots buried in the mud. Two types of mangrove are the stilt or prop-root mangrove and the periscope-root mangrove.

In arid, semi-arid, tropical and sub-tropical regions shoreline vegetation is nearly absent. The high evaporation gives rise to Sabkhas, mudflats and algal mats. Sea-grasses or marine angiosperms grow below or around the low-tide mark. In tropical coasts they are important sediment traps (coastal accretion). Beach and dune plants live in a less stable environment than plants in saltmarshes and mangrove swamps. Dune plants are important in influencing the accretion and stabilisation of dune sands. The trailing vine Ipomoea and grasses like the Australian Spinifex are important in tropical regions. The best known plant on sandy tropical shores is the coconut palm. Another broad-leafed evergreen tree is Barringlonia on the Indo-Pacific shorelines.

1.3.3 Environments of the coastal zone

The global distribution of the major deltas and their location is presented in Table 1.3.

Coleman and Wright (1971) stress the point that the major components of a river form an integrated system. Changes in the catchment area and in the alluvial valley will influence the sedimentation and the hydrology of the deltaic plain and the stability of the shoreline. Galloway (1975) published a subdivision of delta types based on the regime of the delta front. In a ternary diagram, a subdivision is made in fluvial dominated, tide dominated and wave dominated deltas. In Figure 1.13, fifteen of the most important deltas are classified according to this system.

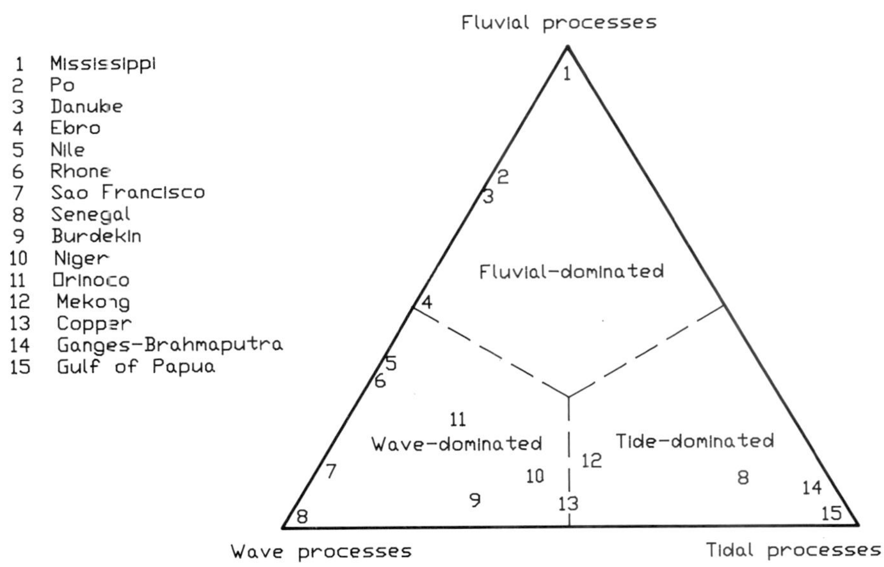

Fig. 1.13 Ternary diagram of delta types.

River	Continent	Receiving sea or ocean	*
Amazon	South America	Atlantic Ocean	31
Burdekin	Australia	Coral Sea	36
Chao Phya	Asia	Gulf of Siam	28
Colville	North America	Beaufort Sea	1
Danube	Europe	Black Sea	7
Dneiper	Asia	Black Sea	8
Ebro	Europe	Mediterranean	11
Ganges-Brahmaputra	Asia	Bay of Bengal	21
Grijalva	North America	Gulf of Mexico	18
Huang Ho	Asia	Yellow Sea	12
Indus	Asia	Arabian Sea	20
Irrawaddy	Asia	Bay of Bengal	24
Klang	Asia	Malacca Straits	30
Lena	Asia	Laptev Sea	5
Mackenzie	North America	Beaufort Sea	3
Magdalena	South America	Caribbean Sea	25
Mekong	Asia	South China Sea	29
Mississippi	North America	Gulf of Mexico	15
Niger	Africa	Gulf of Guinea	27
Nile	Africa	Mediterranean	16
Ord	Australia	Timor Sea	34
Orinoco	South America	Atlantic Ocean	26
Parana	South America	Atlantic Ocean	35
Pec	Europe	Barents Sea	4
Po	Europe	Adriatic Sea	10
Red	Asia	Gulf of Tonkin	22
Rhine	Europe	North Sea	6
Sagavanirktok	North America	Beaufort Sea	2
Sao Francisco	South America	Atlantic Ocean	33
Senegal	Africa	Atlantic Ocean	19
Shatt-el-Arab	Asia	Persian Gulf	17
Tana	Africa	Indian Ocean	32
Volga	Europe	Caspian Sea	9
Yangtze-Kiang	Asia	East China Sea	13

Table 1.3 Major deltas and their locations.

* - the number in column 4 refers to the map in Figure 1.2

Tide dominated deltas have an extensive tidal plain cut by deep tidal channels. The morphology can be described as lobate. Offshore tidal current ridges may be present. Examples are the Irrawaddy delta (Bay of Bengal) and the Yangtze Kiang (East China Sea).

River dominated deltas are able to propagate seawards. Their morphology resembles a *bird foot* as in the Mississippi delta. Due to the dominant fluvial input the tidal plain is narrow (see Figure 1.15).

A wave dominated delta is characterised by the occurrence of cheniers and beach ridges. Good examples are the Niger delta, the Grijalva delta and the Hwang Ho delta. Most deltas, however, are a mixture of the above mentioned types. The delta of the Mekong with high tides and intermediate wave energy is an example of a mixed type.

Coleman and Wright (1971) divided the components of the delta and river plain into:
- alluvial valley; braided or meandering river plain, natural levees, swamps and floods plains.
- deltaic plain; distributary levees and inter distributary and delta flank marshes and backswamps.
- tidal plain; tidal deposits and salt marshes, in the humid tropics mangrove swamps and a network of interconnected tidal channels.

Alluvial valley

This is a more or less balanced system in which the river flows over and through its own deposits. Wolman and Leopold (1957) concluded that the volume of material deposited tends to be about equal to the volume eroded. Erosion occurs during periods of vigorous flow while at other places deposition takes place. The alluvial valley is a system through which water and sediment are transported from the drainage basin to the deltaic plain.

A river channel with point-bars and levees is situated in the floodplain. The type of channel can be braided, straight or meandering as a result of changes in sediment load and large seasonal changes in discharge. The point-bars, situated at the high water level of the river, have the coarsest sediment of the system. In the non-tidal river reach, the natural levees build up to around the level of the mean annual flood. In the tidal reach, the elevation is slightly above the high tide level. These levees are built up by sediments deposited during overbank flow and crevassing. Crevassing consists of overflow and cutting of small channels through the levees resulting in the formation of fan-shaped sheets of sediments in the backswamps or flood plains. The highest elevation and the greatest width (more than 300 m) of natural levees are found at places where river stage fluctuations are greatest. Also, upstream in the alluvial valley, the elevation of levees is generally higher than downstream. A levee has an asymmetrical form. It has a steep slope to the distributary and a very gentle slope into the inter-distributary (Figure 1.14).

Broad and high levees are used for building villages and towns. In general, the elevation and the firmness of the levees are ideal for house building, road building, garden plots and orchard sites. The flood plain itself is only flooded by very high river discharges. The sediments are generally fine-grained clays and peats. In the flood plain itself, abandoned stream courses can be found at a small elevation above the plain and containing coarser sediments. In humid, tropical regions seasonal high water discharges and subsequent flooding of the flood plain results in heavy swamp vegetation in large parts of the alluvial plain. On the other hand, in arid to semi-arid, warm climates the alluvial valley is devoid of vegetation and the depressions of the flood plain are built up of salt-flats and aeolian dune fields during periods of low river discharge.

Deltaic plains and tidal plains

The apex is the point at which the alluvial valley branches into multiple distributaries. Figure 1.15 shows the different components of a deltaic plain. The lower tidal plain is influenced by daily tidal fluctuations and the upper deltaic plain is influenced by the river discharges. The most important areas are those parts of the upper and lower deltaic plain that are abandoned. These areas are more susceptible to flooding by the sea, by the river or by precipitation.

In general, the relief of the upper deltaic plain is higher than the lower deltaic plain. The levees of the distributaries, made of coarse materials, provide the high levels and the inter-distributary depressions, where organic debris and clays accumulate, the low levels. Some of these depressions may become poorly drained swamps.

The height of the levees is related to flood levels (2 to 5 years return period). The levees of the upper deltaic plain are lower than those in the alluvial valley and they are rarely more than 200 metres wide. In the lower deltaic plain, natural levees are even lower and very poorly defined, merging almost imperceptibly into the adjacent basins. Brammer (1990) describes the levees of the Ganges-Brahmaputra delta. In the lower alluvial valley the levees are 2 to 5 metres above the level of the backswamps. In the upper deltaic plain they are 2 metres and in the tidal plain less than 1 meter above the level of the backswamps.

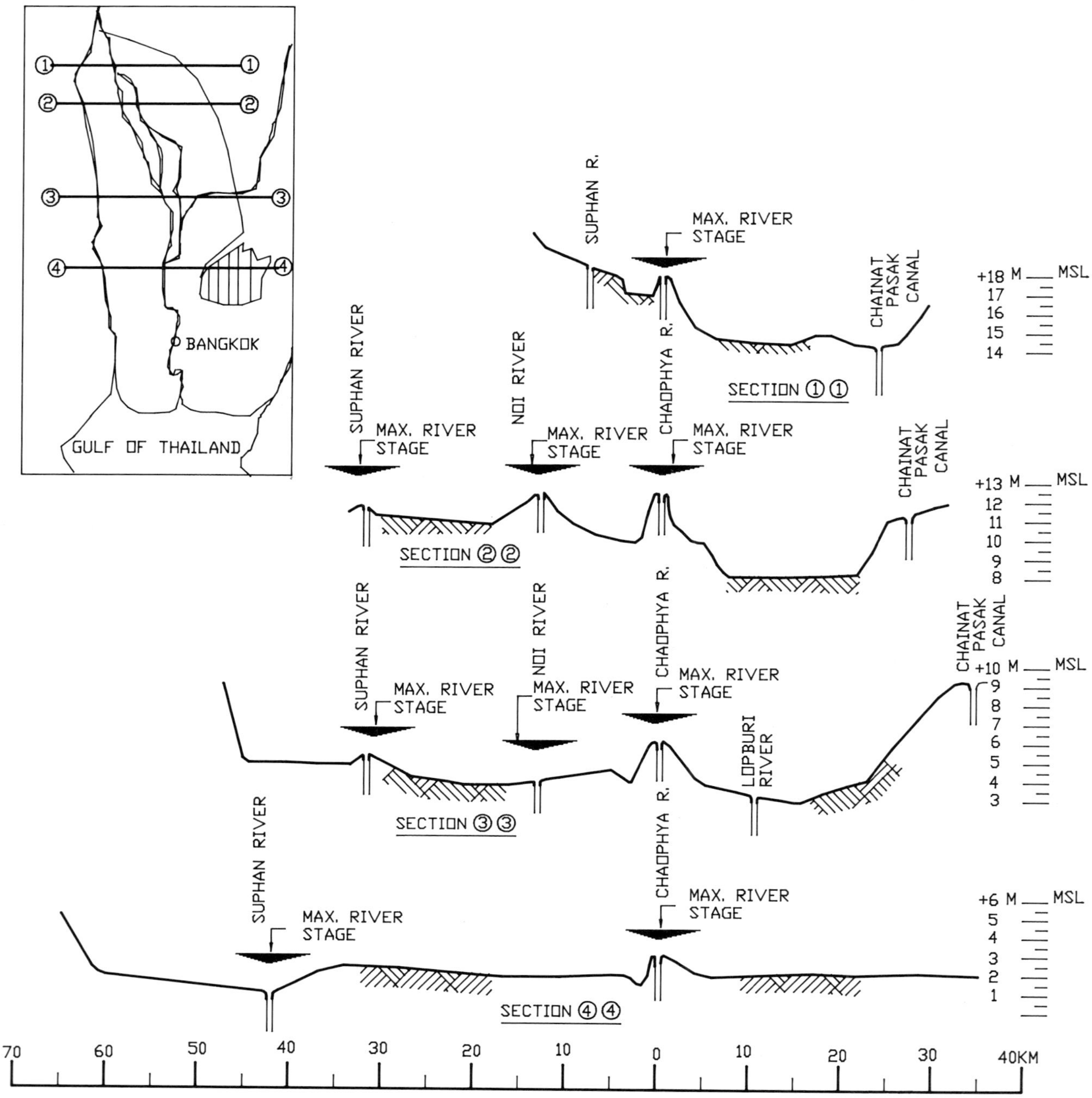

Fig. 1.14
Cross-sections of the Chao Phya detaic area.

When the tidal range is high, a strong tidal current may occur in the tidal plain. This results in the forming of interconnected channels which can be quite deep (approximately 30 metres in the Ganges-Brahmaputra delta).

Beach ridges and cheniers

In wave dominated deltas, beach ridges can develop. On the Nayarit coast in Mexico, 280 beach ridges can be found north and south of the Rio de Santiago. Other examples are the Mekong, the Grijalva and several small tropical deltas on the east coast of Western Malaysia. In many places the vertical thickness of the sand bodies in unknown. Some of them could be cheniers.

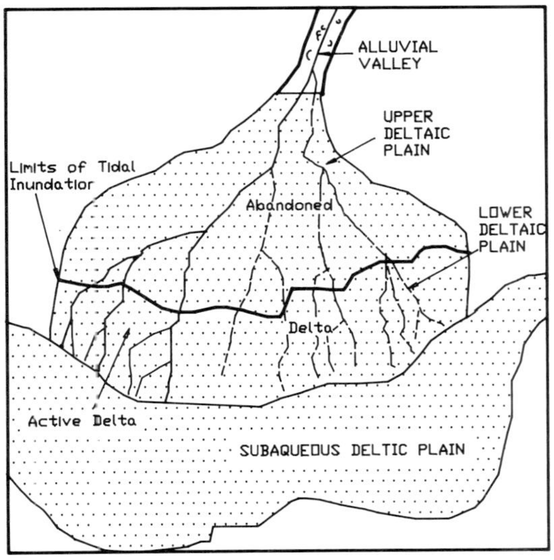

Fig. 1.15
Components of a detaic plain.

The best example of cheniers is on the plain of south-western Louisiana. These cheniers are thought to have been formed during periods of reduced river sediment influx when wave attack slowed or halted the advance of the coast and the shore retreated locally. During these periods the eroded material of the shore was sorted by wave activity. The mud fraction was suspended and the sand fraction was sorted and deposited on the shoreline. In contrast to beach ridges and barriers, cheniers are thin bodies of sand situated on mudflats. Cheniers are developed on micro and mesotidal shorelines.

Barrier islands

Barrier islands are most prevalent in coastal settings that have the following characteristics:

- a low-gradient continental shelf adjacent to a low relief coastal plain;
- an abundant sediment supply;
- moderate to low tidal ranges (micro and mesotidal);
- wave dominated.

The three main components of a barrier-island system are the barrier beach, a lagoon and a tidal channel delta complex (ebb and flood delta). A good example are the barrier islands in the north of the Netherlands. On the seaward side of the barrier islands, an extensive dune area can develop. The main tidal channels between the barrier islands is bordered by an ebb delta on the seaward side and by a flood delta on the landward side. The lagoon environment can be divided into the subaqueous lagoon, the tidal flats, the marsh and the fresh and brackish water swamp. In tropical regions there is a very distinct zonation of the mangrove swamps according to the salinity of the water.

Estuaries

Estuaries can be described as semi-enclosed bodies of water which have an open connection to the sea. Within an estuary, the salt water is diluted with fresh water originating from land drainage. In most cases, they are the tidally influenced, lower region of a river. Tidal processes are of great importance as this controls the movement and interaction of the fresh and salt water bodies in the estuary. Sedimentation and sediment transport are also governed by the tidal regime. Micro-, meso- and macrotidal estuaries can be distinguished. Macrotidal estuaries are characterised by a funnel shaped form connected on the landward side to a tidal influenced channel. The funnel shaped channel is bordered by extensive tidal flats and in the centre of the channel elongated tidal current ridges are present. Several examples of mesotidal estuaries can be found on the East Coast of the United States. The mesotidal estuaries of the Eastern Scheldt, Rhine and Meuse rivers in the South-West Netherlands are closed by dams in order to protect the hinterland from storm surges. In the German Bight, there is an extensive tidal flat which is dissected by mesotidal estuaries.

Offshore sandbars

When the tidal range is of importance (meso- or macrotidal) and sand is available, offshore sandbars can develop. Examples are the Southern Ocean, the Gulf of Papua and the Gulf of Korea. Offshore from the Ganges-Brahmaputra delta there are no sandbars present. The material required to build up sandbars disappears into an undersea canyon. Similar canyons can be found at the Niger delta and Indus delta. These canyons act as sinks for the sediment which is consequently lost from the near-shore and shoreline.

Coastal dunes

The importance of eolian deposition in coastal areas is clearly demonstrated by the size and bulk of sand dunes in many areas. Sand dunes occur where there is a large supply of sand, wind to move it, and a place in which it can accumulate. The coastline is an ideal location for these criteria to be met. Longshore transport supplies the sand and waves accumulate the sand on a beach. Onshore winds transport the sand from the beach and it accumulates in dunes. Extensive coastal dunes can be found in arid, temperate and humid areas. In some places, the heights of the coastal dunes can exceed 100 metres (e.g. South West France and the Oregon coast of the USA), but in general they are between 10 and 30 metres in elevation. The only climatic areas lacking extensive coastal dunes are the humid-tropical. This may be related to a lack of sand supply because of intense chemical weathering, higher shear stresses required to move almost continuously wet sand, and dense vegetation adjacent to the beach.

Two main types of coastal sand dunes are:
- vegetated dunes (i.e. fixed);
- transverse dune ridges (generally migrating and bare of vegetation).

This basic coastal dune classification can be refined into four types: Vegetated, Artificially induced, Madanos (transverse dune ridges) and Parabolic dunes. This is called the VAMP classification. Vegetated dunes commonly consist of stabilised parabolic dunes. Transverse dune ridges or migrating dunes are characterised by a lack of anchoring vegetation. They move generally landwards in response to the prevailing winds. Precipitation dunes, a form of transverse ridges, deposit sand in front of and often migrate over houses, roads, forests and marshes. Large transverse ridges can take the form of isolated birchen dunes. Parabolic dunes form behind the fore dune (or transverse dunes) under the increasingly stabilizing effects of vegetation. The morphology of parabolic dunes differs from birchen dunes in that the former are open to the upwind direction and the latter are open to the downwind direction.

Under tropical conditions, sand composed of calcium carbonate, instead of quartz, is common. Calcium carbonate sand, acted on by the wind, will produce a distinct type of coastal dune referred to as aeolianite, or aeolian calceranite. These types of dune sand easily become lithophyte under tropical conditions. They are called dune limestones.

The coastal dune area is also an important source of fresh water. In most dune areas a fresh water lens occurs which floats on the brackish and saline water. Another important function of coastal dunes is the natural defence against the sea for the low-lying backswamps and hinterland.

1.3.4 Subsidence

In many deltaic and other low-lying coastal areas, the land is sinking with respect to the sea-level. This phenomenon is due to a rise of the sea-level (section 1.2.1) and a subsidence of the land surface. Jelgersma (1988) gives a comprehensive description of the phenomenon of land subsidence in coastal lowlands. A summary of the most important natural and man-induced factors in this process is given below.

Sea-level rise and land subsidence have serious consequences for the protection against the sea floods and for the water-management of the low-lying coastal areas. These aspects will be dealt with in later.

Natural subsidence

Natural subsidence can result from tectonic movement, compaction by sediment during loading, oxidation of peat, vibration (earthquakes) and lowering of the groundwater table (dry periods or a drop in the sea-level).

Tectonic subsidence consists of basement subsidence enlarged by subsidence caused by isostatic loading due to the sediments deposited in the basin.
Not much is known about the amount of tectonic subsidence in sedimentary basins. One method that can be used to give an impression of the subsidence is to compare the level of a given coastal deposit in the basin with the same deposit outside the basins. The difference in depth indicates the differences in tectonic subsidence during the given period. By doing so the tectonic subsidence in the basin generally is not more than a few centimetres per century. In the Netherlands the depth of the marine deposits from the last interglacial (Eemian) are 8 m below mean sea-level and in North West France they are 8 m above mean sea-level. The latter areas are thought to be rather stable. Apparently a change in level of 16 metres has occurred during the last 120,000 years between The Netherlands and France. This indicates a tectonic subsidence of 1.4 centimetres per century in the Dutch part of the North-Sea basin (Zagwijn, 1983). This method assumes that tectonic subsidence is a continuous process which may be not true. Evidence from the sedimentary basin of The Netherlands indicates that there are areas of maximum subsidence, the so-called depot centres (Zagwijn 1989). Some sedimentary basins show more tectonic activity. An example is the Ganges-Brahmaputra basin, where faults are active and earthquakes occur.
Fine-grained sediments, particularly clays, can show the effects of natural loading. Recent mud deposits with a clay content of more than 40% have porosities of about 85%. The effect of loading reduces the porosity to 40-50%. Subsequently, the density increases and the volume decreases. An example of compaction due to loading is the site of New York's La Guardia Airport. The natural compaction of an 18 kilometre thick water saturated organic silt and clay deposit was accelerated by placing 7.5 metres of fill material on top of the soft sediment. After 25 years the land subsided 2.5 metres due to compaction of the underlying clay deposits. Vibration of soft fine-grained sediment during earthquakes can cause serious compaction due to the decrease in porosity and corresponding increase in density.

Lowering of the groundwater table during dry periods or lowering of sea-level will reduce the porosity of the drained sediment and subsequently cause compaction and surface lowering. The surface peat layers may also oxidise when the swamp dries out. This oxidation results in surface lowering.

In the delta of the Rhine and Meuse well-founded data are available with respect to the so-called "relative land subsidence" (RLS). This is defined as the subsidence of the top of the Pleistocene with respect to the Mean Sea Level. The RLS is a natural phenomenon explained for this region by the eustatic sea-level rise, the geosynclinal subsidence, and the isostatic compensation after the last glaciation period. The rate of RLS is not constant with time. The present rate along the coasts of the Netherlands can best be ascertained from observations of the tidal levels on 10 tidal gauges. The recording started around 1860. The annual averages of the mean sea-level show a rising trend in all stations but there are significant differences between the variations in time which cannot be fully explained. The overall average is a rate of RLS of 15 to 20 centimetres per 100 years. This figure is only valid for the coastal areas of the Netherlands along the North Sea as explained in Section 1.2.2.

Man-induced subsidence

Man-induced subsidence can result from irrigation, land reclamation and extraction of gas, oil and water. When water-saturated, fine-grained sediments are irrigated or reclaimed the lowering of the water-table stimulates compaction of the sediments with subsequent subsidence of the surface. In the Netherlands, Bennema et al (1954) have done extensive research on compaction of reclaimed land. Clays with 35% finer than 2 µm have a porosity of more than 80% and they will be compressed to about half of their original thickness after reclamation over a period of one year.

Peat will be compressed much more; to about one ninth of its original thickness. Holocene sands are less porous and are not subject to much compaction.

Striking differences in compaction can be seen in land which was reclaimed centuries ago, where peat layers have been eroded by creeks. The latter have been filled up by sand and silt and now emerge up to 2 metres above the adjacent part of the landscape due to compaction of the peat and clay (Figure 1.16).

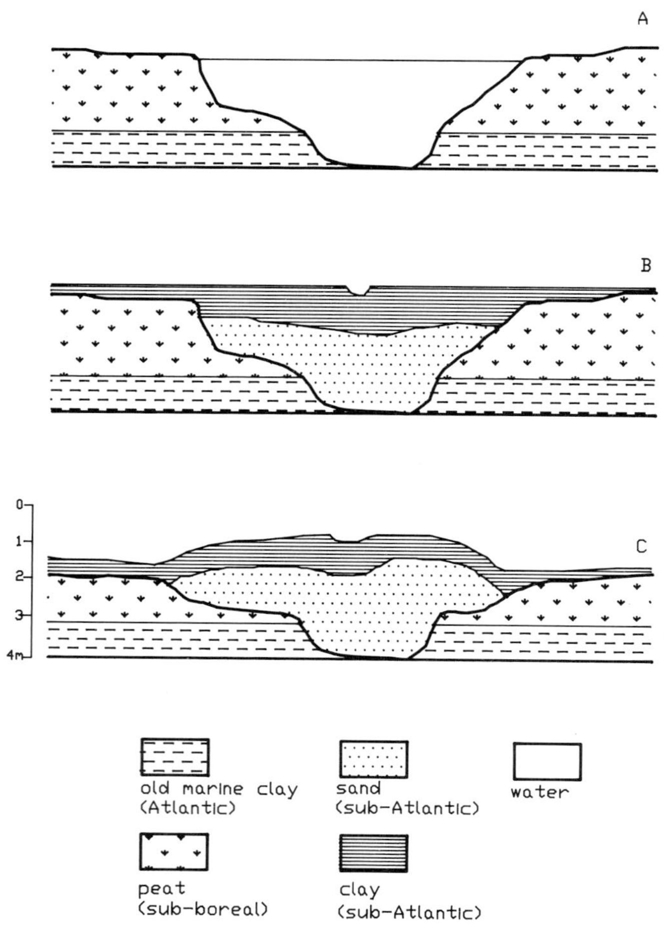

Fig. 1.16
Formation of a gully ridge,
a) during marine ingression
b) after filling of gully and
c) after compaction.

After World War II, new polders were reclaimed in The Netherlands. To forecast the subsidence of the surface due to reclamation, consolidation constants were derived from the pore space and the thickness of the sediment before reclamation (de Glopper, 1969, 1973). Before the polder was pumped dry, permanent measuring sites were installed to observe the regional pattern of subsidence after reclamation. In 25 years, the observed subsidence varied between 10 and 100 centimetres. This can be related to the thickness of the Holocene layer and its composition. Due to poor drainage conditions compaction during the first 10 years was much lower than in the following 15 years.

A serious, man-induced geological hazard is land subsidence due to the extraction of groundwater in areas underlain by Cenozoic basins. Detailed information about this problem can be obtained from the proceedings of various international symposia on land subsidence (IAHS, 1969, 1976 and 1984) and the UNESCO Guidebook to studies of land subsidence due to groundwater withdrawal

(1984). Surface and sub-surface layers in coastal areas consist of young unconsolidated material deposited in alluvial, lacustrine and shallow marine environments. The permeabilities may vary considerably depending on the lithology of the layers. Subsequently, sub-surface layers generally consist of confined or semi-confined aquifers of gravel and sand with high permeability and low compressibility, alternating with aquitards of clay with low permeability and high compressibility. Groundwater extraction will cause a drop in the artesian head which results in compaction due to an increase in the effective stress. The compaction of these aquifers is immediate and is mostly recoverable if fluid pressure is restored. This compaction, however, is small compared to the compaction that occurs in the aquitards due to pore-pressure reduction in these fine-grained sediments. A drop in head in the aquifer creates a hydraulic gradient from the aquitards to the aquifers. The fine-grained, clay beds have low hydraulic conductivity and high compressibility. Therefore, the vertical escape of water and the consequent decrease in pore pressure and increase in effective stress are slow and time dependent. The resulting compaction is large and irreversible. The stresses at the interface of the aquifer and aquitard are shown in Figure 1.17.

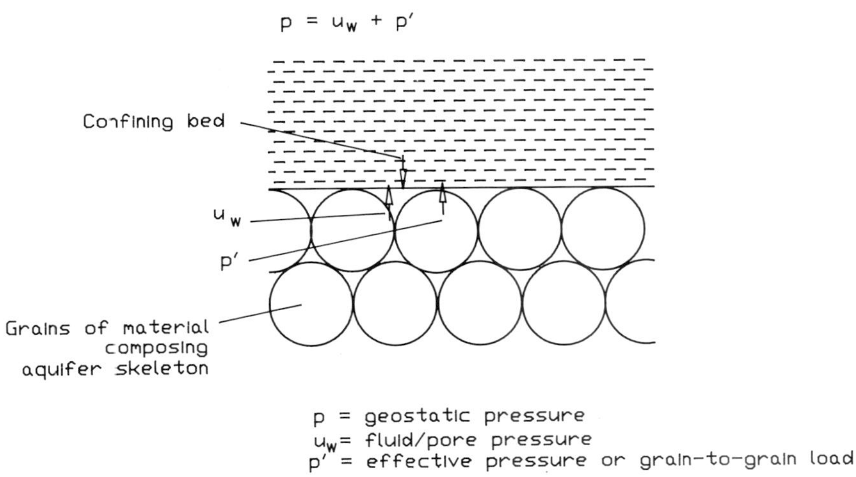

Fig. 1.17
Stresses at the interface between an aquifer and aquitard.

The biggest problem in coastal areas is that of tidal encroachment associated with subsidence due to groundwater extraction. This can cause salt-water intrusion in the aquifer and flooding of settlements. Other problems are differential changes in elevation and gradients of channels and drains and failure of pipelines. In water wells, casings may fail due to compressive stresses caused by compaction of aquifers. Over-extraction of groundwater can occur in deltaic areas with heavy population and intense industrial development. Increased subsidence may then necessitate the construction of an extensive system of dykes, flood-walls, locks and pumping stations for protection from river and marine inundation. The costs of such hydraulic engineering works can be very high. In the USA, subsidence in the Houston-Galveston Region has amounted to as much as 2.8 metres since 1915 in the area of highest groundwater extraction. As shown in Figure 1.18, there is a close relation between the observed drop of the water-table in the wells and the observed subsidence of the bench marks.

Due to increased subsidence, some areas are now subject to inundation by normal tides and an even larger area is subject to catastrophic flooding by hurricane tides. In Japan, the Nobi Fluvial and deltaic plain is underlain by more than 2000 metres of young clastic sediments. Groundwater extraction has caused a lowering of the land surface of between 20 centimetres and 140 metres in 15 years. Due to this subsidence, 248 square kilometres of the area is now below mean sea-level. To stop this disastrous situation withdrawal of groundwater had to be severely regulated. In Taiwan, land subsidence in the Taipei river basin amounted to 2 metres during 22 years of heavy

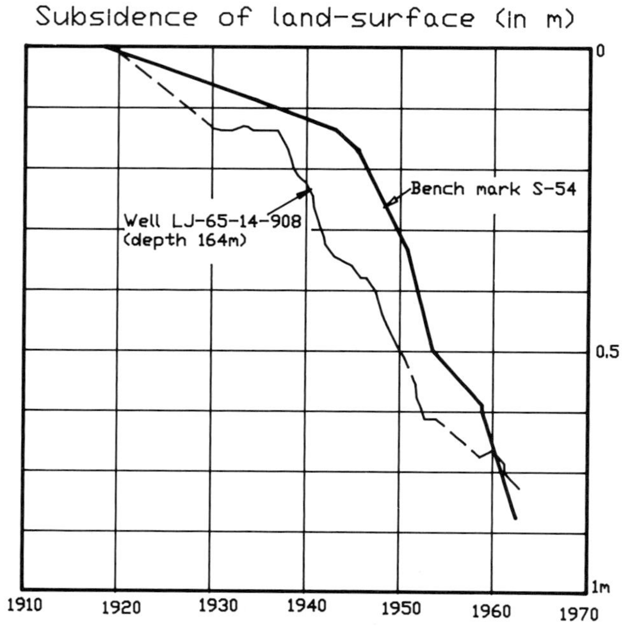

Fig. 1.18
The relation between water-level decline and land subsidence, Houston, Texas (UNESCO,1984).

groundwater extraction. This situation has hampered the normal drainage of heavy rainfalls which now cause severe flooding (Figure 1.19).

Controlling man-induced subsidence

In coastal areas with heavy groundwater extraction it is advisable to have an annual geodetic survey to monitor ground movements and to install a network of piezometers to check pressure heads in the exploited aquifers. Measurements of compaction in the aquifer itself can be done by extensometers in boreholes. Investigations of permeability, grain size, clay chemistry and compressibility of aquifers and aquitards should be made to identify the physical properties of the surface formations. These data can be used to build a numerical model for the prediction of compaction of the groundwater reservoir caused by the extraction. Based on the results of these investigations withdrawal of groundwater in coastal plain aquifers should be controlled by the authorities concerned.

Methods to arrest or control subsidence are generally quite costly and require the development of organisational and legal instruments. Whether any of the methods mentioned below can be economically justified will depend upon its implementation costs compared with the costs of continued or even increased land subsidence.

One method to control subsidence is to reduce the amount of pumping. Groundwater extraction can be reduced by regulations. Water demands can be met by importing surface water. Examples are the major aqueduct that carries Colorado River water to South-Central Arizona, and the aqueduct near Venice (Italy) that supplies the industrial area with river water. Reduction of pumping also can be facilitated by using surface water for irrigation and reusing treated water in industrial plants.

In many places groundwater extraction is concentrated in certain areas. It is advisable to spread the pumping over a much larger area in order to reduce land subsidence. Compaction due to a drop in the hydraulic head in an aquifer can be reduced by restoring this head through artificial recharge in the outcrop area of the aquifer. Generally this is located at the edge of the sedimentary basin. Repressuring of confined aquifer systems by artificial recharge directly through wells, although expensive, is another possibility to slow down or stop land subsidence. An example is the Wilmington oil field in Southern California, U.S.A. where subsidence was controlled by injection

Land subsidence in Taipei Basin

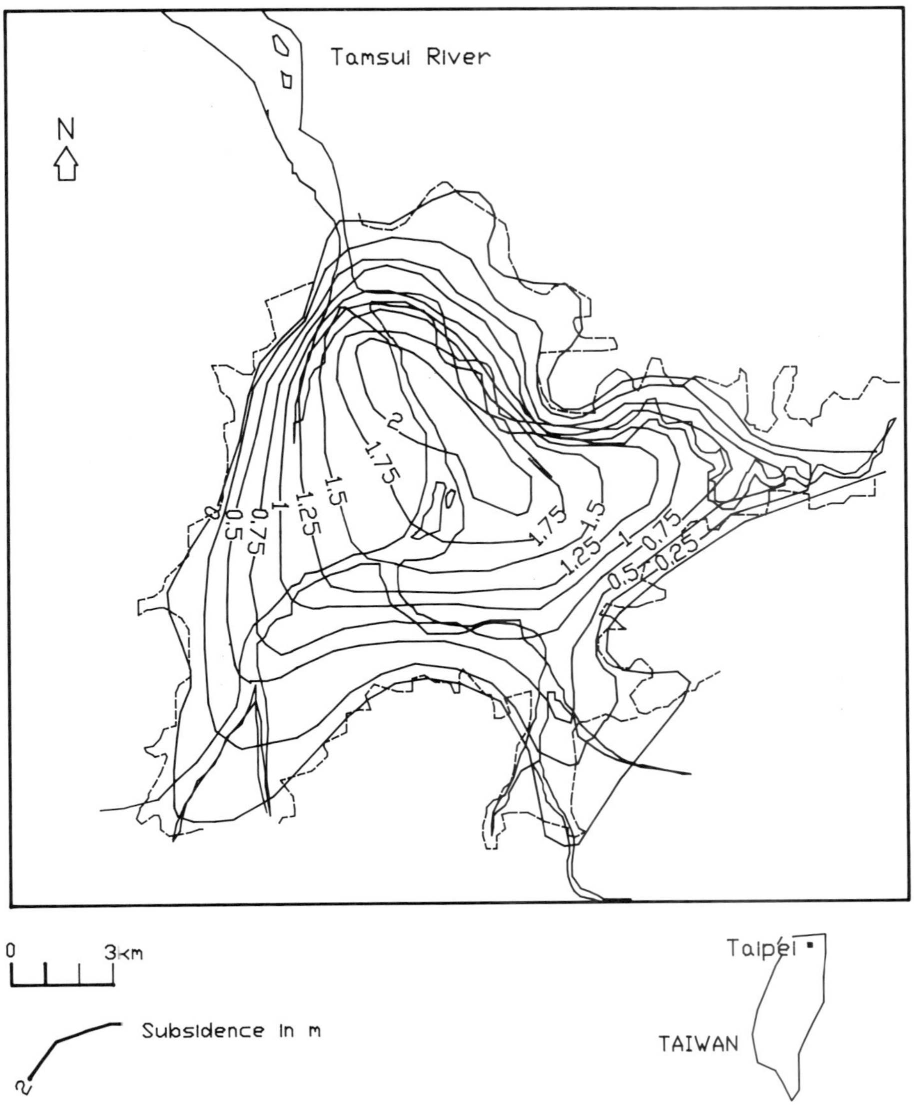

Fig. 1.19
Land subsidence in the Taipei Basin due to groundwater extraction (UNESCO, 1984).

of water through wells. In this case, the area of subsidence was reduced from 58 to 8 square kilometres. In some places, the land surface has also rebounded by as much as 30 centimetres (Mayuga and Allem, 1969). One problem with the artificial recharge of aquifers is that the water must be very clear, otherwise clogging of wells and aquifers may occur.

1.4 Soils of Deltaic Areas and their Agricultural Potential

1.4.1 Soil forming factors

Soils, forming the uppermost part of the earth's crust, are the result of interactions between parent material, climate, relief, hydrology and organisms (including man), over time. The interactions between these soil forming factors are complex and their relation to soil properties is indirect, being governed by many chemical, physical and biological processes. The various soil forming processes to be considered here are:

- parent material;
- climate;
- relief;
- hydrology;
- organisms;
- man;
- time.

In the following sections, the soil forming factors and processes are described in relation to their effect on the properties of the soils of deltaic areas.

Parent material

Parent material is, by definition, the material from which the present soil is derived. Generally, the parent material of soils formed in situ is the underlying unweathered rock. In deltaic areas however, this is not the case. Here, the soils have been formed mainly from clastic sediments, i.e. materials which have been disrupted, weathered and transported individually from their places of origin. These materials include marine sediments (tidal sea deposits), fluviatile sediments (tidally-influenced or non-tidal river deposits) and eolian sediments (wind deposits). In addition, some soils in deltaic areas may have developed from organic materials which accumulated in poorly drained or submerged areas (peats and mucks).

The mineralogical composition of the sediments in deltaic areas is largely determined by the composition of the soils and the weathering materials in their catchment areas. These, in turn, are determined by the lithology of the catchment and, through its effect on weathering, by the climate. Sediments originating from highly weathered soils (e.g. Ferralsols or Acrisols; FAO/ UNESCO/ ISRIC, 1988) as formed under conditions of high rainfall and temperatures are relatively rich in quartz, kaolinite and some resistant minerals (orthoclase, muscovite, epidote, zircon), but poor in terms of chemical content. In arid and semi-arid climates chemical weathering and leaching have generally not been as intense and sediments tend to be richer. Depending on the lithology of the catchment they may contain variable amounts of feldspars, micas, hornblende and even biotite, an easily weathered mineral. Lime and gypsum may also be present.

The simplified picture outlined above does not strictly apply, of course, to the complex situation actually existing in catchments of large river systems. Such systems (e.g. Mississippi, Nile, Niger, Ganges and Mekong), flow through lithologically widely varying regions and/or through rather contrasting climatic zones.

The parent material of soils in beach ridges and sand bars at the shoreline of deltaic areas may originate from other coastal deposits brought in by longshore drift. Wind deposits (dunes) are mainly of a local origin having been transported over short distances only. Transport occurs in various ways depending on the velocity of the transporting agent (water or wind) and on the size and specific gravity of the particles. The largest particles (boulders, gravel and sand) are transported either by rolling or dragging along the riverbed or landsurface, in the case of water and wind respectively. Smaller particles (sand and silt) move in saltation, while the smallest particles (fine silt, clay and organic substances) are carried in suspension. In addition water can transport various compounds in solution (e.g. salts, organic acids, etc.). Obviously, higher stream velocities are required for the movement of larger and heavier particles. For example, in water a velocity of approximately 20 cm/s is required to initiate movement of fine sand (diameter 0.1 mm), whereas almost 100 cm/s is necessary for the displacement of gravel of 1 cm diameter. Conversely, at decreasing velocities the transporting capacity will also decrease and consequently the particles will be deposited according to their size and weight. In general, this selective process causes a differential sedimentation, or *sorting*, of the particles according to their size. This explains why in a river system, generally speaking, coarser sediments are found upstream, where higher stream velocities prevail, whereas finer sediments occur in the downstream areas.

Sorting also occurs in local sedimentary systems, giving rise to the formation of lighter textured sediments adjacent to the river bed (levees). Here, stream velocities are usually high in comparison with those of the basins and backswamps, which form at a greater distance. The latter are comprised of fine textured sediments due to the deposition of smaller particles from slowly moving or even stagnant water. Sedimentation is also affected by vegetation, which reduces the stream velocity. In general more clayey materials are deposited in overgrown areas in comparison with bare sites.

Not only does the particle size of sedimentation vary from place to place in river-plains, deltas, etc., it may also vary significantly in the vertical direction. This stratification is caused by periodic changes in water movement like seasonal and tidal floods, changes in the river course, etc. In tidal areas, the water is most restless at low places. With the exception of the turn of the tide, the water here is always moving. With an increasing height of deposition, the duration of flooding becomes shorter and the flow becomes more still. In such a system, the particle size of the deposits gradually decreases in an upward direction through the profile.

Climate

Climate directly influences soil formation. The rate of weathering, i.e. the chemical and physical processes whereby rocks on exposure to the weather change into soil, is directly determined by the soil temperature. Leaching, i.e. the vertical removal of weathering products and other soil constituents, is mainly determined by the volume of water passing through a profile which, in turn, is mainly a function of rainfall. In deltaic areas however, climate has mainly an indirect effect on the parent material through its influence on weathering preceding the transportation and sedimentation. A direct effect of rainfall on soil formation is that of leaching (decarbonisation and desalinisation) in the young soils. In dry and warm climates salinity can build up at or near the soil surface as a consequence of evaporation and the capillary rise of saline groundwater.

Soil ripening, i.e. the irreversible extraction of water from fresh, unripe sediment, is greatly affected by climate. High evaporation in dry conditions causes stronger ripening and at a faster pace than in humid zones.

Another indirect effect of climate is through its influence upon organisms, and particularly upon vegetation. Biological processes are generally dependent on temperature. This applies to the synthesis of organisms (plant growth), as well as to decomposition and subsequent mineralisation of organic matter. Climate, or rather seasonality, affects the discharge pattern of rivers and, as such, it has an effect on physiography. In per-humid climates river discharges are generally continuously high, creating less pronounced physiographic differentation than under alternating wet and dry seasons.

Relief

Relief, or the shape of the land surface, is a major soil forming factor, even in deltaic areas which have only little relief. Level or low-lying areas may be flooded during heavy rainfall, when adjacent rivers overflow their banks, or at high tides. The higher parts of the land may remain above flood levels. Such differences in relief cause different wetness regimes in the soil profile. These in turn, give rise to different chemical processes in the soil, in particular the oxidation and reduction of iron components. Under saturated conditions reduced iron compounds, which are easily soluble, prevail in the soil. They give rise to the blueish and greenish grey colours characteristic of reduced soils. Under aerated conditions, ferric iron compounds are formed. These are non-soluble and they cause yellowish brown and reddish soil colours. Alternating conditions for oxidation and reduction cause both, grey and brown/red colours which are usually mottled. Relief in deltaic areas is the natural expression of the geomorphologic processes which have taken place. The following land facets or geomorphologic components can be distinguished in deltaic areas: river levees, point bars, alluvial flats, river basins, oxbow lakes, creek ridges, tidal flats, salt marshes, sandbars, beach ridges and dunes, and terraces. These are all distinct geomorphologic units with their own characteristic soil conditions.

Hydrology

In deltaic areas, where many soils are subject to frequent flooding or waterlogging, hydrology is the dominant soil forming factor. Gleying, the alternating reduction and oxidation of iron into ferrous and ferric forms, is not particularly dependent on temperature. Therefore, soils with impeded drainage look alike in the tropics and in temperate regions. Temporarily water-logged horizons are mottled, and permanently saturated layers are uniformly bluish or greenish grey. In the seasonally wet tropics however, soils dry out to a greater extent and to much greater depths than in temperate areas, while ferric iron tends to concentrate into hard concretions rather than in mottles as is common in the cooler climates. In poorly and imperfectly drained soils in arid and semi-arid conditions salinity can build up at or near the soil surface due to evaporation and the capillary rise of saline groundwater.

Soil drainage classes, a concept widely used in soil survey, describe the moisture condition of the soil as a whole, i.e. the net effect of internal and external drainage (Table 1.4). A distinction can be made between external and internal drainage. External drainage refers to the depth and fluctuation of the groundwater table and the capacity of the site to remove excess surface water. Internal drainage is the capacity of the soil to remove excess water vertically through the profile, which is determined by the permeability of the soil. The occurrence of impeding layers in a soil may give rise to perched water tables. External drainage as defined above is largely determined by the topographic position of the soil in the landscape. Soils of deltaic areas are obviously low-lying and to a large extent they are subject to seasonal flooding by rainwater and/or bank overflow. Soils near the coast are subject to diurnal (or semi-diurnal) tidal flooding. These tidal floods can be freshwater floods occurring as a result of the river water being pushed up by the incoming tidal water. Closer to the sea, however, the tidal floodwater is saline.

Drainage class	Description	Occurrence
Very poorly drained	Water table at the soil surface for most of the year; frequent standing water; grey soil colour.	Low-lying sites; depressions; infilled oxbows; swamps; marshes.
Poorly drained	Water table at or near the soil surface for much of the year; gley mottles in the topsoil.	Level and slightly depressed sites; river basins; tidal flats.
Imperfectly drained	Water is removed slowly from the profile; soil wet for significant periods; gley mottles in the B horizon.	Lower transitional sites; soils with moderately deep impermeable layer; alluvial and tidal flats.
Moderately well drained	Water is removed somewhat slowly from the profile; soil wet for short periods; gley mottles in the subsoil.	Higher transitional sites; soils with deep impermeable layer; lower parts of levees; tidal flats.
Well drained	Water is removed freely from the profile but not rapidly; no mottles in the profile.	Relatively high sites; river levees; creek ridges.
Excessively drainend	Water is removed freely and rapidly from the profile.	High sites; sandy deposits; dunes; point bars; beach ridges; sand bars.

Table 1.4 Soil drainage classes - description and occurrence in deltaic areas.

Organisms

Plants and animals affect soil formation in various ways. Their remains are decomposed microbiologically to humus. Through a mineralisation process, nitrogen (as ammonia), phosphorus (as phosphates), sulphur (as sulphates), potassium, calcium and magnesium are eventually released from the humus. The rates of decomposition and mineralisation largely depend on the climate and hydrologic conditions, the pH of the soil, and the stability of the organic material. Under conditions of excess wetness and poor aeration, as common to deltaic areas, decomposition and mineralisation are slow, which may result in the accumulation of organic material and humus. The latter give topsoils their characteristic dark colours. Under permanently saturated conditions, decomposition of organic matter may be so slow that it accumulates as peat or muck. Humus, consisting of colloidal humic substances, has a large specific surface which is able to adsorb water molecules and ions. The cation exchange capacity of humus is particularly high, considerably higher than that of clay. Together with clay, the humic substances in soils largely determine its structure and its water holding capacity.

Many soil animals (e.g. earthworms, millipedes, termites, etc.) live on plant remnants and other organic matter. Their movement through the soil disturbs the soil matrix in many ways. It disrupts the stratification in fresh alluvium, it creates holes which allow transport of water, air and soil material, and it mixes mineral particles with organic matter. A particular form of disturbance by animals in the coastal environment is the formation of mounds and tunnels by mud lobsters and crayfish.

Man

Human activity plays an important role in the soil formation in deltaic areas. Natural vegetation is widely removed by man to make way for crops. This not only occurs in deltaic areas where it considerably reduces the nutrient cycle, but also in the upland areas of the river catchment. Here, it causes accelerated erosion which results in increasingly severe flooding and sedimentation downstream in the floodplains, deltas and estuaries. Increased stream velocities also lead to an increase in bank erosion along major rivers.
Cultivation has a pronounced influence on the physical properties of soils. Tillage changes the structure and porosity of the topsoil. Ploughing, especially for wetland rice *(puddling)*, forms a ploughpan near the soil surface. Large scale intervention by man may alter soil properties even further. Drainage, for example, will generally cause deeper development (ripening and oxidation) of the soils, resulting in a high permeability. Drainage of soils that contain pyrite, as is common to many coastal swamps under reed vegetation (temperate climates), or mangroves (tropics), may cause extreme acidification due to the formation of sulphuric acid *(acid sulphate soils)*. Inadequate drainage in the irrigation schemes of drier climates may give rise to salinity and alkalinity problems due to accumulation in the soil, of salts present in the irrigation water.

A specific use by man of soils in deltaic areas is the excavation of clay for making bricks or fine ceramics, and peat for fuel.

Time

Time acts on soil formation differently from the other soil forming factors. It affects soil properties through its effect on the stage of development rather than on the type of processes. Soils of deltaic areas are relatively young, or *immature*, soils. Insufficient time has passed since their deposition for distinct horizon differentiation by translocation processes to be made. Also, because of ongoing flooding and sedimentation, they are rejuvenated periodically over large areas. Their relatively young age does not imply that these soils invariably have large mineral reserves. To the contrary, the fact that their parent material has gone through at least one weathering and erosion cycle renders them only relatively fertile, especially in humid tropical areas. Exceptions to this very general rule exist of course, where recent volcanic material or other fresh and rich mineral compounds are mixed with the sediments.

1.4.2 Soil forming processes

The soils of deltaic areas are relatively young soils with little profile development. Although most of these soils are medium to fine textured (silt loams to clays), textural variations are large. Heavy clays may occur in low-lying physiographic positions such as basins and tidal flats whereas coarse sands may occupy point bars and beach ridges. Drainage conditions also vary significantly according to physiographic position. Profile development, reflecting the soil forming processes which have taken place in the soils, is mainly confined to:

- physical processes like soil ripening and subsequent structure formation, shrinkage and subsidence;
- chemical processes like oxidation (change of colour and formation of acid sulphate soils), and leaching (decalcification and desalinisation);
- biological processes like homogenisation.

Physical processes

Physical ripening is the process of water extraction from fresh, soft sediments (Pons and Zonneveld, 1965). This dehydration can be due to drainage under gravity, evaporation of water from the soil, and water uptake by plant roots. The uptake of water by plants is particularly important. In fact, physical ripening is hardly possible without vegetation, especially in temperate climates. This is because of the very impermeable nature of the unripe layers, even if the clay contents are low. The water extraction causes the collapse of the initial open structure formed by the laminar shaped clay minerals. In fresh sediments these clay minerals stick together due to the electrical forces at the contact points (Figure 1.20).

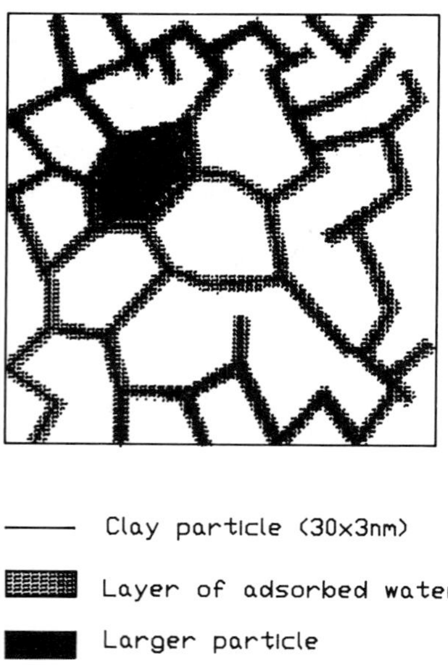

Fig. 1.20 Clay minerals in fresh (unpire) sediment: a schematic cross section (Schothorst and Broekhuizen, 1990).

Ripening is an irreversible process because once the initial open structure has collapsed, it cannot be restored. The ripeness stage of the soil can be expressed by its n-value, i.e. the quantity of water absorbed by the clay (Pons and Zonneveld, 1965). This index can be calculated as follows:

$$n = (A-0.2R)/(L+bH)$$

where:

A	is the water content (expressed as a percentage of the dry soil);
R	is the sand and silt content;
L	is the clay content;
H	is the content of organic matter;
b	is the ratio of water absorption of organic matter to clay (usually in the range 3 to 5, depending on humification status of the organic matter).

The n-value of soil material can be assessed in the field by means of a simple squeezing test (Table 1.5).

Ripeness class	n-value	Consistency*, Squeeze test	Water content(%)
ripe	< 0.7	firm(wet), hard(dry), not sticky; cannot be squeezed through the fingers.	< 50
nearly ripe	0.7 - 1.0	fairly firm, slightly sticky; can be squeezed with difficulty through the fingers.	50 - 60
half ripe	1.0 - 1.4	fairly soft and sticky; can be squeezed through the fingers.	60 - 70
practically unripe	1.4 - 2.0	soft, very sticky; can easily be squeezed through the fingers.	70 - 80
unripe	> 2.0	very soft to liquid; flows between the fingers without squeezing.	> 80

Table 1.5 Classification of soil material according to physical ripening (Pons and Zonneveld, 1965).

* - descriptions of consistency refer to the average clay content (20-50% clay) and mixed mineralogy.

An important effect of soil ripening is the formation of structure elements through irreversible vertical and horizontal cracking of the sediment. This effect is most pronounced in clayey sediments and in materials rich in organic matter. The process enhances soil permeability and water storage capacity. The extent and kind of cracking also depend on the clay mineralogy and on the cations adsorbed. High clay contents, especially of expanding clay types (e.g. montmorillonite and smectite) and high sodium adsorption, give a coarse cracking pattern (prismatic and columnar structures), low contents of non-expanding clays (kaolinite) and calcium adsorption give finer elements (blocky and granular structures).

Mainly as a result of the loss of water through ripening, but also because of settlement (i.e. the compression of soil material due to its increasing specific weight upon water loss and, to a lesser extent, because of oxidation of organic matter) the soil volume decreases and the land subsides upon drainage. Clayey sediments subside more than sandy deposits, due to the greater amounts of water which the former contain initially. This results in the enhancement of height differences between the lower-lying clayey basins of the river flood-plains and the more elevated, coarser textured levees and creek ridges.

Chemical processes

The entrance of oxygen into the soil upon drainage initiates several chemical processes. Organic matter is rapidly oxidised and disappears partly. In the process, inorganic compounds are liberated which either are leached, or they react with other compounds. Oxidation of ferrous-iron present in the organic material in reduced conditions results in the formation of ferric (hydr-)oxides which, due to slight differences in water content or amounts of iron and organic matter present, have yellow, brown or reddish colours. These appear as mottles in the soil profile and reflect the hydrological condition of the soil such as fluctuations of the groundwater table.

A specific chemical process that may occur in marine sediments, is the oxidation of pyrite (FeS_2) to sulphates and sulphuric acid (H_2SO_4). This may cause extreme acidification in what is referred to as acid sulphate soils. Pyrite can be formed microbiologically in sediments, under conditions of flooding with saline or brackish water, in the presence of large amounts of organic matter. Such flooding regimes occur in tidal swamps. The required high contents of organic matter are associated with specific types of vegetation like reeds *(Phragmites spp.)* in temperate climates, and mangroves *(e.g. Rhizophora and Avicennia spp.)* in the tropics.

Pons et al. (1982) present the overall reaction equation for pyrite formation as follows:

$$Fe_2O_3 + 4SO_4^{2-} + 8(CH_2O)_n + 1/2O_2 \rightarrow 2FeS_2 + 8HCO_3^- + 4H_2O$$

The oxidation process of pyrite is complex and takes place in several stages, involving chemical as well as microbiological processes. Van Breemen (1982) presents the following overall oxidation reaction:

$$FeS_2 + \frac{15}{4}O_2 + \frac{7}{2}H_2O \rightarrow Fe(OH)_3 + 2SO_4^{2-} + 4H^+$$

Most of the colloidal $Fe(OH)_3$ formed ultimately crystallises as reddish brown iron oxide (goethite) in mottles, coatings and nodules in the soil. Under strongly oxidising, severely acid conditions, jarosite is formed from oxidising pyrite (Dent, 1986), according to the following reaction:

$$FeS_2 + \frac{15}{4}O_2 + \frac{5}{2}H_2O + \frac{1}{3}K^+ \rightarrow \frac{1}{3}KFe_3(SO_4)_2(OH)_6 + \frac{4}{3}SO_4^{2-} + 3H^+$$

Jarosite ($KFe_3(SO4)_2(OH)_6$) precipitates as straw-yellow mottles, pore fillings and coatings, characteristic of young acid sulphate soils. Acid sulphate soils are problem soils. If oxidised, toxic amounts of aluminium, hydrogen, manganese and possibly ferric-iron prevail at low soil pH. This, together with the resulting low availability of phosphorous, restricts and inhibits plant growth. Root development may also be affected if acid layers occur at shallow depths in the soil. This may not only cause water stress to crops but also restricts further ripening of the soil. Under flooded conditions (e.g. in ricefields or in fishponds), toxic levels of ferrous-iron, hydrogen sulphide and carbon dioxide prevail.

Leaching of calcium carbonate (decalcification) takes place under the influence of rainwater or as a result of seasonal flooding. Rain dissolves carbon dioxide in the atmosphere and, once in the soil, it may take up organic acids. The dilute acid solution thus formed can dissolve lime present in the profile. This process can be important on the higher parts of the coastal environment. In the lower parts, seasonal floods may temporarily cause reduced conditions in the topsoil. Anaerobic decomposition of organic matter may then result in accumulation of carbon dioxide. Soluble calcium bicarbonate is formed and subsequently removed by the retreating floodwater.

Decalcification is a rapid process, to the extent that calcareous sediments may be immediately decalcified upon deposition or upon incorporation of new deposits in the reduced topsoil, for example by ploughing or biological mixing. These mechanisms are held responsible for the decalcification of young sediments, for example, in the flood-plains of the Ganges river (Brammer, 1968) and in the Rhine delta (Zonneveld, 1960). Desalinisation, or the removal of salts, takes place by means of a normal dissolution process by rain and/or floodwater. Its occurrence is strongly related to climatological conditions or more specifically, rainfall. A net surplus of rainfall over evapotranspiration is required.

Biological processes

Homogenisation is the process by which animals, insects and plant roots disturb and mix the soil. During growth, plant roots exert pressure on the soil matrix and disturb the original alluvial stratification of young sediments. On decay, the roots leave pores behind. By burrowing in the soil, animals and insects further disturb the soil structure. When feeding on soil material or on plant roots, they transport parts of it, vertically as well as laterally. Important burrowing animals include rats and mice, moles, earthworms, beetles and, under relatively well-drained conditions, termites. Mudlobsters and crayfish are of particular importance in saline or brackish coastal areas.

Microbiological processes play an important role in soils of deltaic areas as well, for instance in the formation of pyrite and, upon aeration, its subsequent oxidation to sulphuric acid. Other important microbiological transformations include those of nitrogen compounds in wet and/or periodically wet soils (Figure 1.21). In soils that become anaerobic, nitrate (NO_3^-) is lost as gaseous N_2 and N_2O in what is called the denitrification process. On the other hand, microbiological decomposition of organic matter produces NH_4^+ (ammonification). Under reduced conditions more organic-N is mineralised than in oxidised environments at a given rate of microbial activity. Therefore, especially during the first week of soil reduction (ponding or flooding), if oxygen has disappeared, there is a flush of microbial activity and ammonium production (Brinkman and Van Diepen, 1990). Other anaerobic micro-organisms can capture N_2 from the atmosphere and transform it to NH_4^+, ready for uptake by plants (N fixation).

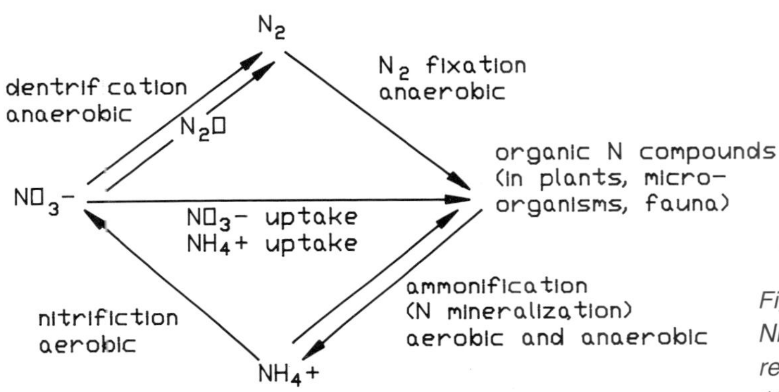

Fig. 1.21
Nitrogen Transformations in periodically reduced soils (Brinkman and Van Diepen, 1990).

1.4.3 Typical soils of deltaic areas

As discussed above, the soils of deltaic areas are typically relatively young soils of varying texture, with little profile development and, initially, strong stratification. Many of these soils have restricted drainage. Soil profile development is mainly confined to physical processes like ripening and subsequent shrinking and subsidence, chemical processes like oxidation, decalcification, leaching and exhange of sodium by calcium, and biological processes including homogenisation by soil organisms, root penetration, and (evapo)transpiration.
Other common soils in deltaic environments are peat soils, acid sulphate soils, heavy clay soils with strongly swelling and shrinking clays (in the lowest physiographic positions) and, on sand bars and dunes, coarse sands. Saline and/or alkaline, or *sodic*, soils may develop where, in arid and semi-arid climates, groundwater of poor quality is at or close to the surface of the soil. FAO/UNESCO/ISRIC (1988) provides a Soil Map of the World with an overview of the main soils in deltaic areas.

1.4.4 Management of acid sulphate soils

With the danger of acidification lurking, it will be no surprise that many acid sulphate soils have been wasted by thoughtless drainage or, as is the case locally in West Africa, as a result of the lowering of the groundwater table, due to extended periods of drought.

Because of the overall effect of water management on the behaviour of acid sulphate soils, any attempt to reclaim them must take into account the interaction between soil, groundwater and floodwater (Dent, 1992). In this respect, seasonality of rainfall and availability of irrigation/leaching water largely determine the use possibilities of these soils. If a large water surplus exists, the main management problem is drainage. Maintenance of a high watertable (i.e. above the pyritic layer) will prevent, or at least limit, soil acidification. Dry periods, even as short as some 2-3 weeks, create increasingly severe acid conditions. In such conditions, water availability for crop growth depends on the thickness of the non-acid/non-pyritic topsoil. In even drier climates, salinity becomes an additional, if not the sole, constraint. Here, the situation is aggravated by the fact that water for leaching of the salts and acids is scarce.

In southern Kalimantan in Indonesia, Kselik (1990) subdivided a deltaic area with acid sulphate soils into four hydrological land classes (Figure 1.22). The classification is based on the (tidal) flooding regime of the land as well as on its drainage characterics. Type A' land comprises the areas between mean low tide and mean neap tide, which are under the influence of daily flooding and drainage. Type B' covers the areas between mean neap tide and mean spring tide. These areas are flooded during spring tide only but they are subjected to daily drainage. Type C' is land above spring tide, but still under the influence of the tide. There is no tidal flooding in these areas but they are permanently drained. Type D' land is beyond the influence of the tide, there is no tidal flooding and limited drainage. The watertable in these areas drops only during the dry season. Monthly average (ground) waterlevels for the various hydrological land types are shown in Figure 1.23. In the research project concerned, recommendations as to water management practices are being developed specifically for each of these hydrological land types (Kselik, 1990 and AARD/LAWOO, 1993).

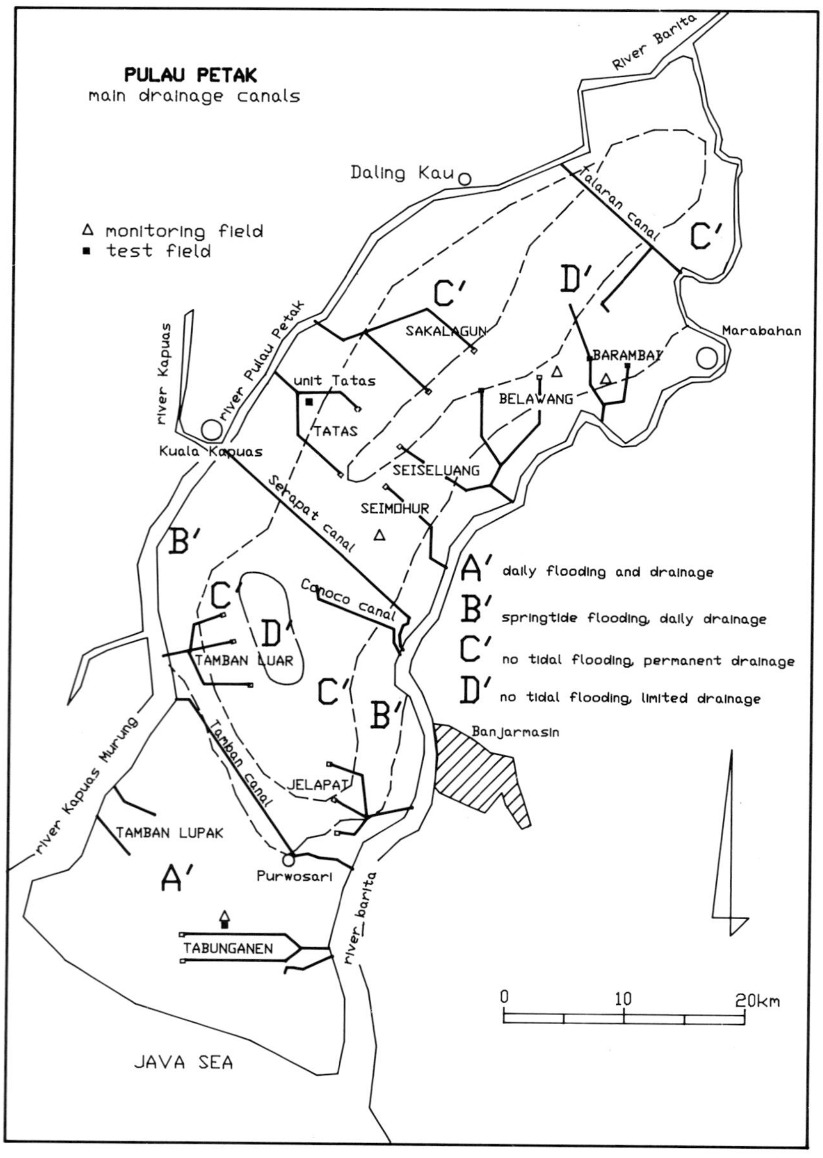

Fig. 1. 22
Hydrological land classes for Pulau Petak, southern Kalimantan, Indonesia (wet season).

Broadly, there are three options for the use of acid sulphate soils:

- leave as is, especially if alternative land is available for development: mangroves swamps, salt marshes and mud flats have essential ecological functions for example as natural coastal defense barriers, as breeding grounds for fish and wintering grounds for migratory birds and for the gathering of natural products (FAO, 1982; Maltby, 1986);

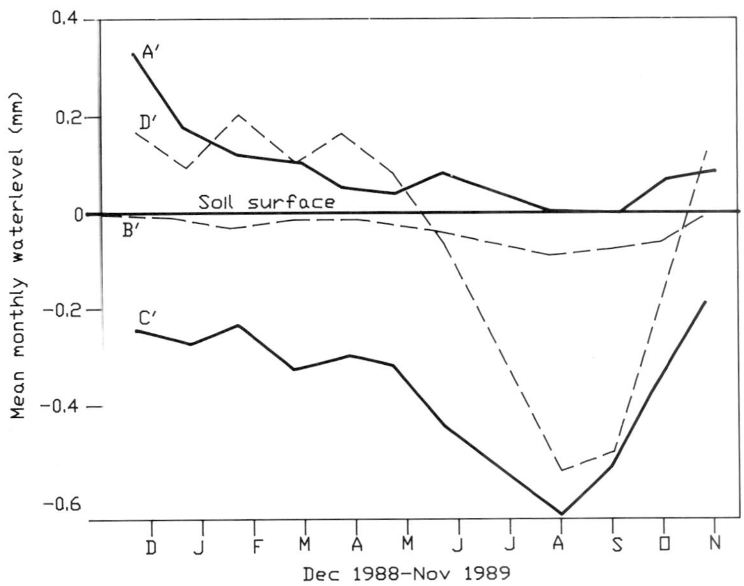

Fig. 1.23
Monthly average waterlevels (in cm) in four hydrological land classed in Pulau Petak, southern Kalimantan, Indonesia.

- reclaim, but with minimum disturbance in order to prevent, as much as possible, the oxidation of pyrite: flooded rice cultivation, shallow fishponds (shrimps if tidal water is available), dryland cropping under (controlled) high water table, forestry and grazing are possible options (Janssen et al., 1992);
- total reclamation aimed at complete drainage and ripening of the land and flushing or neutralisation of the acidity formed.

An example of traditional land use in deltaic areas with acid sulphate soils is described by Sarwani et al. (1993) for southern Kalimantan. Here, the Banjarese and Buginese population groups are able to reclaim (potential) acid sulphate soils gradually, for the cultivation of rice as well as other crops such as coconut, citrus, rambutan and pineapple. This is done by applying escape mechanisms in their cropping systems, starting from the water management system, the land preparation, and the subsequent transplantation sequence, as well as in the choice of crops and crop varieties grown. Central in this reclamation process is the gradual building of raised beds (local name: *baluran*). Acid formed in these beds is either flushed by (tidal) floods, or it is washed out with the rainwater percolating through them.

1.5 References and Bibliography

AARD/LAWOO, 1993,Water management and soil fertility research on acid sulphate soils in Kalimantan, Indonesia. Final Report, Vol. III, Agency for Agricultural Research and Development, Jakarta, Indonesia and Land and Water Research Group, Wageningen, The Netherlands.

Bakker, H. de, 1979. Major soils and soil regions in the Netherlands. Junk Publishers, The Hague, The Netherlands.

Bennema, J., Geuse, E.C.W.A., Smits, H. and Wiggers, A.J., 1954, Soil compaction in relation to Quaternary movements of sea-level and subsidence on the land, especially in The Netherlands, Geol. & Mijnb. N.S. 16, pp. 173-178.

Brammer, H., 1968. Decalcification of soils developed in calcareous Gangetic alluvium in East Pakistan. Pak. J. Soil Sci. IV (2), pp. 8-20.

Brammer, H., 1990a, Floods in Bangladesh - Part 1, Geographical background to the 1987 and 1988 floods, Geographical Journal 156, No. 1, pp. 12-22

Brammer, H., 1990b, Floods in Bangladesh - Part 2, Flood mitigation and environmental aspects, Geographical Journal 156, No. 2, pp. 158-165

Breemen, N. van , 1982, Genesis, morphology and classification of acid sulfate soils in coastal plains, Acid Sulfate Weathering. Soil Sci. Soc. Am. Special Publ. 10. Madison, Wisc., USA. pp. 95-108.

Brinkman, R. and van Diepen, C.A., 1990, Mineral soils, Wetlands and Shallow Continental Water Bodies, Patten, B.C. (ed), Vol. 1. SPB Ac. Publ., The Hague, The Netherlands. pp. 37-59.

Clark, J.A., Farrell, W.E. and Peltier, W.R., 1978, Global changes in post-glacial sea level: a numerical calculation, Quaternary Research 9, pp. 265-287.

Coleman, J.M., 1981, Deltas: Processes of Deposition and Models for Exploration. Burgess Publ. Co., U.S.A.

Coleman, J.M. and Wright L.D., 1971, Analysis of major river systems and their deltas, Technical Report No. 95, Coastal Studies Institute, Louisiana State Univ, Baton Rouge.

Davies, J.L., 1964, A morphogenic approach to the world's shorelines, Journ. Geomorphology 8, pp. 127-142

Dent, D., 1986, Acid sulphate soils: a baseline for research and development. ILRI Publication 39. Wageningen, The Netherlands.

Dent, D., 1992, Reclamation of acid sulphate soils, Advances in Soil Science Vol. 17., Lal, R. and Stewart, B.A. (eds), Springer Verlag, New York, USA. pp. 79-122.

FAO, 1982, Management and Utilization of Mangroves in Asia and the Pacific. FAO Environment Paper 3., Rome, Italy.

FAO/UNESCO/ISRIC, 1988, Soil Map of the World. World Soil Resources Report 60. Rome, Italy.

Galloway, W.E., 1975, Process framework for describing the morphologic and stratigraphic evolution of deltaic depositional systems, in Deltas, Bronssard, M.L., (ed), Houston Geological Society, pp. 87-98

Glopper, R.J. de, 1969, Shrinkage of subaqueous sediments of Lake IJssel (The Netherlands) after reclamation, in I.A.H.S. (1969), pp. 192-201.

Glopper, R.J. de, 1973, Subsidence after drainage of the deposits in the former Zuyder Zee and in the brackish and marine forelands in the Netherlands, Van Zee tot Land, Staatsuitgeverij, 's Gravenhage.

Hayes, M.O., 1975, Morphology of sand accumulation in estuaries, in Estuarine Research, Cronin, L.E. (ed), Vol. 2, Academic Press, New York, pp. 3-22

IAHS, 1969, Publications 88 and 89, Proceedings of the First International Symposium on Land Subsidence, Tokyo, Japan.

IAHS, 1976, Publication 121, Proceedings of the Second International Symposium on Land Subsidence, Anaheim, California.

IAHS, 1984, Publication 151, Proceedings of the Third International Symposium on Land Subsidence, Venice, Italy.

ICID, 1966, Proc. 6th Congress, Development of Deltaic Areas, New Delhi.

Jelgersma, S., 1966, Sea level changes during the last 10,000 years, Proc. Int. Symp. on World Climate 8000 to 0 B.C., Royal Meteorological Society, London, pp. 54-69

Jelgersma, S., 1988, Land subsidence in coastal lowlands, Proc. SCOPE Workshop on Rising Sea Level and Subsiding Coastal Areas, November 9-13, 1988, Bangkok

Janssen, J.A.M., Andriesse, W., Prasetyo, H. and Bregt, A.K., 1992, Guidelines for soil surveys in acid sulphate soils in the humid tropics; the main problems considered. Agency for Agricultural Research and Development, Bogor, Indonesia and Land and Water Research Group, Wageningen, The Netherlands.

Köppen, W., 1931, Grunddriss der Klimakunde. De Gruyter Publ. Co., Berlin (in German).

Kselik, R.A.L., 1990, Water management on acid sulphate soils at Pulau Petak, Kalimantan, Wkshp on Acid Sulphate Soils in the Humid Tropics. Bogor, Indonesia. pp. 249-276.

Maltby, E., 1986, Waterlogged wealth. Why waste the world's wet places? Earthscan, International Institute for Environment and Development, London, UK.

Mayuga, M.N. and Allem, D.R., 1969, Subsidence in the Wilmington oil field, IAHS-AJSH Pub. 88, pp. 66-79

Pirazolli, P.A., 1991, World Atlas of Holocene Sea-level Changes, Elsevier Sc. Publ., Amsterdam.

Pons, L.J., van Breemen, N. and Driessen, P.M., 1982. Physiography of coastal sediments and development of potential soil acidity, Acid Sulfate Weathering, Soil Sci. Soc. Am. Special Publ. 10. Madison, Wisc., USA. pp. 1-18.

Pons, L.J. and Zonneveld, I.S., 1965, Soil ripening and soil classification. Initial soil formation in alluvial deposits and a classification of the resulting soils. ILRI Publication 13. Wageningen, The Netherlands.

Sarwani, M., Lande M. and Andriesse, W., 1993, Farmers' experiences in using acid sulphate soils: some examples from tidal swampland of southern Kalimantan, Proc. Symp. on Acid Sulphate Soils, February 1992, Can Tho, Vietnam.

Schothorst, C.J. and Broekhuizen, J., 1990, Zakking van grond, Bodemkunde van Nederland Locher, W.P and de Bakker, H. (eds), Malmberg, Den Bosch, The Netherlands. pp. 303-320. (in Dutch).

Thornthwaite, C.W., 1933, The Climates of the Earth, Geographical Review no. 23, pp. 433-440.

UN-ECAFE, 1963, Proc. Regional Symposium on Flood Control, Reclamation, Utilisation and Development of Deltaic Areas, Bangkok. Water Resources series no. 25, New York

UN-ECAFE, 1969, Proc. 2nd Regional Symposium on Deltas, Bangkok-Tokyo. Water Resources series no. 39, New York

UN-ECAFE, 1979, Proc. 3rd Regional Symposium on the Development of Deltaic Areas, Bangkok. Water Resources series no. 50, New York

UNESCO, 1969b, Proc. Symposium on the Scientific Problems of the Humid Tropical Zone Deltas and their Implications, Dacca.

UNESCO, 1969a, Proc. Symposium on the Hydrology of Deltas, Bucharest.

UNESCO, 1984, Guidebook to studies of land subsidence due to groundwater withdrawal. Prepared for the I.H.P. working group, ed. Joseph Poland.

United States Department of the Interior, Bureau of Reclamation, 1965, Design of Small Dams, U.S. Government Printing Office, Washington D.C., U.S.A.

Way, D.S., 1973, Terrain Analysis: A guide to site selection using aerial photographic interpretation. Dowden, Hutchinson & Ross Inc., Pennsylvania, U.S.A.

Wolman, M.G. and Leopold, L.B., 1957, River channel patterns - braided, meandering and straight, U.S. Geological Survey Prof. Paper 282-B

Zagwijn, W.H., 1983, Sea-level changes in The Netherlands during the Eemian, Geol. & Mijnb. Vol. 62, Nr.3, pp. 437-450.

Zagwijn, W.H., 1989, The Netherlands during the Tertiary and Quaternary - a case history of coastal lowland evolution, Geol. & Mijnb. Vol. 68, pp. 107-120.

Zonneveld, I.S., 1960, The Brabantse Biesbosch; a study of soil and vegetation of a fresh water tidal area. STIBOKA Soil Studies No. 4., Wageningen, The Netherlands.

2.1 Surface Water

2.1.1 General

The hydrologic conditions of a delta are entirely different from those of other parts of the river basin to which the delta belongs. This applies both to the surface water and groundwater. In this section the system of surface waters will be discussed. At the upstream boundary of the system, at or near the apex, the water levels and discharges are supposed to be given. These depend on the rainfall and the physiographic conditions of the upstream part of the catchment. The recipient basin (e.g. ocean, sea or lake) is the downstream boundary. Here the water levels are governed by the instantaneous sea-levels which in turn depend on the mean sea or lake level, the astronomical tides, storm surges and tsunamis, and, in some embayments, oscillations of the water level. The sea-level, which is independent of the flow in the deltaic river branches, exerts a back-water effect on the flows and water stages. The variations of the sea-level propagate into the open estuaries and river branches. The mechanism of this propagation is an hydraulic process and is considered to be a special part of tidal hydraulics with the propagation of tides in shallow waters. The same applies to coastal and estuarine currents, with respect to the movement of sediment and dissolved matter, which are very important factors in determining the morphology of coasts, estuaries and river branches, and for the utilisation and management of water and land resources.

2.1.2 Coastal and estuarine dynamics

Coasts and estuaries form the aquatic boundaries of deltaic areas along about half their perimeter (Bird & Schwartz, 1985). The interaction between the water and the land is intense and of great importance. The main characteristic is a multi-faceted dynamism in space and time. Variations of the water-level continuously shift the water-line. Water flows back and forth over large distances carrying salt and other dissolved substances, useful nutrients as well as toxic pollutants. Sediments are also transported by the water, causing continuous changes of the geometry of the area. In fact, these are the processes which gradually created the neighbouring low-lying land. Water is essential for all life in the area, including man. It is useful for irrigation and for navigation, but can also be dangerous, in the form of floods and saline water. No wonder man has, for many centuries, tried to bend and harness its natural forces for his own safety and profit.

Mean sea-level

The water-level along a coast exhibits a continuous variation, with time scales varying from many centuries (e.g. eustatic variations), to a few seconds (e.g. waves). The short term variations are generally considered as fluctuations around a long term average. The definition of mean sea-level (MSL) involves the adoption of a period of time over which the average is taken. For phenomena of a short period this MSL seems to be fixed, but on a longer time scale it may show a variation. The attractive forces of the earth, the moon and the sun directly influence the motions of water on our planet. Indirectly, they act via seasonal and daily variations of meteorological phenomena such as air pressure and wind. These give rise to variations in sea-level of shorter periods ranging from several years to fractions of a day. The longest period of this nature is the rotation of the ascending node along the lunar orbit which takes 18.6 years. This causes a variation of about four percent around the mean of some important constituents of the tidal motions, and indirectly results in a slight variation of the mean sea-level in shallow coastal waters. Shorter and more pronounced cycles are the annual period of the earth in its orbit (365.25 days), the monthly orbital cycle of the moon (29.53 days), the daily rotation of the earth with respect to the sun and moon (24 hours and 24 hours 50 minutes respectively), and combinations and higher harmonics of these oscillations. All these periods must be taken into account in the observation and analysis of water-levels in coastal areas.

Astronomical tides

Tidal variations of sea-level show a very regular pattern with cycles directly linked to the relevant astronomical phenomena from which they originate. The rotation of the earth and the attraction of the moon and the sun lead to tidal waves, mainly originating from the Arctic Ocean and propagating into other oceans, adjacent seas and estuaries. Reflection, interference and frictional damping lead to an intricate pattern of phases and amplitudes throughout the world.
Theoretical considerations lead to the concept of tides (level variations as well as streams), as

being the sum of a number of harmonic constituents (Schureman, 1958):

$h = h_0 + \Sigma f_n a_n \cos(\omega_n t + \psi_n)$

where:

h_0 = mean sea level
f_n = node factor of constituent n
a_n = amplitude of constituent n
ω_n = angular frequency of constituent n
ψ_n = phase of constituent n
t = time

Of these f_n and ω_n can be predicted theoretically. The other harmonic constants (h_0, a_n and ψ_n), can be derived from analyses of observations at any location on earth. The phases (ψ), are linked to the relevant astronomical phenomena. Once the constants are known, they can be used for the prediction of future tidal phenomena. The number of constituents that can be detected and the accuracy of the constants depend greatly on the length of the record that has been analysed. The more that is known, the more accurate are the predictions. Table 2.1 gives the codes, angular frequencies, amplitudes and phases of the most usual constituents for two locations, Hook of Holland, The Netherlands and the Bangkok Bar, Thailand. The most important constituents are generally the M_2 and S_2 which give rise to a semi-diurnal tide of which the combined amplitude varies during a fortnight (14 days), between $M_2 + S_2$ (spring tide), and $M_2 - S_2$ (neap tide). The main diurnal constituents K_1 and O_1 give rise to a difference between the two high waters on the same day. In some cases, the diurnal constituents predominate, resulting in a diurnal tide. The long term constituents (S_a, S_{sa}) are generally caused by meteorological phenomena. The monthly and fortnightly variations are caused by the propagation of the tidal wave into the shallow waters of continental seas and estuaries. Higher harmonics and combinations of constituents arise from the non-linear interaction and distortion of the tidal wave in shallow water. Consequently, the prediction of tides in shallow coastal waters and estuaries requires the inclusion of more constituents than in the deep open sea.

Code	Frequency	Hook of Holland		Bangkok Bar	
	ω (deg/h)	a (cm)	ψ (deg)	a (cm)	ψ (deg)
h_o	-	120	-	260	-
S_a	0.041	9	218	24	284
S_{sa}	0.082	-	-	9	47
M_m	0.54	4.5	8	1.0	13
M_{sf}	1.02	3.6	62	1.8	244
K_1	15.041	7	1	67	161
O_1	13.943	10	189	45	168
P_1	14.959	3	344	21	169
Q_1	13.399	4	143	9	95
M_2	28.984	78	86	55	138
S_2	30.000	19	146	28	209
N_2	28.440	11	59	9	106
K_2	30.082	5	149	10	188
M_4	57.968	17	162	1	164
MS_4	58.984	10	219	1	233

Table 2.1 Tidal constants for Hook of Holland and Bangkok Bar.

Tidal predictions are published by many local authorities, mainly for navigational purposes. A compilation is published annually by the British Admiralty together with some information about the main tidal constants and other characteristics. The local amplitudes and phases of the tidal constituents depend greatly on the geometry of the area. Geo-morphological developments cause gradual changes of the constants as do artificial changes of the geometry by the construction of dams, canals and bunds. Experiments with tidal (mathematical) models can provide information about the effects which can be expected from human intervention (Dronkers, 1964).

Surges

The orderly and predictable tidal phenomena are more or less disturbed by stochastic phenomena called surges. These are occasional single deviations from the normal pattern, with a duration ranging from a few hours to several days. Surges are caused by storms, earthquakes, landslides, calving glaciers and areas of high and low pressure, and result in a set-up of water against the coast. Tropical cyclones (hurricanes, typhoons) and the depressions of higher latitudes are known to cause severe surges. A rise of the water surface moves along with the low pressure centre of the system. Resonance can occur when its speed equals the celerity of a tidal wave, and the effect is amplified when it enters shallow coastal waters. Waves and gale force winds add to the destructive power of the phenomenon. Surges can leave the source area and traverse the oceans and seas over very long distances before they strike a coast. Well known storm surge areas are the Gulf of Mexico (hurricanes), the North Sea (storms), the Bay of Bengal (cyclones) and Japan (tsunamis caused by earthquakes). The temporary rise of the sea-level is superimposed on the tidal variation. Extreme water levels occur when the peak of a surge coincides with High Tide, or (in the case of a negative surge) when the minimum of the surge meets Low Tide.

The levels reached during surges have to be taken into account in the design of structures such as sea defence (positive), and navigation channels (negative). For these purposes design levels are estimated using a frequency analysis of the extreme high and low waters. Observed frequencies of exceedance (n), are related to the levels, using log-normal paper or the method of Gumbel (1945).

Waves

Waves are generally generated by wind and propagate along the surface of the water until their power is dissipated by internal friction or by breaking against a coast (Wiegel, 1964 and Ippen, 1966). As a result of their common origin, periods of high waves can coincide with storm surges. However, as both phenomena can travel great distances, this is not usually the case. Waves are characterised by a height, H, a period, T, and a wave length, L. The period and the length are related to the celerity, c, in the following way:

$$L = c\,T$$

The celerity, in turn, is related to the wave length and the water depth, h, according to:

$$c^2 = \frac{g}{2\pi} * L * \tanh\left[\frac{2\pi h}{L}\right]$$

This relationship leads to the definition of two distinct domains:

- short waves, deep water ($h > L/2$) giving:

$$c^2 = \frac{g}{2\pi} * L = 1.56L$$

- long waves, shallow water ($h < L/20$) giving:

$$c^2 = gh$$

The energy per unit width, contained in a wave of length, L, and height, H, is given by:

$$E = 1/8\, \rho\, g\, H^2\, L$$

This energy propagates with a speed, c/2, in deep water, and c in shallow water. Apart from some loss by friction, the energy remains constant in a shoaling wave. As L decreases in shallow water ($h > L/20$), the height increases until the wave becomes too steep and breaks. Refraction also plays a role in waves which approach obliquely to the depth contours, because of the dependence of the celerity on depth. Diffraction occurs around relatively sharp corners in the geometry of a coast or a structure. During the generation of waves, energy is transferred from the wind to the waves until an equilibrium wave height is reached. The process takes time during which the waves grow and travel a certain distance referred to as the fetch. The relationship is shown in Figure 2.1 (Groen & Dorrestein, 1976). If the wind blows over a shorter distance and/or a shorter period, the waves do not reach the equilibrium height. Water depth also limits the growth of waves, the equilibrium wave height being about 0.45 times the depth.

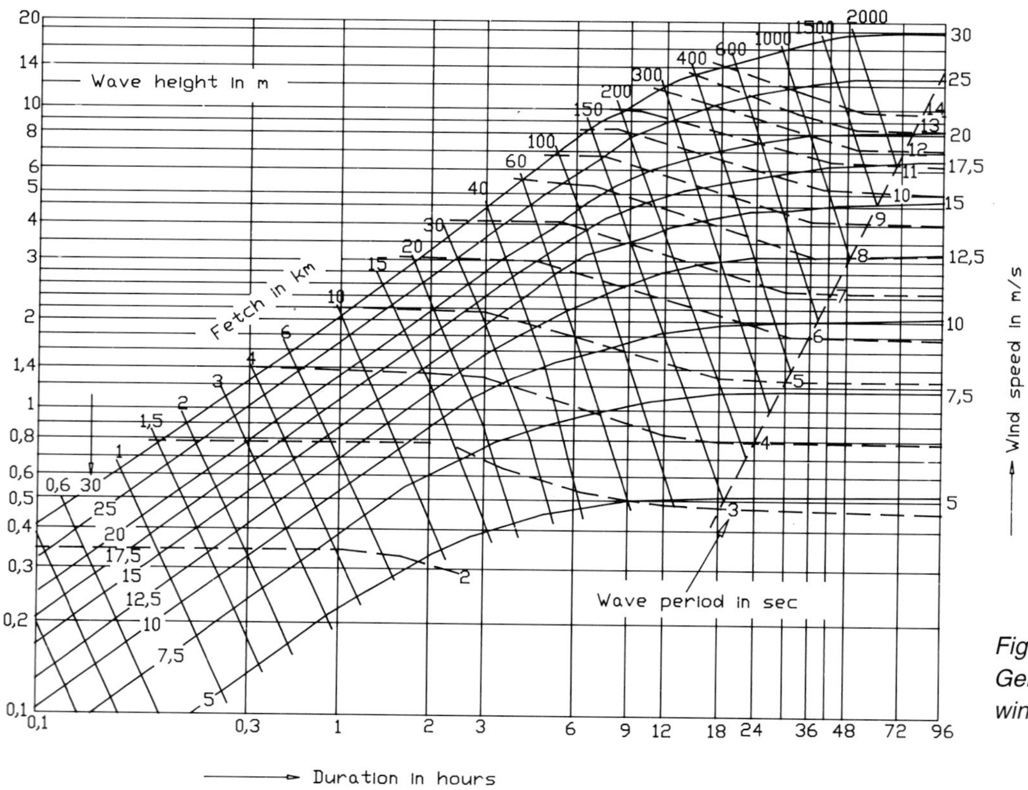

Fig. 2.1 Generation of waves by wind.

A wave field is by no means regular. The height and the period vary in space and time around a significant value. The phenomenon can be described by a spectrum. The significant wave height, H_s, is defined as the average of the highest one third of the waves in a record, and is the height generally given by observations. It is exceeded by about one in every eight waves (12.5 %) and is approximately 1.6 times the average wave height. A height of 1.5 H_s is exceeded by about 1%, 1.8 H_s by 0.1 % and 1.95 H_s by about 0.05 % of the waves. Waves with a narrow spectrum (i.e small variation in the period) generally arrive in groups of about 7 waves between the maxima.

Surges as well as waves pose a threat to coastal defence and navigation. Methods have been developed for the prediction of both phenomena (Dronkers (1960), Delta Committee (1960) and Groen & Dorrestein (1976)). All are based on a prediction of the relevant meteorological phenomena, in combination with observed water-levels and wave-heights previous to the time of prediction. The tides and the fact that both phenomena propagate through the area, are also taken into account. By nature, these predictions are limited in accuracy and time span, but for many purposes, a warning time of 6 to 12 hours is quite acceptable.

Coastal currents

The wave phenomena discussed above are usually accompanied by horizontal movements of the water in the coastal area. These flows carry dissolved and suspended matter and induce movements of the material on the bed. The transport of sediments is a major factor in the geomorphology of the coastal area, especially in the case of low accretion plains. The flows involved in the main oceanic circulations rarely penetrate significantly into the near shore area. Much more prominent are tidal currents and drifts caused by wind and waves (Ippen, 1966). Many of these currents are oscillatory with a resultant displacement because of some asymmetry. The same holds for the transport phenomena. Tidal currents generally run parallel to the coast with maxima around high (flood) and low (ebb) water. Often, this implies a resultant flow, in the direction of the flood, in shallow water. Strong tidal currents perpendicular to the coast occur in and around coastal inlets and the mouths of estuaries. In these cases, flood flows occur during the rising tide, ebb flows during the falling tide with slack water occurring in between.

Waves in deep water are accompanied by circular orbital motions in a vertical plane of which the strength decreases with depth. The effect almost vanishes at a depth of around 0.5 times the wave length. In the case of shallow water waves, the water oscillates in a horizontal direction over the full water depth. The strength of these orbital motions is more or less proportional to the wave height and inversely proportional to the period (Figure 2.2).

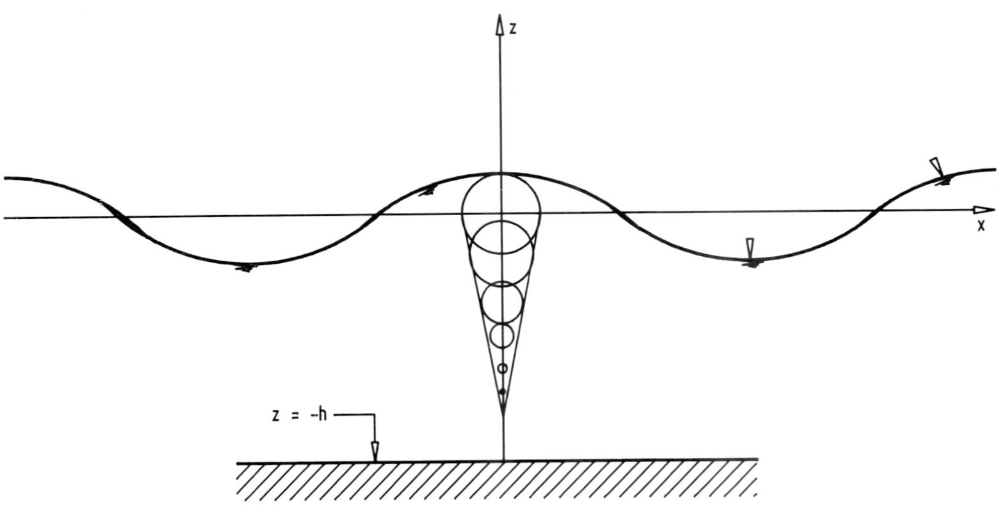

Fig. 2.2 Orbital motion under a deep water wave.

In shallow water waves the motion is almost constant throughout the entire depth (Figure 2.3).

Tidal currents vary with the tidal range and reach up to 0.5 to 1 m/s off the coast, but much higher values (e.g. up to 2 or 3 m/s) can occur in tidal inlets and in the mouths of estuaries. Resultant flows vary between a few centimetres per second offshore in calm weather, to several metres per second in the near shore region with strong winds and waves. The direction of these resultant flows depends on their causes.

Estuarine systems

Hydraulically, an estuary is the area, at the mouth of a river, in which the observed phenomena are

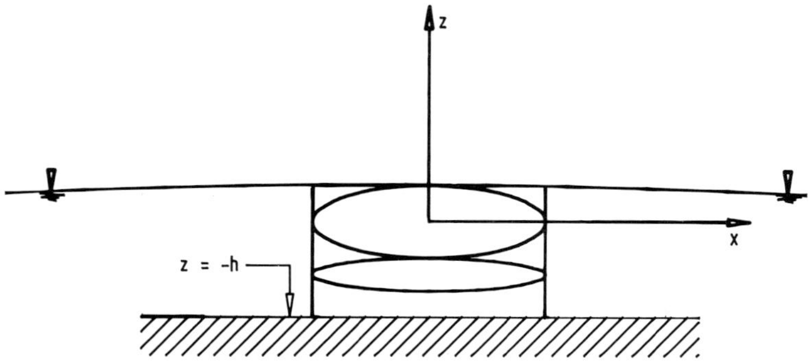

Fig. 2.3
Orbital motion under a shallow water wave.

jointly influenced by the regimes of the river and the sea. From the river comes a discharge of fresh water, dissolved matter and sediments, and the sea intrudes with saline water, tides, storm surges, etc. The geometry of the bottom and the action of the wind on its surface also influences the hydraulic phenomena (Dyer (1973) and McDowell & O'Conner (1977)). The upper boundary of the estuary is the location where the tidal motion ceases to be visible during the dry season, while the lower limit is the location where the fresh water from the river is mixed with the saline seawater. As a consequence, the stage of the river is only a few metres above mean sea-level, at the location where the sea flows into the estuary.

The heavy saline water from the sea tends to intrude into an estuary along its bottom. It mixes with the fresh water from the river and the mixture flows seaward along the surface of the estuary. The mixing process is greatly enhanced by the oscillating tidal motion. As a consequence, the mean intrusion length of the salt depends on the fresh water discharge of the river and the tidal range. The tip of the salt wedge shows a tidal excursion around its mean location. The tidal wave enters the estuary and causes a complex oscillatory motion in combination with the discharge of the river and the density currents described above. Figure 2.4 illustrates the variation of the marine influence in an estuary with the discharge of the river (NEDECO, 1965).

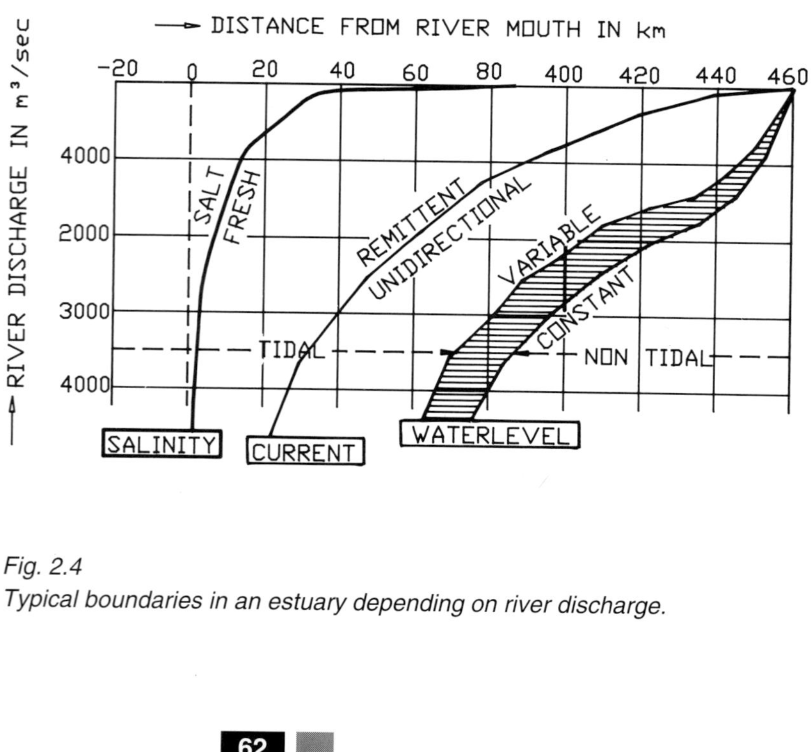

Fig. 2.4
Typical boundaries in an estuary depending on river discharge.

Intrusion of tides

Tidal waves intrude from the sea into the complex geometry of an estuary, against the seaward flow of river water. The shape of the tidal waves is modified by friction along the bottom of the estuary, reflection against its shores and interference with reflected waves. In some cases a local maximum of the tidal range occurs in the estuary. Large areas become dry around low water and are inundated during every high water. The tidal wave fades out at the upper end of the estuary where the bottom and the mean water level rise above the level which can be reached by it. This location depends on the discharge of the river.

The tide is accompanied by horizontal oscillations of the water. The volume of the ebb is always larger than that of the flood, because of the discharge of the river. This asymmetry is more pronounced in the upper parts of the estuary up to the tidal limit i.e. where the variable flow changes into the permanent flow of the river. At the mouth of the estuary, there is a transition from the rather strong inward and outward flows of the estuary, to the generally calmer, coastal water. The tides in estuaries can generally be predicted with reasonable accuracy. The effect of the river discharge can be taken into account, but as with other areas, observations of the phenomena must be the basis of the predictions.

Estuarine transport phenomena

The substances moving in an estuary fall into the four categories: water, salt, other dissolved matter and sediments. Salt substantially increases the density of seawater and influences the hydraulic system. In extreme cases sediments can have a similar effect. In all cases the movement of dissolved matter and sediments, can be regarded as caused by the motions of water without back-feeding. The fresh water discharge of the river flows through any cross section of the estuary. It is modified by the oscillating motion of the tides of which the influence decreases in the upstream direction. This pattern is, in the first place, complicated by the geometry of the estuary. Floods generally appear to occur before and around High Water while the ebb runs before and around Low Water. The effect is that resultant flows across shallow ,sometimes drying, banks are predominantly in a flood direction, while in the deeper channels they are ebbs. The result is a net horizontal circulation of water in a part of the estuary. A similar effect arises from the wind, of which the tractive force is more effective over the shallow banks than in the deeper channels.

Density currents related to the intrusion of seawater give rise to a vertical circulation. The resultant displacement is landwards in the denser saline water and seawards in the upper fresh water zone. Near the tip of the salt wedge there is a convergence of the bottom currents. The resultant flows at the surface caused by this phenomenon are seawards at any location in the estuary. The combination of all these phenomena leads to a complicated three dimensional pattern of resultant flows. Moreover, this pattern varies with the discharge of the river and the wind. It is of great significance for the quality of the water in the estuary.

Dissolved substances, supplied by the river, move through the estuary in a similar pattern to the fresh water. A dilution with saline water takes place. This encounter may also lead to chemical reactions and adsorption or desorption on the particle matter suspended in the water. Dissolved matter introduced at a location within the estuary will move with the water into which it is injected. It is generally dispersed rapidly by the complicated system of oscillating and resultant currents. This also implies that its paths may be intricate and that the time, and location of the introduction, as well as the density of the effluent play important roles.

The main sources of sediments in an estuary are the river and the sea, but local erosion may also be a contributing factor. Parts of the sediment are deposited in the estuary or lagoons, which are generally areas of net deposition. The greater part of the sediments ultimately leaves the estuary through its mouth and are deposited along bars or enters the littoral transport system. The pattern of sediment transport is more complicated than that of dissolved matter. Locally, erosion and deposition temporarily remove and return material to the streams of particle matter. Many of these processes are cyclic with the tides, the spring/neap periods and the seasons. Moreover, there is a distinct difference between the behaviour of coarse material (sand), moving as bed load and fine material (silt), carried in suspension. Sand, with grain sizes of more than 50 μm, moves as individual particles, but the finer silt, consisting substantially of cohesive clay, flocculates in saline water.

The cohesive sediments are eroded during increasing and maximum flows, which increases the concentration. Deposition occurs in quiet areas around slack water, with a resulting decrease in concentration. As a consequence, the maxima and minima of the transport lag behind the extremes of the rate of flow. The excursions of the particles are of the same order of magnitude (i.e. 10 to 20 kilometres) as that of the water. This also leads to spatial lags between the concentrations and the rates of transport. In both modes of transport the rate of transport increases non-linearly with the velocities of the water. Spring tide transports are much higher than neap tide transports.

The resultant transport over a tidal cycle appears as the difference of the two (relatively large) alternating displacements by ebb and flood. Its magnitude depends on the asymmetry of the flow in combination with the non-linearity of the transport mechanism. This dependence on second order aspects of the flow, makes the pattern of the resultant sediment transport in an estuary very sensitive to disturbances, natural as well as artificial. Observations and calculations are therefore very difficult to perform. Very important factors in the resultant sediment transport pattern are the river discharge and the density currents in the saline part of the estuary. The latter cause an inward transport of sediments along the bottom of the estuary towards the tip of the salt wedge, where it meets the seaward transport from the river. The resultant transport of suspended sediment along the surface is always seawards. The system gives rise to a turbidity maximum in the area of the tip of the salt wedge. This phenomenon is significant with respect to sedimentation in harbours and channels. In many cases, the inward transport predominates during the dry season depositing large amounts of sediments as far upstream as the salt travels. Storms at sea may greatly influence the process. The greater part of these deposits are eroded and carried seaward with the increased discharge of the river. This can lead to a large displacement of sediments in a very short period, often involving rapid siltation in channels and harbours. This is also the period of rapid deposition in the bar area and an injection into the littoral transport system. An example is given in Figure 2.5 (NEDECO, 1965).

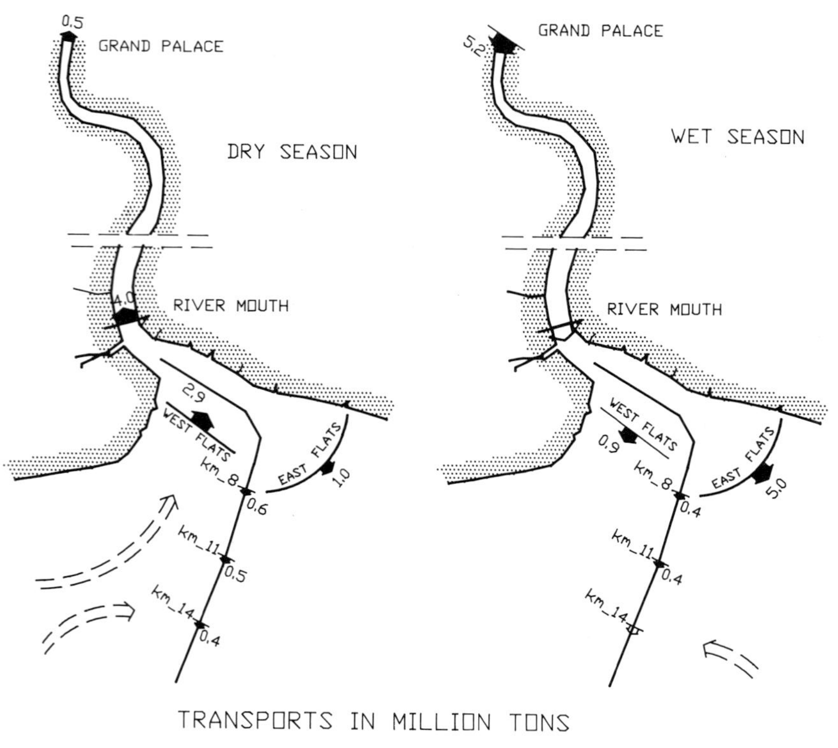

Fig. 2.5
Residual sediment transport per season.

2.1.3 The hydrologic input from the river basin

Upland river discharge

The river discharge and the sediment load at the apex form the input from the river into the delta. These hydrologic characteristics are largely determined by the climate and the geologic structure of the basin. The discharge depends on the amount and distribution of precipitation, and the occurrence of snow and ice. Not all precipitation becomes run-off. A considerable portion enters the atmosphere by evapotranspiration. This depends on meteorological factors such as temperature, humidity, wind and net incoming solar radiation, and is further influenced by the availability and depth of the groundwater.

The many and diverse climatic characteristics of any particular drainage basin combined with its geological and physiographic features lead to a certain rainfall run-off relationship.
The run-off coefficient often used to express the relationship is defined by:

$$r = R/P \times 100$$

where:
r = run-off coefficient (%)
R = annual total run-off (mm)
P = annual precipitation (mm)

This coefficient is characteristic of a given river basin and represents a mean value. It can vary between periods of time and from one basin to another, from less than 1% to as high as 100%. The atlas of the world's water balance (UNESCO, 1977) gives a world map of the annual run-off coefficient.

Water entering streams comes from precipitation falling directly on the surface of the stream and its tributaries, surface run-off, and groundwater flow. This differentiation is important in relation to the variability of the upland discharge. However, this is only one of the many factors determining the variability. The variability of the flow depends, in the first place, on the temporal as well as the spatial variability of the rainfall in the drainage basin of the river. The water supply to the delta is optimal when the river is perennial (i.e. continuous flow). Intermittent and ephemeral streams show periods without any flow. Under these conditions seawater intrudes great distances into the open estuaries creating a serious scarcity of fresh water. Water quality may be seriously affected, biological life may be threatened and, in particular, sedimentation may occur. For these reasons river regulation has sometimes been applied in the upstream portion of river basins. Such regulation consists of storage reservoirs which retain water during the wet season and release it during the dry season, thus augmenting the flow. Figure 2.6 demonstrates a most striking example of stream flow modification for the High Aswan Dam in Egypt (Shahin, 1991).

Deltas are exposed to flooding by storm surges at sea, local rainfall and flood waters from the river. River floods are determined by the storm rainfall pattern and the physiographic characteristics of the river basin. The important parameters, which determine the extent of the flooding and the possible damage in the delta are the volume of the flood, the spill volume (that part of the total flood volume which spills over the banks), the duration of the flood, and the magnitude of the flood peaks. The rate of rise of the flood stages determines whether the flood can be classified as a gentle or flash flood (Figure 2.7). Flash floods are likely to occur in relatively small river basins with steep slopes and rather impervious soils, where most of the flood water originates from surface run-off. Under these conditions, the flood hydrographs consist of a number of instantaneous responses to individual periods of rain. In larger river basins, with a concentration of the flood flow at the outlet resulting from the arrival of run-off from different parts of the basin, at different times, and at a slower rate, the various responses tend to overlap forming a single flood hydrograph during the rainy season. The gentle or flash characteristic of the flood determines the possibility of cultivating rice in the deltas of the humid tropical zone, in the absence of flood protection measures.

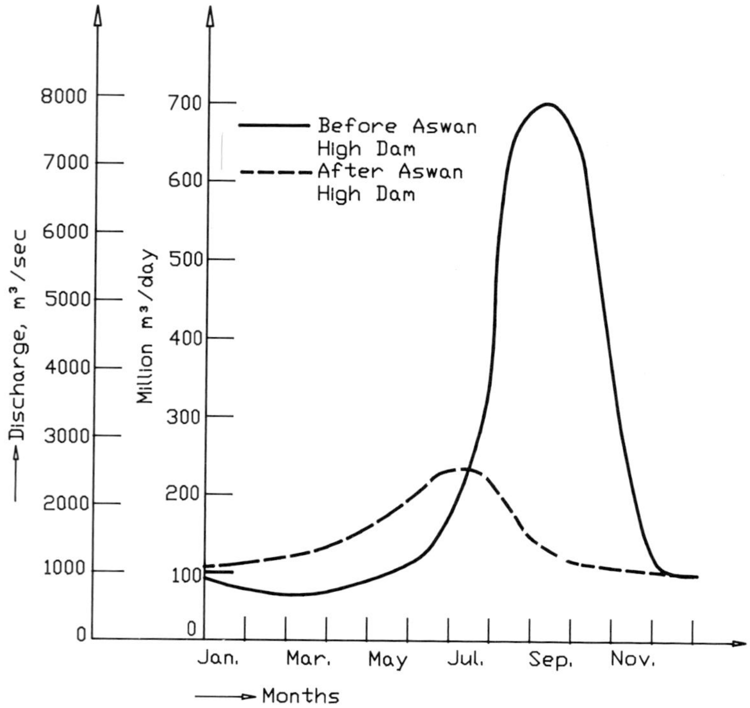

Fig. 2.6
Discharge hydrographs before and after construction of the High Aswan Dam.

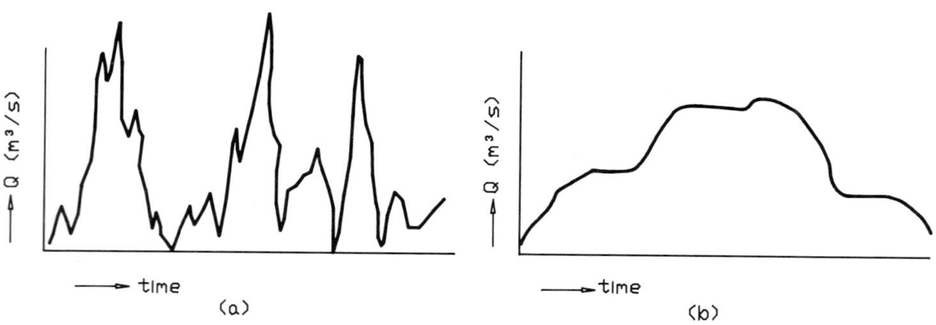

Fig. 2.7
(a) Flash floods and (b) gentle floods.

Sources of sediment

Deltas are formed through the processes of weathering and erosion in the river basin. Sediment derived as a result of these processes are carried by the river branches to the sea, thus contributing to an extension of the delta. During floods, sediments may also be deposited on the land areas, resulting in a further expansion of the delta. A certain part of the sediment originating from other rivers, offshore shoals or from eroding coasts, is redistributed by the sea and deposited at locations far from their places of origin.

The rate at which weathering takes place depends to some extent on the rate at which material is removed by erosion. Similarly, the rate of erosion is controlled by the rate at which material is released by weathering. The rates of denudation are very much greater for high altitude mountainous regions than for low-lying areas, for the same type of climate. Also, the rates are greatest where there is high rainfall distributed throughout the year, and least in arid to semi-arid regions irrespective of climate. Schumm (1965) descibes in greater detail the effect of temperature and mean annual precipitation upon mean annual sediment concentrations and yields. A brief summary of the type of weathering process and the corresponding rate of denudation in each of the four main climatic groups is given in Table 2.2 (Fookes & Vaughan, 1986).

Climate	Type of Weathering	Rate of denudation, (mm/1000 yr)	
		Highland	Lowland
Arctic	Mechanical decomposition and frost wedging	——-	——-
Arid to semi-arid	Chemical decomposition involving water dominates over mechanical disintegration	low latitude desert mountains 100-200	low latitude desert, Sahara 1
Humid to semi-humid	Both mechanical and chemical weathering take place; in cold temperate areas frost wedging causes deep rock disintegration	mid latitudes precipitation all year; e.g. coastal range of NW U.S.A. 3000 - 8000	mid latitudes, e.g.Rhine basin 30
Tropical	Chemical processes are dominant as a result of heavy rainfall and high temperatures	Tropical to low latitudes, e.g. Himalayas, 700	Tropical to low latitudes, e.g. N.India, 20

Table 2.2
Types of weathering processes and denudation rates in each the main climatic regions of the world.

Effects of changes to the upstream catchment

The term delta refers to a very specific downstream portion of a river basin. There is a close relation between the deltaic area and the upstream non-deltaic portions of the basin. This relation exists in several fields such as hydrology, water-management, economic development and trade, culture etc.

The supply of water and sediments to a delta depends on the precipitation and physiographic characteristics of the catchment, and on the effect of changes in these factors. Such changes may be due to natural effects (climatic variations, vegetation, natural erosion etc.) and man-induced effects. The latter are changes in land use, like deforestation, reclamation for agricultural purposes, watershed management and construction of upstream storage reservoirs.

The effect of changes in land use of the catchment is still a controversial matter. Many contradictory statements have been made because so much depends on the specific local conditions. Also, many misconceptions exist, especially with respect to the effects of deforestation. Recent research under field conditions, and compilation and analysis of data from all over the world (UNESCO, 1969) have shown that the hydraulic processes are more complicated than was assumed.

It is now generally admitted that as long as no substantial erosion is occurring and no large scale urbanisation is taking place, the effect of the type of land use on the hydrology of the delta is rather small. It makes little difference whether the upstream reaches, especially the head reaches, are used for agriculture with adequate soil conservation measures, or whether a natural vegetal cover exists like the rain forest in the humid tropics.

There is some difference, though, with respect to the water yield of the catchment and the effect of minor rainstorms. Forests are great water consumers and it has been found that removal of forests in many cases increases the yield by more than 25%. Isolated minor rainstorms are partly absorbed by interception by the trees and retention on the surface and in the weathered top layers. However, this effect decreases with increasing depth of the rainstorm.

An entirely different situation is found when deforestation is carried out without a follow-up of appropriate soil conservation measures, which too often is the case. The weathered top layers with the pervious root zone are eroded by the heavy rains, often down to the solid rocks. Infiltration is greatly reduced and a greater portion of the rainfall drains off along or near the surface. This overland flow is no longer slowed down by the vegetation and the result is a great eroding force and a change of the flood hydrograph (Figure 2.8).

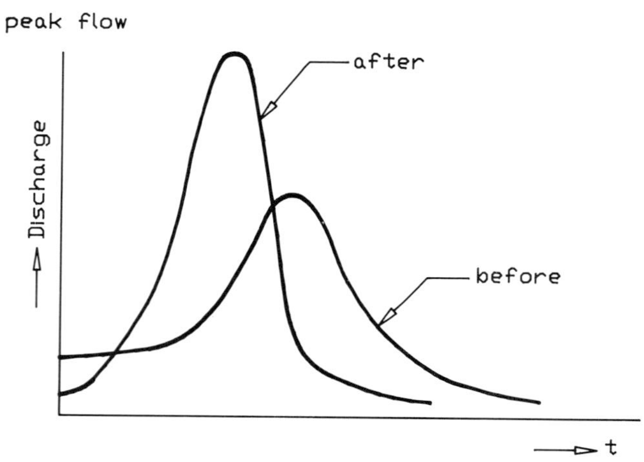

Fig. 2.8
Effect of deforestation on a flood hydrograph.

A similar effect will occur in the case of large scale urbanisation of the catchment. The degree of the flooding in the delta is determined by the areal extent, the depth of flooding and its duration. These parameters do not depend so much on the magnitude of the peak flow, but rather on the spill volume which is a part of the flood volume. As mentioned earlier, deforestation increases the flood volume but the effect is relatively small especially for the floods caused by major rainstorms during the wet season.

The conclusion is that deforestation and subsequent erosion have little effect on the flood phenomena in the delta, and that afforestation or reforestation, however beneficial these are in many respects, cannot be relied upon as a means of reducing the extent, depth of flooding and duration of the flooding in non-protected deltas.

The increase in sediment load (bed load + suspended load), resulting from erosion, will affect the morphology of the river channels upstream and in the delta. As the capacity to carry the bed load is limited, sediments will be deposited on the river bed, firstly near the places of supply and then propagating in the downstream direction. The subsequent rise of the river bed causes a rise of the flood levels and hence more inundations. Reservoirs for flood control are quickly filled up. The increased supply of fine silt to the deltas has a positive effect on the building-up and fertility of the land areas. It also causes a more rapid extension of the delta into the sea. An instructive case is

Cutter dredger in one of the canals of the Mekong Delta. The open canals are exposed to heavy silting up in the delta an anual maintenance dredging of some 10 million m^3 is required.

found on the north coast of Java (Figure 2.9) which is generally rather smooth, but which exhibits bulges protuding into the sea, at the places where rivers are discharging (Bruynzeel, 1983, 1987).

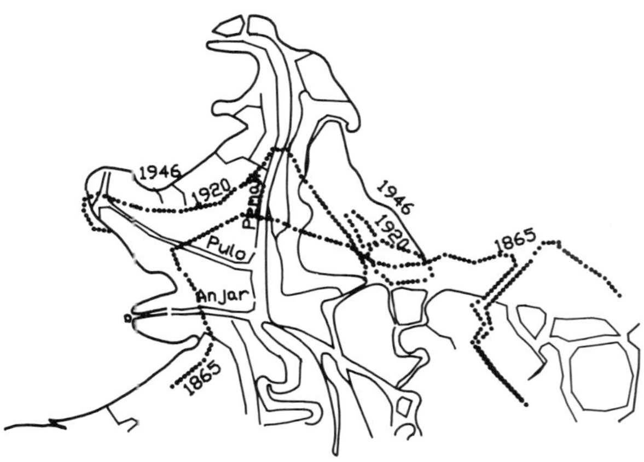

Fig. 2.9
Growth of the Delta of the Pemali (Pekalongan, Central Java, Indonesia). Coastlines of 1865, 1920 and 1946.

These deltas have been formed during the last hundred years or so. The rates of horizontal extension doubled and even quadruped in places, especially after 1940 when deforestation started on a large scale. Extension rates of 400 metres per year have been recorded. A small part of the erosion products is thus recovered in the delta as fine-grained fertile soils, but this is only a poor compensation for the almost irreparable loss of valuable soils and forests upstream.

2.1.4 Hydrology of land areas exposed to flooding

Flooding is a natural and characteristic feature of deltas since it is by this very process that they are created. Distinction should be made between flooding of land in the upstream region of the delta under non-tidal conditions and that in the downstream areas under conditions of astronomical tides and storm surges. In the non-tidal zone, flooding is governed by the flood levels in the river and the duration of the high levels. In this respect the distinction between flash floods and gentle floods is most relevant. In tidal zones where high astronomical tides are limited, flooding depends entirely on the occurrence of storm surges. In the intermediate zone maximum flooding occurs in the case where floods due to the river and sea coincide.

By virtue of the level nature of deltas, and the low elevation of the land with respect to the levels of the open water, the phreatic table is always found at shallow depths below the land surface even during dry periods, and may rise to the flood level under wet conditions. The occurrence of floods has played an important role in the history of human occupation of deltas. Indeed the settlement of deltas occurred long before any flood protection measures were achieved. There are still many densely populated areas throughout the world which are exposed to flooding. In other cases embankments have been constructed to protect the lower areas. This section deals with the hydrology of land areas in the absence of flood protection and considers both the non-tidal and tidal reaches. Flood protection by embankments may have profound impacts on the hydrology and morphology of the channels, and on the water management.

The non-tidal reach

Overtopping of the channel by flood waters in an alluvial river leads to the formation of so-called natural levees along the river channels, encircling depressions or backswamp (Figure 2.10).

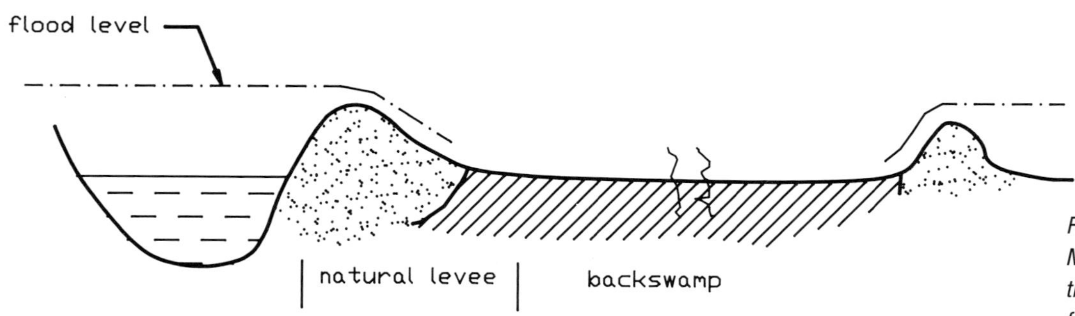

Fig. 2.10 Morphology of the delta floodplain.

In the upstream parts of a delta the height of the natural levee with respect to the backswamp may be as much as three to four metres with widths varying between a few hundred metres up to two or three kilometres. Coarse sediments prevail in the natural levees because of the sorting of the sediments during floods, in a lateral direction to the channel. The backswamps are composed of finer material. As a result of sorting in the longitudinal direction of the channel this configuration changes in the downstream direction. The levees are lower and consist of finer sediments than upstream and the channel also flattens out.

Flooding occurs when bank-full discharge is exceeded and the natural levees are overtopped. In general the natural levees build up to a level between that of the mean annual flood level and the 5-year flood level. This also gives an indication of the frequency of flooding under natural conditions. However, there are lower spots and creeks where flood waters can spill at an earlier phase of the flood. Here, breaching of the natural levee may occur (crevasses), increasing the inflow of water from the river into the depression. In the upstream reaches flooding of the deep backswamps depends on the nature of the floods (i.e. flash or gentle). In the case of flash floods the water levels of the flooded areas will lag behind the river stage, and the maximum water levels will be lower than the peak stage (Figure 2.11a). In the case of gentle floods with flat peaks the flood levels of the land areas rise to about the same height as the river stages (Figure 2.11b).

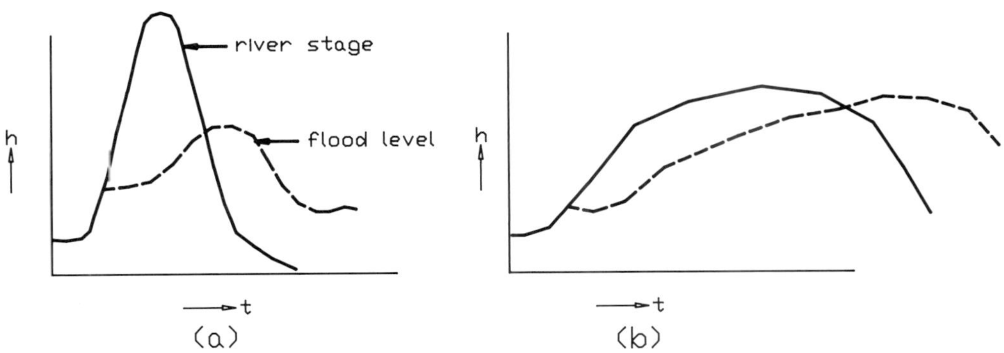

Fig. 2.11
Flood levels of the land and river stages for
a) flash floods and
b) gentle floods.

Further downstream flash floods become less pronounced and filling of the depressions takes water both from the river and from the upstream depressions. Also, a longitudinal floodplain flow occurs carrying flood water downstream and ultimately to the sea. Whereas the flow in the channels has velocities of around 2 to 3 m/s, the longitudinal floodplain flow is generally very slow (0.1 to 0.3 m/s) because of the small depth and the high roughness of the surface due to vegetation, field dykes, houses etc. These obstacles affect the magnitude and direction of the floodplain flow. In particular roads, canals and inner dykes delineate compartments in the floodplain which are reproduced as elements or meshes in physical and mathematical models of the flooding process (Figure 2.12).

Flooding is thus an intricate process where many factors play a role and small differences in topography result in significant differences in the flow pattern. It is not easy to obtain adequate field information on the development of the flood. Satellite pictures are very useful in determining the extent of the inundation with time, and in combination with ground data, the depths of flooding can also be obtained. It is also not easy to obtain data on the direction and velocities of flow at various locations and water levels. Flood maps based on such information in the framework of topography and morphology form an indispensable tool in predicting the effects of flood protection measures such as embanking.

Flooding may not only be caused by the river floods but also by heavy local rainfall which often occurs in the same season as the river floods. It may be aggravated by a rise in the river bed and a subsequent rise of the flood levels. A rise in the river bed may result from human activities as well as from natural effects such as the extension of the delta into the sea and lengthening of the river. A rise in the river bed is not followed by an equal rise of the land areas which are exposed to deeper flooding.

In deltas which are exposed to flooding, the human occupation, the way of life and the agricultural pattern are entirely governed by the flood characteristics. Settlement is concentrated on the natural levees along present and abandoned watercourses and other ridges where the depth of flooding is relatively low. The soils generally have a favourable texture for producing various crops and, under normal conditions, the drainage is better than in the backswamps.
In temperate zones crops are cultivated in the higher parts of the backswamps in summer, when there are no floods. Other areas are used as pasture for cattle. A remarkable adaptation of man and crops to the floods has taken place in the deltas of the humid tropical zone where wetland rice can be produced. Local varieties have been developed which can grow in an increasing depth of water provided that the rate of rise of the water level does not exceed 5 to 10 centimetres per day. An extreme case is that of the so-called floating rice which can grow in water depths of up to 3 or 4 metres, with stem lengths of 4 to 5 metres. The yield of these varieties is of course, much lower than those of the varieties growing in water depths of a few decimetres. Examples of such cases can be found everywhere in the humid tropics especially in the large deltas of Southeast Asia. The

Fig. 2.12
Mekong Delta: meshes in the floodplain.

Market area in Sonng PhiNong (Thailand). During the dry season the shops on the ground floor are accessible.

Market area in Sonng PhiNong (Thailand). During the flood season all goods are moved to the second floor. Shops are accessible by planking and boats.

delta of the Mekong provides an example of a large delta which is very productive, relatively densely populated and unprotected from flooding. In the delta of the Chao Phya (Thailand), many houses have been built on stilts on and outside the natural levees. In urban areas people live on two levels: ground level during the dry season and first floor level during the flood season. In other cases (e.g. Bangladesh) artificial mounds and concrete shelters have been constructed.

The tidal and intermediate zone

At the coast, the river stages are governed by the tidal levels at sea. River floods have little effect. In the intermediate zone, both sea levels and flood levels are relevant. Near the coastline the land areas are raised by the deposition of silt to a level which, under favourable conditions, is slightly above the level of the ordinary high tide, but below the level of the high water spring tide. In the coastal zone, the natural levees are low because of the small variation of the high water levels and the homogeneous composition of the sediment. Schematically the water levels and land elevation in the three reaches can be represented as in Figure 2.13.

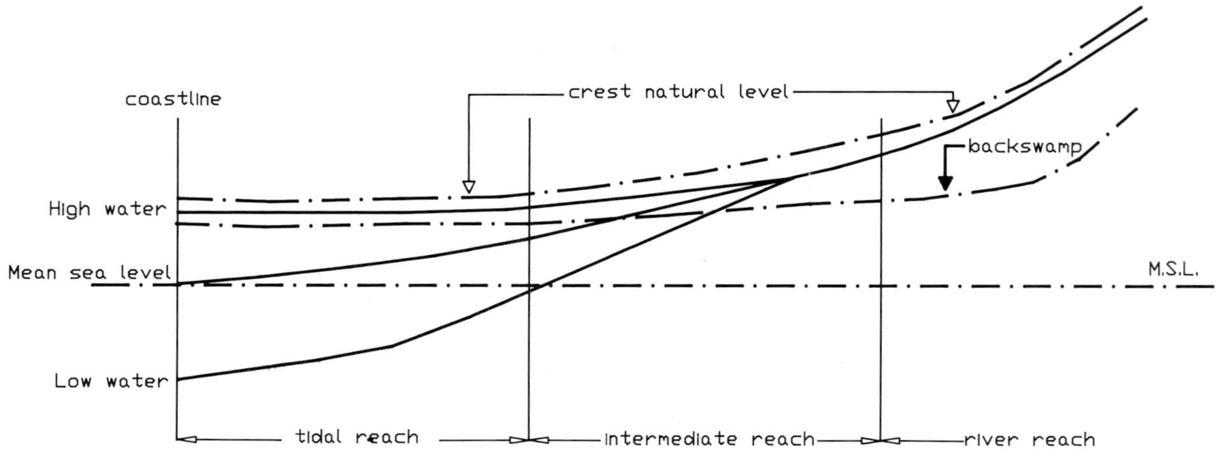

Fig. 2.13
Water levels and land elevation.

During spring tides the land in the coastal zone is below sea-level for part of the tidal cycle resulting in flooding. The short duration of the period with high levels and the time required for the propagation of the water inland restricts the inundation to a narrow strip along the coast. This inundation can be arrested by low coastal embankments (Figure 2.14).

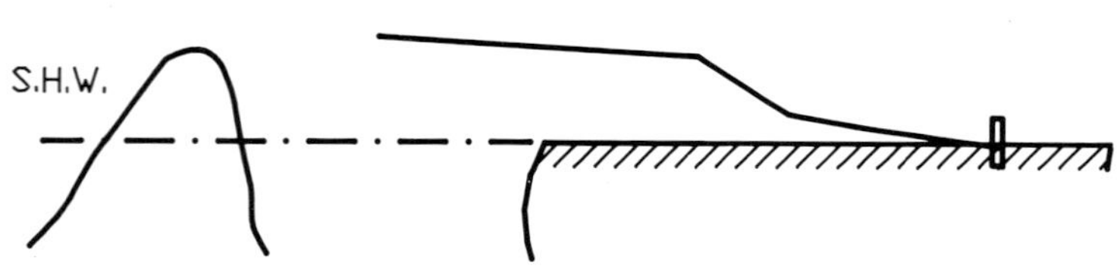

Fig. 2.14
Positioning of low coastal embankments.

Extensive flooding occurs in coastal areas of deltas which are not protected from storm surges. These surges do not contribute to the building up of the land areas, but produce deep flooding, extending to areas in the intermediate zone. The extent of the flooding in the upstream direction depends on the duration of the storm surge since the filling of the unprotected land areas takes some time. Maximum flooding of the intermediate zone occurs when the storm surge coincides with a river flood.

Coastal embankment in the delta of the Mekong (Golong).

An example of a deltaic area that is regularly subjected to flooding, both from rivers and the sea is in Bangladesh. Every year, one third to one half of the country is affected by floods, causing damage to 40 to 60% of the agricultural areas in the flooded zones (Figure 2.15).

Most of the flooding originates from the main rivers, which are the Ganges-Padma, Brahmaputra-Jamuna and Meghna, all of which have gentle floods. Flash floods occur in the smaller streams crossing the northern and eastern borders between April and October. Generally all the rivers in Bangladesh begin to rise in April or May. By mid or late June the low and moderately low lands are normally flooded by the rising river stages. The Brahmaputra rises in March and April due to melting snow and usually peaks in June, followed by another peak in August. The Ganges, on the other hand, begins to rise in June and July and peaks in late August or early September. When the second peak flow of the Brahmaputra coincides with that of the Ganges, heavy flooding occurs at their confluence and in downstream areas. The drainage of excess water to the sea during the wet season is hampered by the rise in mean sea-level of about 0.5 m due to the southwesterly monsoon winds.

The flooding in the coastal zone is caused by the set-up of the sea-level above the high water of the astronomical tides which occurs during cyclones in the Bay of Bengal. During the cyclone of 11th November, 1970, more than 300,000 people lost their lives in a relatively small area. The maximum sea-level was around 6.3 metres above mean sea-level. There are no reliable data on water levels during historic floods but it is likely that even higher water levels have actually occurred, perhaps reaching up to 9 metres.

2.1.5 Hydraulic and morphologic effects of embanking

Embankments (e.g. dykes and levees) constitute a commonly applied means of protecting land areas from flooding by rivers or the sea. They are generally easy to construct manually using spades and wheelbarrows, or with simple mechanical equipment like draglines and scrapers. In most cases, embankments can be built with material available at the site (i.e. clay, sand, gravel etc.). In the Netherlands, compressed seaweed and peat have also been used, while in Japan, India and Kerala walls consisting of natural rocks or bricks have been constructed to prevent flooding. Legend has it that King Menes, the first Pharaoh, built his capital around 3000 B.C. at Memphis, and protected it from the inundations of the Nile by a wall. This wall was constructed at a bend in the river where the Nile was most likely to break out during periods of excessive flow. In China

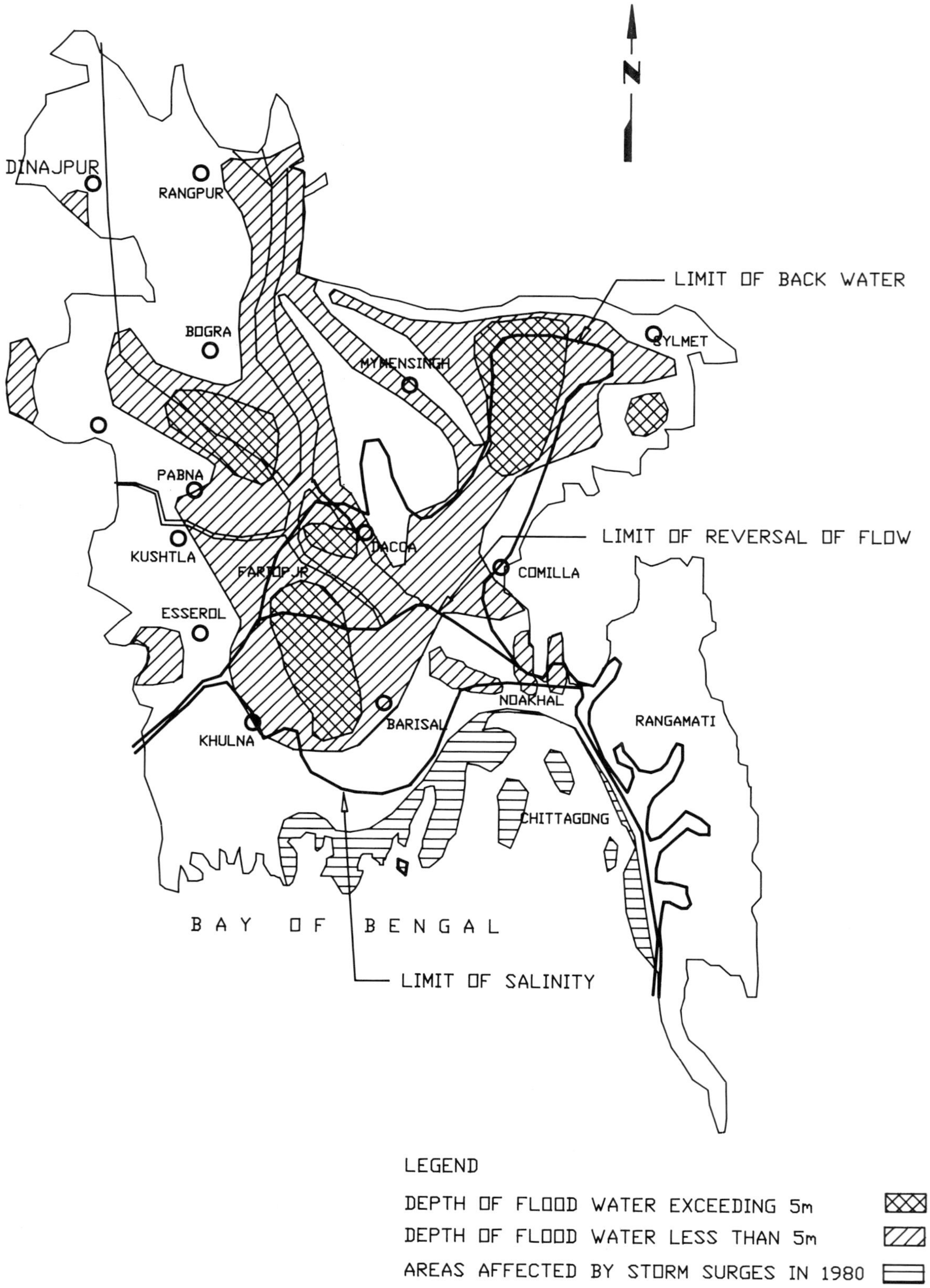

Fig. 2.15
Flooding in Bangladesh.

the legendary hero-emperor Yü the Great, who lived around 2280 B.C.. is reputed to have built great water works, dams and dykes. Later he became the patron of all hydraulic, irrigation and water resource engineers.

In modern times, extensive embankment systems, in some cases several thousands of kilometres in length, have been constructed along rivers. e.g. the Mississippi (U.S.A.), the Yellow River and Yangtze Kiang (China), the Rhine and Meuse (the Netherlands), the Red River and Song Cau (Vietnam), the Po (Italy), the Indus (Pakistan), the Loire (France) and many others. Embankments have also been constructed along coasts to protect low-lying areas from flooding by high astronomical tides and storm surges at sea. Examples of such coastal embankments and sea dykes can be found in many countries especially in Bangladesh, Japan, China, Korea, Vietnam, U.S.A., Guyana, Germany and the Netherlands. There is no doubt that embankments are, and will for a long time continue to be, an effective and economical means of flood protection. However, they present a number of side effects that, under certain conditions, may render their application less attractive. The various side effects of embanking may be considered in more detail by examining the effects upon:
- hydraulics;
- morphology;
- water management.

Side effects differ in relation to the particular reach of the river under consideration. In the non-tidal reach the river stage and discharge are only affected by the rain on the upstream catchment. Without embankments any flow exceeding bank-full discharge overtops the natural levees and gradually fills the backswamps or floodplain. At the same time a floodplain flow starts in downstream direction (Figure 2.16).

Hydraulic side effects

Embanking has two hydraulic side effects which occur simultaneously. These are the elimination of the floodplain storage, and the confinement of the total flow to the cross-section in between the embankments. The combined effect is a rise in the flood level for a given flood from the upstream catchment. The magnitude of this effect can be assessed accurately using scale or mathematical models. Scale models, where the river, the floodplains and the embankments are represented on a small scale in an hydraulic laboratory and where water is conveyed through the model, have become almost obsolete and have been replaced by more convenient mathematical models. Mathematical models are based upon the fundamental laws of conservation of continuity and momentum which are solved for a given topographical configuration at discrete moments in time. The topograhical schematisation of deltaic areas is determined by special features like canals, embankments, reaches of river channels, roads, ridges etc.

The application of both scale and mathematical models requires detailed data about recorded historical floods. These data refer to discharges, river stages and depths of flooding. The models are calibrated by routing a recorded flood through the model and adjusting coefficients (e.g. Chezy, etc.) in such a way that a good reproduction of the recorded flood is simulated. The embankments are then included in the model and the same flood is routed through the model. The difference between the computed river levels in each case indicates the magnitude of the hydraulic effect of embankments.

The two effects of floodplain filling and channel confinement can be considered separately. The filling of the floodplain area requires a lateral flow Q_L from the channel. Hence the discharge at section B is smaller than that at section A (Figure 2.16).

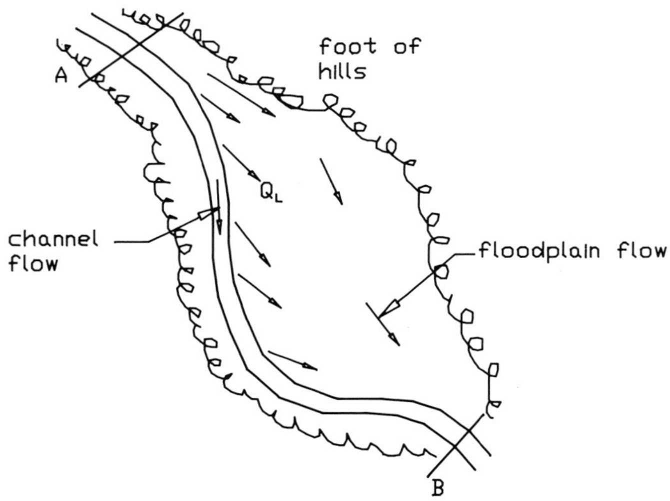

Fig. 2.16
Channel flow and floodplain flow.

Figure 2.17 shows the hydrographs in sections A and B before and after embanking. After embanking, the flood hydrograph in section B tends to become similar to that of section A. This means that the peak flow increases.

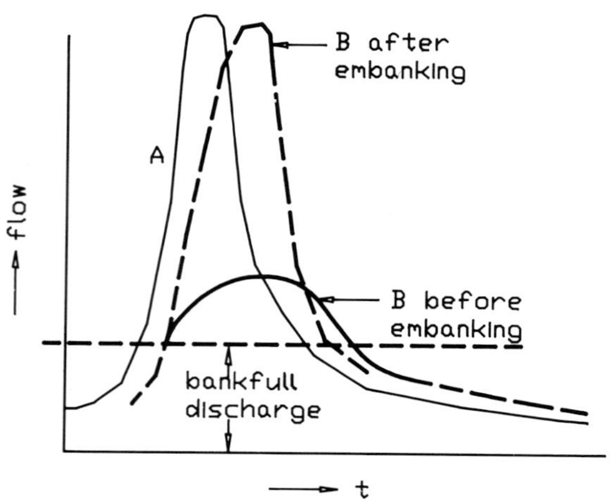

Fig. 2.17
Hydrographs at A and B, before and after embanking.

After embanking, the floodplain flow is conveyed by the channel confined between the embankments. This results in an increase in the channel flow which can be estimated if data on the velocity and depth of flows are available. The computations start at the coastline assuming mean sea-level, and gradients are computed for the situation prior to embanking (as recorded), and that after embanking (Figure 2.18).

Effects from flow confinement occur for both flash and gentle floods. The effect of embanking on the flood levels may be considerable with the rise amounting to several metres. In the 1960's the effect of completely embanking the Pampang, a floodplain in the Philippines (Luzon), where flash floods occur, was investigated using a physical model. A rise in flood levels of 2.5 to 3.5 metres was found for the middle reaches and thus the idea was rejected. The first low embankments along the Mississippi were built along one bank around 1840. After 1880 the embankment system was extended to both banks, and the levees were heightened. The result was a substantial rise in the

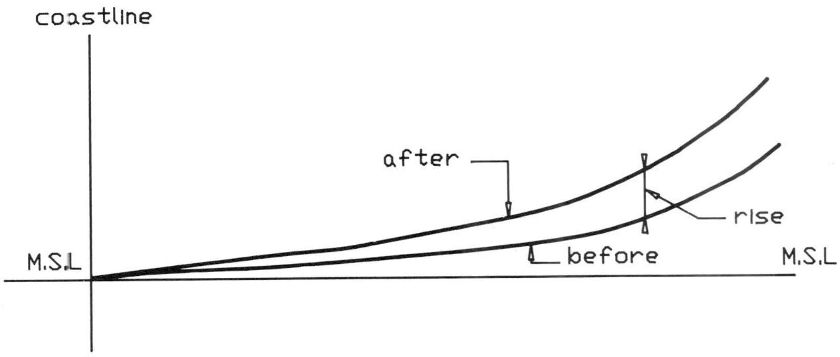

Fig. 2.18 Longitudinal flood profiles before and after embanking.

flood level in the order of 2 metres. The devastating effect of the flood of 1927 when the levees were breached was attributed to this double embanking. The rise was compensated by the implementation of a programme of cut-offs in the period 1932 to 1942. The hydraulic and other side effects of embanking have led to a long controversy on the desirability of a complete flood

Pampanga flood plain and delta (Philippines). Embankment consisting of a stone wall and house on stilts.

protection of the lowlands in the delta of the Irrawaddy river in Burma. A compromise was adopted consisting of horseshoe shaped embankments open at the downstream ends. During high floods part of the protected area is flooded thus maintaining the storage function of the floodplain. Recently a mathematical model of the floods in the delta tributaries was elaborated based on extensive field measurements. With this model it was possible to conclude that the complete flood protection by closed embankments would have increased the water levels of major floods by 1 to 1.5 metres in the middle reaches.

The water levels in estuaries and the lower river reaches are governed by those in the sea. These in turn vary according to the astronomical tides. Along certain coasts abnormally high levels, above the high water levels of astronomical tides, may occur under the effect of landward winds and differences in barometric pressure which causes a set-up of the sea-level. Between the zone where the sea levels prevail and the non-tidal reach, an intermediate zone is found where the maximum levels depend both on the sea-level and the river-stages upstream.
In the absence of surges the effect of embanking tidal marshes in deltaic areas is small, if the land

has been silted up to a level slightly above mean high tide. Flooding will only occur during high spring tides and the elimination of this flooding has little effect on the propagation of the tides in the estuaries and tidal creeks.

The mangrove belts along the coasts of Bangladesh and West-Bengal (India) reduce the propagation of storm surges into the non-embanked deltaic river branches. Here cyclones generally of short duration (less that 12 hours), generate surges of 6 metres or more. The mangrove belt which extends 120 kilometres inland offers a high hydraulic resistance to the flow and the water levels inland remain considerably lower than at sea. It thus offers natural protection to the city of Khulna and the low-lying areas behind the belt. Entire or partial reclamation of the mangrove belt would expose the area to possible flooding from the surges.

Effect on morphology

Sediment moves in streams as suspended load in the flowing water and as bed load along the channel bottom. The two processes are not totally independent as material which travels as bed load at one section may become suspended load at another, and vice versa. A transition zone between the two is known as saltation, which is the movement of particles by bouncing along the bed. The movement of the particles and the range of particle size depend on the physical properties of the flowing water and of the particles themselves. The forces acting on a particle in water are the drag between the solid and the water,the buoyancy of the particle and the gravitational force. The first two forces tend to keep the particle in suspension while gravity causes it to settle. Consequently, when a mixture of grain sizes is involved the coarser material is deposited first while the finer material is carried further downstream. This leads to sorting of the original mixture by size. Sorting, together with varying sediment loads and water velocities lead to deposition of layered sediments. The relationship between the current velocity and the size of particles that may be transported has been determined experimentally and is presented in Figure 2.19 (Sawkins, 1978).

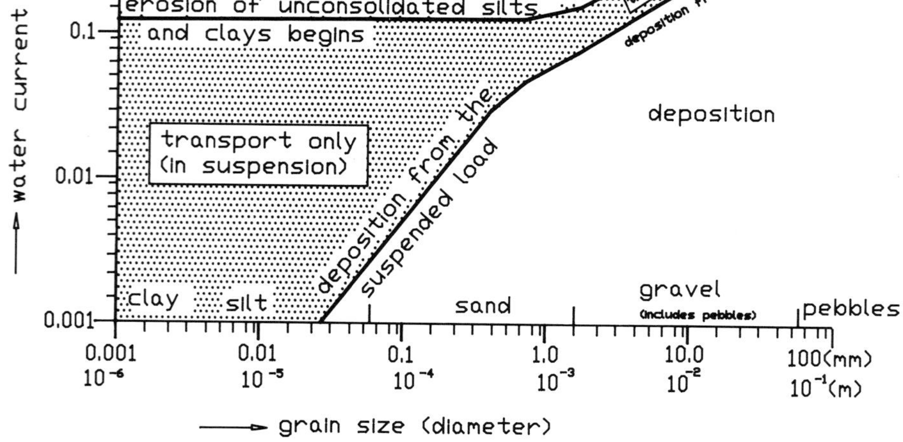

Fig. 2.19
Relationship between current velocity and the erosion and deposition of sediment.

Delta of the Irrawaddy river Myanmar. Bank Erosion near Henzada threatening an embankment which protects the town. To be prepared for a possible failure, a set-back embankment has been built to serve as a second line of defence.

Embanking may affect the morphology of river channels in several different ways. In some cases, after embanking, a rise of the river bed is observed, while in other cases, a deepening has occurred. This may be explained by the fact that, after embanking, the river channel between the dikes has to carry more sediment than before, while at the same time, the sediment-carrying capacity of the river is increased because of the rise in the flood levels. Whether silting up or scouring occurs depends on which of the two factors is predominant. The Yellow river in China and the Red River in Vietnam are examples of rivers where the river bed has risen several metres above the level of the land area. The Rhine in its deltaic reach in the Netherlands exhibited this feature until large scale dredging was introduced. Around 1800 it was seriously considered to abandon certain embanked areas so that these could again function as storage basins for the retention of flood water and silt. The suggestion was never put into effect because the farmers did not want to be drowned out for the sake of future generations.

River training leads to an increase in the sediment-carrying capacity of the river. When more sediment is carried to the sea the extension of the delta into the sea is accelerated and in the long term a further rise of the river bed can be expected. Embanking often entails increased meandering and subsequent bank erosion because of increased velocities. Local bank protection only provides temporary relief, and protection has to be extended to upstream reaches which ultimately results in complete river training. This is much more costly than embanking and suitable material, like heavy stone, is often difficult to obtain. Embanking halts the natural process of building up of the land areas, especially in deltas. It also stops the deposition of the fertilizing silt. Experience in Burma, Cambodia and Rumania with the operation of sluices for the controlled admission of silt-laden water, have shown that the sediment tends to remain near the site of the sluices, little being deposited further away in the lowest parts of the backswamps.

Effect on water management

After embanking a natural land area in a delta, excess water from local rainfall cannot drain away. Sluices in the encircling dyke are not sufficient to maintain an adequate control of the water level in the entire area. An artificial drainage system, with field drainage and conveyance canals (i.e. primary, secondary and tertiary canals), has to be constructed. Embanking eliminates the beneficial effects of moderate floods. In many tropical deltas, where rice is cultivated, floods often supplement irrigation water when local rainfall is low. There are cases where farmers make breaches in the dikes in order to admit water during dry periods. To prevent this action, an irrigation system, including sluices for the admission of water, may be necessary. Drainage may also be necessary to alleviate the effect of over-irrigation. Embanking eliminates the flushing effect of the floods, which remove waste products (e.g. human disposal, salt etc.) that accumulate during dry periods. This is a serious problem in warm tropical areas with long dry seasons. Flood protection by embanking, accompanied by artificial drainage and irrigation, creates new possibilities for intensive agriculture. However, it requires a completely new attitude and adaptation of the farming population and the agricultural pattern, if the full benefits of the new situation are to be realised.

2.1.6 Salt water intrusion

Estuaries are, by definition, areas where the fresh river water meets the salt water of the sea, leading to a transition zone between a riverine and a marine ecosystem. From a physical point of view, interesting phenomena are involved, like the intrusion of sea water into the river, the mixing between fresh and salt water and the effects on sedimentation.

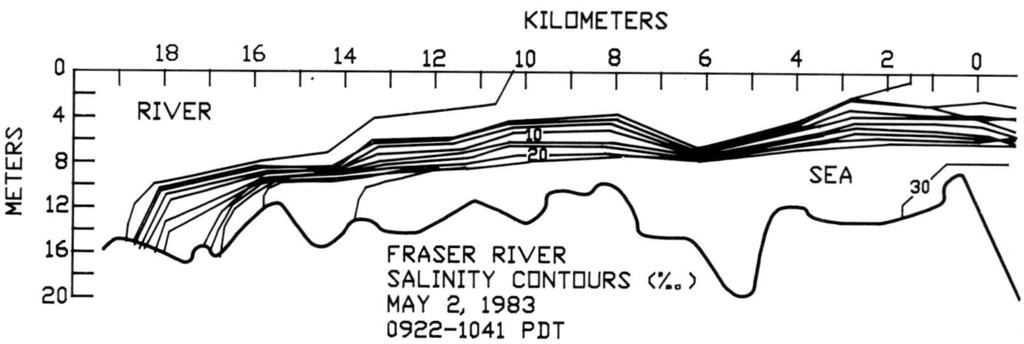

Fig. 2.20
Salinity wedge in Fraser River.

Figure 2.20 shows a longitudinal and vertical distribution of the salt contents in the mouth of the Fraser river in Western Canada (Rockwell Geyer, 1987). The heavier sea water penetrates about 20 km under the river water. There is hardly any mixing so the upper part of the salinity wedge is practically pure river water and the lower part is pure sea water.

Figure 2.21 gives the salinity for average conditions in the Western Scheldt estuary on the Dutch-Belgian border. The vertical distribution is not given, since the water is completely mixed and there is only a longitudinal salinity gradient. The (diluted) sea water reaches as far as Antwerp which is about 90 km from the sea.

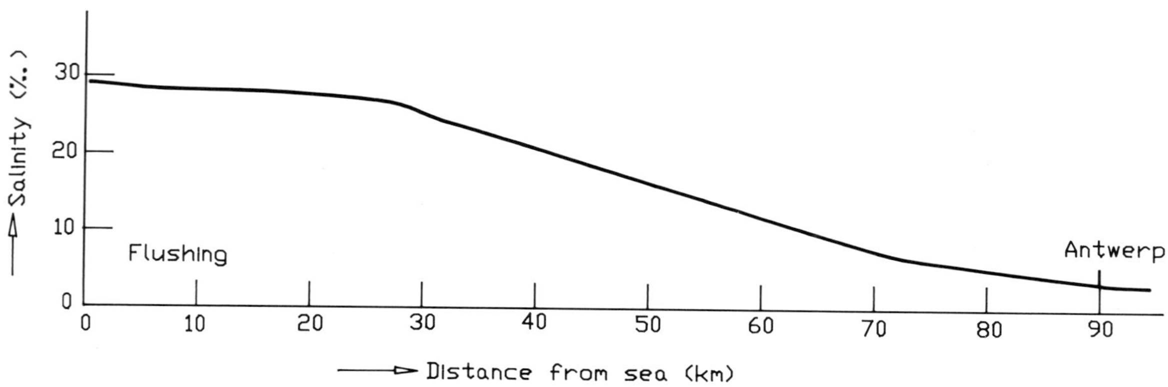

Fig. 2.21
Salinity in the Western Scheldt.

These differences in the characteristics of the salinity intrusion are above all governed by the river discharges and the tidal motion in relation to the geometry of the estuary. This difference in character can even exist in the same river. Figure 2.22 shows the salinity distribution in the Mesuji river in Sumatera, Indonesia during the dry and the wet monsoon (with less than two months in between). With low river discharge, the salinity is almost vertically mixed. With high discharges, there is a sharp interface and the intrusion, at least in the upper layer, is pushed back over 10-20 km.

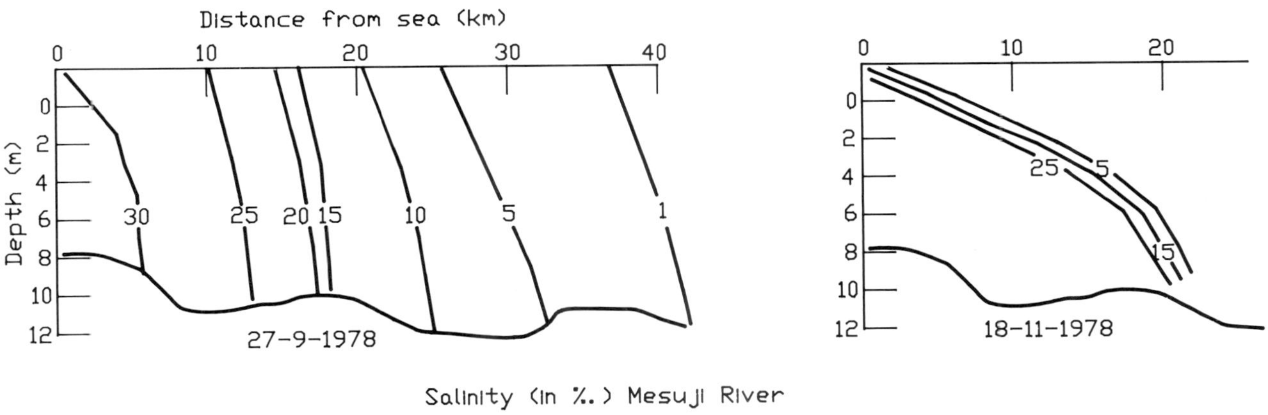

Fig. 2.22
Salt insrrusion in Mesuji River, Indonesia.

Sea water intrusion can endanger the supply of drinking or irrigation water in deltaic areas. Sea water intrusion also influences sedimentation in estuaries, which can cause shoals in navigation channels. Deepening of these channels will cause the sea to intrude further inland, giving again more difficulties with fresh water supply. These examples will make it clear that a good understanding of the mentioned phenomena, including the influence of human interference, is of paramount importance to people dealing with water management in river deltas.

Even in a simple estuary with one branch, sea water intrusion is a rather complicated phenomenon. In this section we will analyse this phenomenon in separate parts, in order to get a better understanding. In the introduction, river discharge and tidal motion in relation to geometry were already mentioned. The influence of geometry is easy to imagine when we consider the depth and the width, determining the effectiveness of the river discharge in pushing back the sea. There are other geometrical factors which are less evident at first glance, like harbours and confluences of branches. They can, through phase-shifts in the tidal motion, disperse sea water in the estuary. Other factors are the density difference (not every sea is equally salty) and wind. We will discuss these factors briefly, with the main focus on the understanding of the phenomenon For formulae and computational methods, references are given for further studies.

Saline wedge

The difference in density between sea water and fresh water is only a few percent, but it strongly influences the flow in river mouths. To understand how it works, think of a lake, separated from the sea by a dam.

Removing this dam leads to the exchange flow as shown in Figure 2.23a. Salt water flows under the fresh water, resulting in a situation where the fresh water floats on top of the salt water. When there is a flowing river, instead of a stagnant lake, there is a friction force at the interface that works against the exchange flow. The intrusion stops when there is equilibrium between both forces. This is then an *arrested saline wedge* (Figure 2.23b).

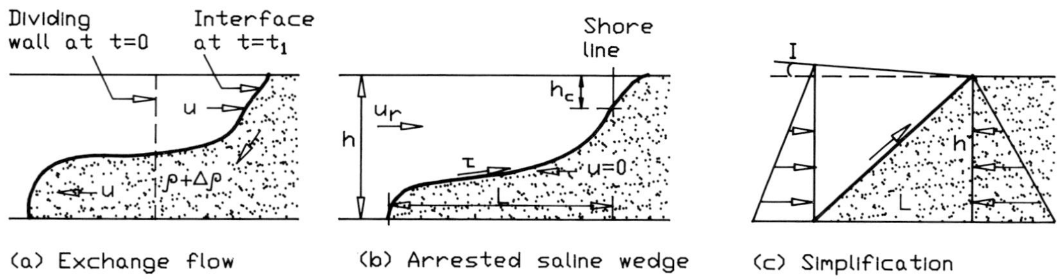

Fig. 2.23 Exchange flow and arrested saline wedge.

At the place where the river enters the sea, the flow of the fresh water over the salt water is critical, meaning that the velocity there is equal to the celerity of a wave at the interface, expressed by means of the internal Froude number:

$$Fr_i = \frac{u_r}{\sqrt{\frac{\Delta \rho}{\rho} g h_c}} = 1$$

The first description of the shape of a saline wedge was given by Schijf and Schönfeld (1953), experimentally confirmed by Keulegan (1957) and summarised by Harleman (1961). Here, a simplification will suffice to get insight in the relevant parameters. Assume that the shape of the wedge is as given in Figure 2.23c. The total force acting on the wedge to the right is then $1/2\rho g(h+IL)^2$ and to the left is $1/2(\rho+\Delta\rho)gh^2$. Equilibrium between the two forces leads to:

$$\frac{1}{2}\rho g h^2 + \rho g h I L + \frac{1}{2}\rho g I^2 L^2 = \frac{1}{2}\rho g h^2 + \frac{1}{2}\Delta\rho g h^2$$

With representative values of $h \approx 10$m, $I \approx 10^{-5}$ and $L \approx 10{,}000$m, the third term at the left hand side can be neglected, leading to:

$$L = \frac{\Delta\rho}{\rho}\frac{h^2}{2hI} = K\frac{\Delta\rho}{\rho}\frac{h^2}{u_r^2}$$

Where K is a constant of proportionality since, in uniform flow, hI is proportional to u^2 (Chezy's law). From this we can see that salt intrusion is proportional to the square of the river velocity, indicating that a decrease of the river discharge and widening or deepening will lead to more intrusion. The effect of deepening is especially pronounced through its influence upon both h and u_r.

Tidal excursion

With flood tide, sea water flows into the river, flowing back again during ebb (Figure 2.24). The distance travelled is called the tidal excursion. A larger difference between high-water (HW) and low-water (LW) leads to a larger excursion into the river and to more intrusion. A larger river discharge gives a smaller excursion and gives less intrusion.

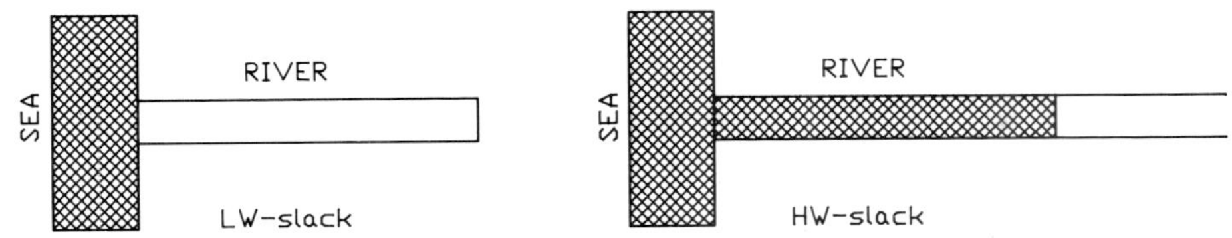

Fig. 2.24
Sea water intrusion due to tidal motion.

Local mixing

Consider the situation shown in Figure 2.25a for stratified flow and Figure 2.25b for fully mixed flow. The potential energy of the stratified flow can be expressed as:

$$PE_{stratified} = \frac{h}{2}(\rho + \Delta\rho)\frac{gh}{4} + \frac{h}{2}\rho g\frac{3h}{4} = \frac{h^2}{2}\rho g = \frac{h^2}{8}\Delta\rho g$$

The potential energy of the mixed flow can be expressed as:

$$PE_{mixed} = h(\rho + \Delta\rho)\frac{gh}{2} = \frac{h^2}{2}\rho g + \frac{h^2}{2}\Delta\rho g$$

This shows that mixing leads to a rise in the center of gravity of the flow and hence to a higher potential energy. Energy is therefore needed for mixing. Sources of this energy can be the friction at the interface of the saline wedge, the wind or the bottom friction of the flow during the tidal excursion in the river. The latter is usually the most important effect.

Fig. 2.25
Rise in potential energy due to mixing.

Large scale mixing

Figure 2.26 shows the situation of a harbour along a tidal river. The phenomenon described here is similar to that in side branches, flood plains, confluences etc. The flow from the river to the harbour is completely governed by the waterlevel in point A. When the waterlevel rises, the water flows into the harbour and when it falls, the water flows out. In the river, the situation is somewhat different. Due to inertia, the water still flows inland even after the water level has started to fall. Because of this phase shift, a cloud of salt water is divided into two separate clouds. At time 2, the cloud is separated and the parts do not rejoin. Other phase differences are possible, but they all have in common a large scale dispersion of clouds of water. This will be discussed later in more detail in the section on diffusion and dispersion.

In most estuaries, all of the above phenomena can be seen. They can work together, like the tide that intrudes and mixes at the same time, and they can influence each other. For example, the exchange flow due to density differences and large scale mixing effects may both be present in a

harbour that is situated in a river stretch with a saline wedge. For more information on these effects see Roelfzema et al. (1978) and Abraham et al. (1986).
Figure 2.27 shows the relation between the tidal flow and the chlorinity in Rotterdam. The

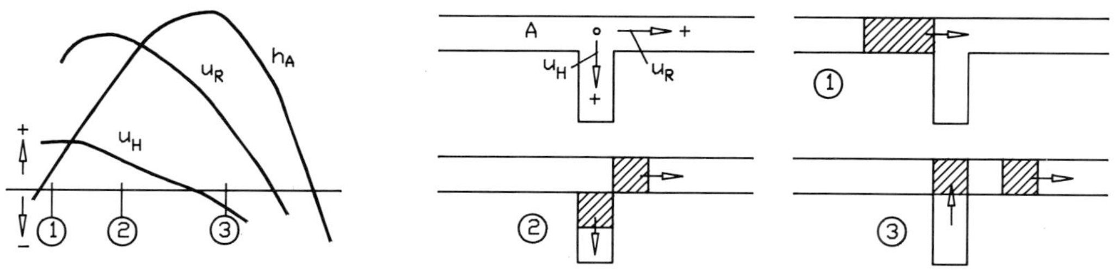

Fig. 2.26 Influence of tidal phase differences and geometry.

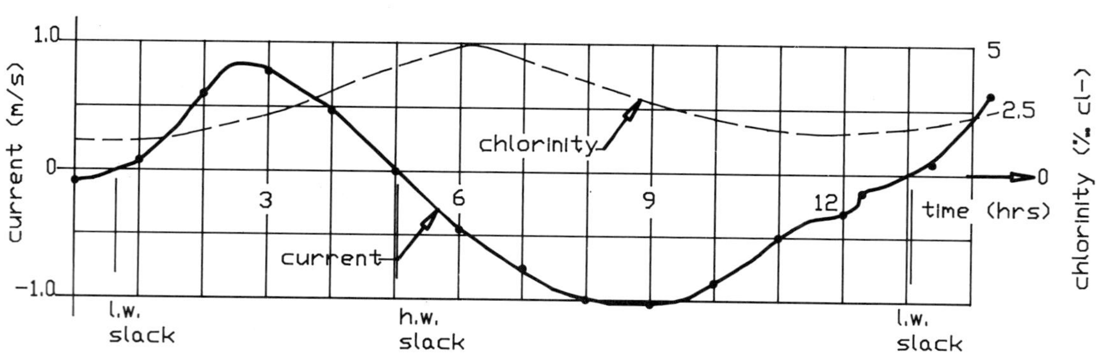

Fig. 2.27 Current and chlorinity in Rotterdam Harbour.

maximum is reached shortly after HW-slack, due to the combination of tidal excursion, density differences and large scale mixing.

Classification

With regard to sea water intrusion, estuaries are usually divided into three types: stratified, partly mixed and mixed (Figure 2.28).

There are several parameters that are used to indicate the type, of which two will be presented here. As mentioned before, tides are very important in the mixing process. A simple parameter therefore is the tidal volume related to the river discharge. The *flood number*, α is defined as:

$$\alpha = \frac{Q_r \, . \, T}{V}$$

where:

Q_r	= river discharge	(m^3/s)
T	= tidal period	(s)
V	= 1/2 (ebb volume + flood volume)	(m^3)

Another parameter was presented by Thatcher and Harleman (1972), including the internal Froude number. The *internal estuary number*, E_D is defined as:

$$E_D = \frac{V}{Q_r \cdot T} Fr_i^2$$

The internal estuary number also contains the inverse flood number, so a small flood number indicates the same type of estuary as a large internal estuary number. Table 2.3 gives ranges of these numbers for the types of estuaries shown in Figure 2.28.

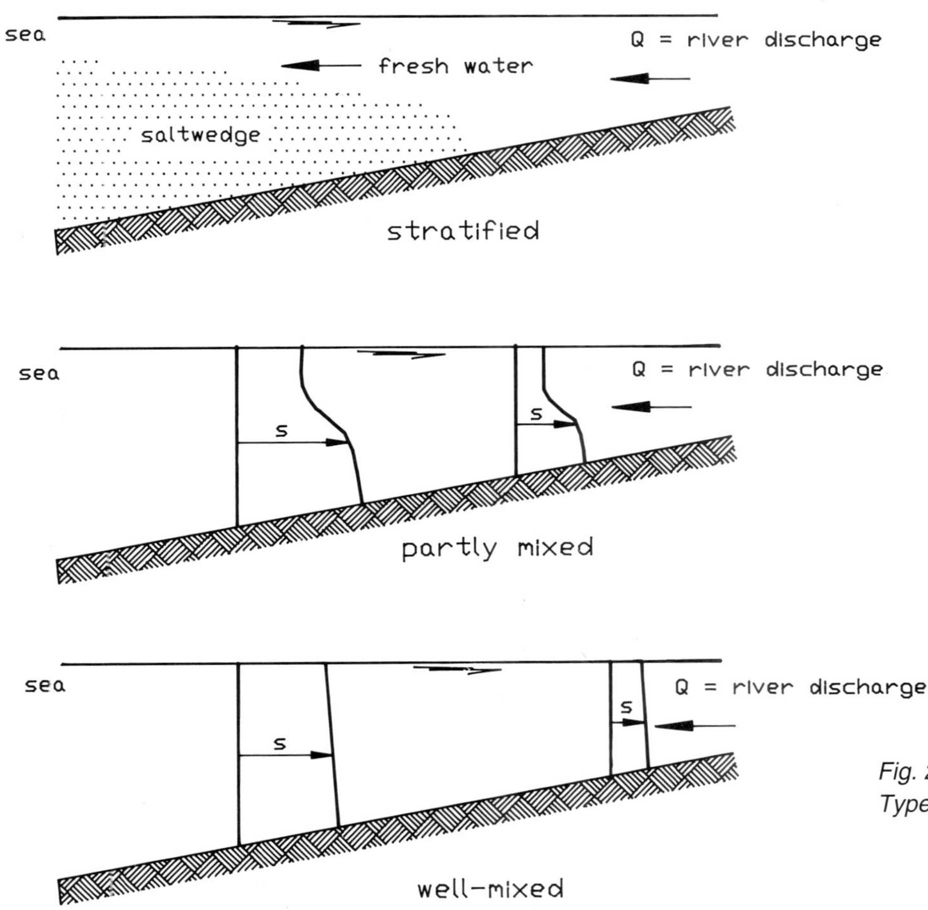

Fig. 2.28
Types of estuaries.

Type of estuary	Well mixed	Partly mixed	Stratified
Flood number α	< 0.1	$0.1 < \alpha < 1$	> 1
Int. est. number E_D	> 8	$0.2 < E_D < 8$	< 0.2

Table 2.3
Classification of estuaries and their parameters.

This classification makes it possible to get an idea of the type of sea water intrusion in an estuary. Savenije (1992) also presents a new method to obtain an impression of the degree of salt intrusion in a (mixed) estuary.

Diffusion and dispersion

There is a lot of confusion about diffusion and dispersion. Many 1-dimensional dispersion models are being used, in which the basic equation looks very similar to the one used for diffusion. Nevertheless, they are different and much depends on the scale on which the problem is viewed and averaged.

Diffusion is the molecular process of mixing at an interface where some property (e.g. temperature or concentration of salt) has a gradient. The influence of this diffusion is limited to the vicinity of the interface. The term diffusion is also used for turbulence, where water particles, containing some property, are exchanged due to the erratic movement of the water in turbulent flow. This process enlarges the interface tremendously, thus stimulating the molecular diffusion, leading to real mixing.

Dispersion is the process of scattering and has nothing to do with real mixing, but when viewed on a larger scale, averaging of values makes it look almost the same. This can be demonstrated by considering the example of a concrete beam reinforced with steel. For the whole beam it is possible to express the average steel content of the beam. But, there are areas of pure steel in the beam and other areas with pure concrete. In this case, there is no mixing at all!

The same can be seen in Figure 2.29 where the vertical averaging of the salt content in a saline wedge (Fig. 2.29a), the averaging of salt and fresh water in time (Fig. 2.29b) or the salinity distribution for a well-mixed estuary (Fig. 2.29c), all lead to the same result for the average salinity level (Fig. 2.29d). This explains the similarity of many equations, while the processes behind it can be completely different.

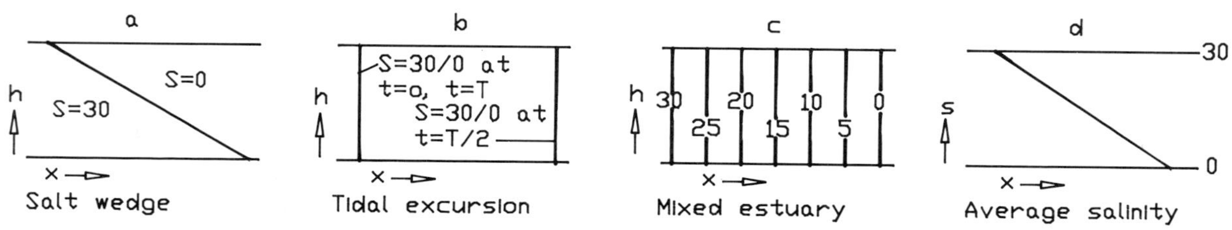

Fig. 2.29
Longitudinal dispersion due to averaging.

The use of 1-dimensional dispersion models is widespread. It is therefore important to have at least some idea of the backgrounds and limitations of these models. In the following sections we will look somewhat more in detail at the processes. The emphasis is again on insight. For more information the reader is referred to Fischer et al. (1979).

Any substance present in flowing water moves with the current and the result is called *advective transport*. The (turbulent) *diffusive transport* is the exchange of fluid particles with the substance and is proportional to its concentration gradient. The result can be expressed per unit area perpendicular to the flow as:

$$T_{tot} = T_{adv} + T_{dif} = uc + D\frac{\partial c}{\partial x}$$

where:

u	= area averaged velocity	(m/s)
c	= area averaged concentration	(kg/m^3)
D	= turbulent diffusion coefficient	(m^2/s)

The result is that a cloud in a flow, with equal velocity everywhere in the cross-section of the flow, is moved with that velocity and diluted at the same time, because of the advective and diffusive transport (Figure 2.30).

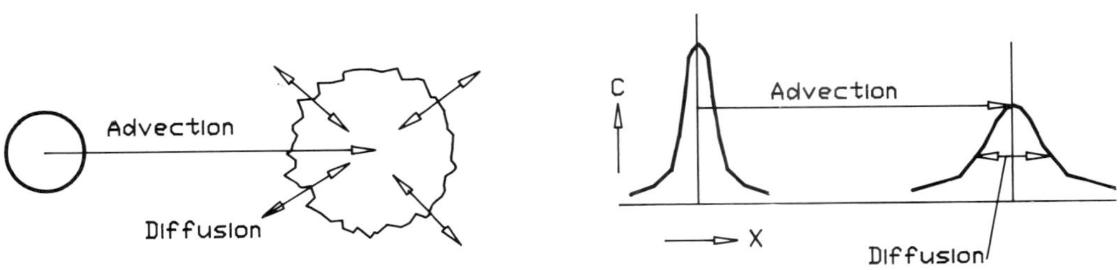

Fig. 2.30 Advection and diffusion.

In a real flow, however, the velocity is not equal over the cross-section. The result of the advective transport is a scattering of the particles. After some time and averaged over some area (for a 1-dimensional model this is the cross-sectional area, A), this result looks very similar to diffusion (Figure 2.31).

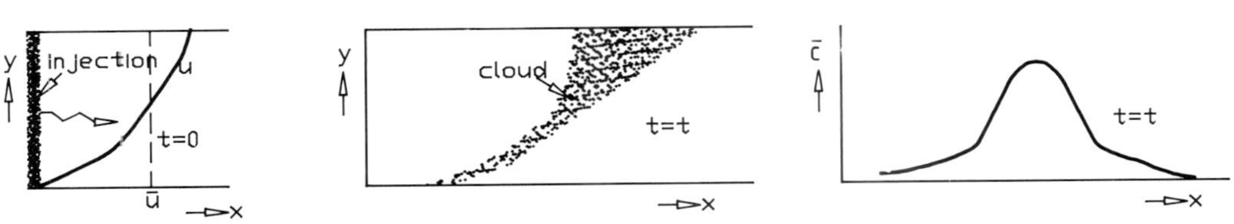

Fig. 2.31 Advection, dispersion and diffusion.

Dispersive transport follows from averaging a non-uniformly distributed velocity and concentration profile. The magnitude is the total transport minus the product of the average velocity and average concentration in the cross-section. The dispersive transport is all the transport that is not expressed by the average velocity and average concentration in a cross-section or, in other words, it is the transport as observed when moving with the average velocity. Relative to this average velocity, areas with lower velocity move backwards and areas with higher velocity move forward, leading to an apparent diffusion.

In order to make 1-dimensional computations possible, the dispersion transport is assumed to be proportional to the gradient of the vertically averaged concentration, in a similar way to diffusion. This then leads to the familiar 1-dimensional dispersion equation:

$$\frac{\partial (Ac)}{\partial t} + \frac{\partial (Qc)}{\partial x} + \frac{\partial}{\partial x}\left[AD_x \frac{\partial c}{\partial x}\right] = 0$$

where:

Q	= flow	(m^3/s)
A	= cross-sectional area	(m^2)
c	= concentration averaged over cross-section	(kg/m^3)
D_x	= longitudinal dispersion coefficient	(m^2/s)

For a steady state or a tidal-averaged sea water intrusion, the transport equation reduces to:

$$uc = -D_x \frac{dc}{d_x}$$

which, after integration and with $c = c_z$ at the seaward boundary, gives:

$$c = c_z e^{-\left(\frac{u}{D_x} x\right)}$$

which is the simple 1-dimensional, steady-state equation for salt intrusion. Many empirical models have been based on this equation, the main problem being to find an expression for D_x (see Rijkswaterstaat, 1976, and Savenije ,1992). For an overview of the relations between flow phenomena and the dispersion coefficient see Fischer et al. (1979) and Winterwerp (1983). Only two aspects of the dispersion coefficient will be discussed here.

The dispersion in Figure 2.31 is caused by the velocity distribution which is, in turn, caused by shear due to the bottom friction. From assumptions on turbulent exchange coefficients, it is possible to calculate the dispersion coefficient, which is proportional to the water depth, h and the shear velocity, $u_* = \tau/\rho$. In an estuary, horizontal velocity differences also play a role. For Dx in this case, a typical value is:

$$D_{x\text{-}flow} = 100\, hu_*$$

but values up to 5 times higher or lower can be found.

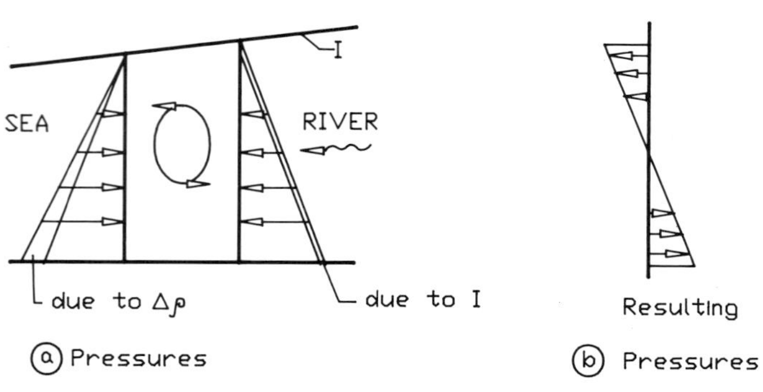

Fig. 2.32 Gravitational circulation.

The pressure distribution shown in Figure 2.32 causes a circulation referred to as *gravitational circulation*. Again, with assumptions about turbulent exchange, it is possible to calculate the dispersion coefficient. More stratification will give a higher dispersion coefficient. D_x can then be expressed as:

$$D_{x\text{-}grav.circ.} = K\left[\frac{dc}{dx}\right]^2$$

where the constant K is a function of turbulent exchange parameters and the water depth. Typical values of D_x in this case can be 10 to 100 times as high as the values calculated for the flow dispersion, which has, in turn, an order of magnitude 100 times higher than the turbulent diffusion coefficient, D.

A widely used approximation for the total dispersion in a non-stationary model, is given by Thatcher and Harleman (1972):

$$D_{x\text{-}T.H.} = 20Ru_* + K\frac{\partial c}{\partial x}$$

The first part of this expression represents the flow dispersion and the second part the gravitational circulation. K expresses the degree of stratification and is a function of the estuary number, E_D.

Effects on sedimentation

Salt intrusion also influences the sedimentation pattern in estuaries. Beside physics, also chemical factors play a role. Sediment transport is a function of flow velocity. Where the velocity decreases, sedimentation can be expected. This is the case where the salt intrusion meets the river water (see Figure 2.33).

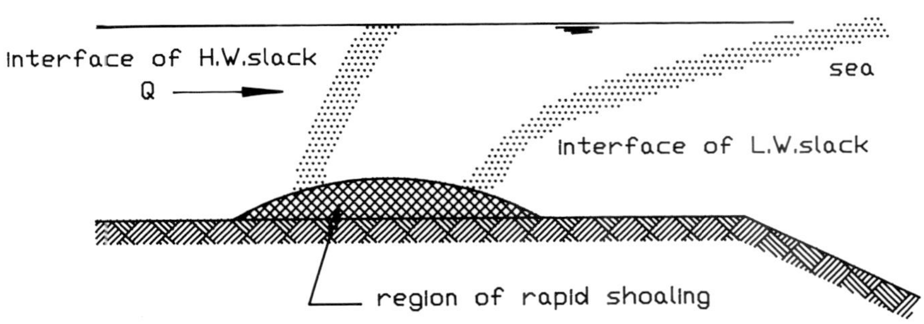

Fig. 2.33
Shoaling and salt intrusion.

The origin of the sediment can be both riverine and marine. With a sharp saline wedge and a constant river discharge, this shoaling will be very pronounced, while in a well-mixed estuary with a large tidal excursion and a varying discharge, the sedimentation will be spread over a large area. Sediment in the lower river regions and deltas is very fine and often consists of clay minerals. In that case, not only gravity but also electrostatic forces play a role. In fresh water, the clay particles are negatively charged on the surface. The positive ions in sea water (Na^+, Ca^{++} and others) neutralize the electrostatic forces and the particles flocculate. This happens already at salinity values of about 3 ‰, resulting in large flocs with a settling velocity that exceeds the value in fresh water by a factor 10 or more (see Massie et al. 1986).

Figure 2.34 shows the relation between the current velocity, the chlorinity and the concentration of suspended sediment at Rotterdam. The increase in settling velocity, due to flocculation when the chlorinity increases, seems to result in lower sediment concentrations during maximum salt intrusion. The minimum concentration at 6-7 hours goes together with an increasing velocity and maximum chlorinity.

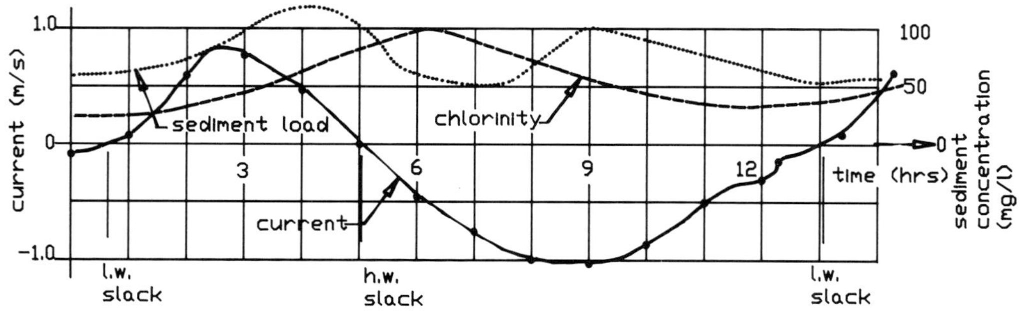

Fig. 2.34
Current, chlorinity and sediment at Rotterdam.

Other sources of salt

Seawater intrusion into open estuaries is only one of the sources of saline or brackish water which produce a high salinity of the surface water in canals and other water courses in deltaic areas. It is a source which is found in all deltas regardless of their state of technical development. The other sources of salt are especially found in deltas with a high degree of development. In the first place mention should be made of the navigation locks between the sea or an estuary with saline water and a freshwater canal. This process can be explained by considering a situation where the canal and the lock chamber contain fresh water and the outer harbour saline water. The water levels are the same everywhere.

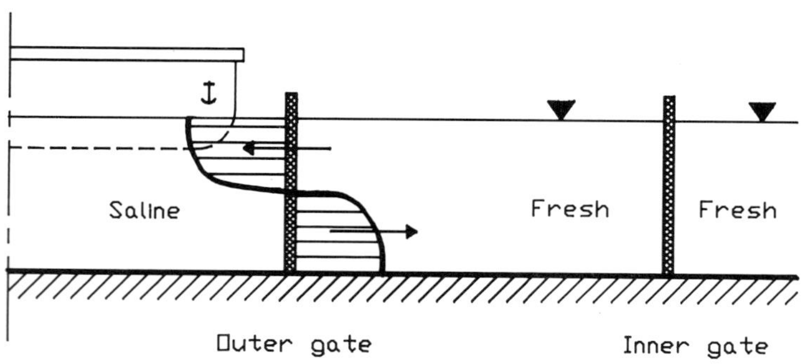

Fig. 2.35
Filling of a lock chamber with saline water.

When opening the outer gates to admit the ship into the lock chamber, the saline water, by virtue of its greater density, slides over the lock floor and drives the fresh water out as indicated in Figure 2.35. Finally, all the fresh water in the lock chamber is replaced by the saline water from the outer harbour. The duration of the exchange depends on the dimensions of the chamber and is of the order of 10 to 30 minutes. The velocity of the outflowing fresh water on the surface may be as high as 20 cm/sec. When, after the exchange, the inner gates are opened to give the ship access to the canal, a volume of saline water corresponding to the entire volume of the chamber will spread over the canal. When the sea level outside is higher than the inside water levels the levels have to be equalized and an additional volume of saline water will enter corresponding to the difference of the levels and the area of the lock chamber.

It is of interest to note that the navigation locks in the delta of the Rhine and Meuse in the Netherlands were identified as early as 1667 as places where high amounts of seawater were entering the canals. It was an engineer, named Hendrik, who considered these locks as the only cause of the high salinity of the inland waters in the low-lying part of the Netherlands. He also indicated how to reduce the amount of saline water entering and how to improve the general water management system. More information about this will be given in Section 3.3.

Salt water may also penetrate into the inland water courses by leakage at the gates of sluices and other barriers. Such structures should be provided with rubber slabs to achieve water tightness. Another and sometimes quite important inflow of saline water into the surface waters consists of seepage of brackish or saline groundwater in areas where differences in land elevation have been created by reclamation. The inflow depends on the hydraulic head and the geotechnical constants of the subsoil (e.g. resistance, c and transmissivity, k/D). For further information reference is made to Section 2.2.3 on groundwater flow.

Finally, the river supplying water from the catchment to the delta may carry water with a high salinity. The salt may be of natural origin like in (semi)-arid zones where the groundwater which supplies the baseflow to the river is often quite saline. In the humid temperate zone the natural salinity of the river is nearly always less than 10 or 20 parts per million (p.p.m.) of Chloride (Cl^-). In the arid zones it may amount to several thousands p.p.m. Cl^-. The salinity of riverwater may be increased by human activities such as drainage of saline soils and disposal of industrial effluents containing waste salts. A striking example of the latter is provided by the river Rhine which drains waste water (mainly with sodium chloride) from potassium mines in France. The present load from this and other sources is such that, at low river discharges, the chlorinity of the river water where it enters the delta can be as high as 600 to 700 p.p.m., corresponding to more than 1200 p.p.m. Total Dissolved Solids. This water is unfit for most uses and agreements have been made to reduce the salt load and to temporarily retain salt during periods of low flow.

2.1.7 Surface water modelling

Every human interference in the water flow of a delta has consequences. Deepening, widening, closing off river branches, water withdrawal or waste water discharges influence the water distribution, sea water intrusion or water quality. To predict the consequences, some model of the phenomena, applied for the local geometry, is indispensable. Such a model can be simple, like a measured relation between river discharge and salinity at a certain point, or it can give a detailed description of the physical phenomena in more dimensions. It can be small, like a river stretch of a few kilometers or cover a whole delta with many branches. Often, there is not just one model for a particular problem and a tailor-made tool is required. In the last decades, there have been major advances into the understanding of physical backgrounds and into the possibilities of hydrodynamic mathematical modelling. However, even the most sophisticated model can not work without local knowledge and field measurements. So, field data are needed whenever you want to make any statement about a certain estuary.

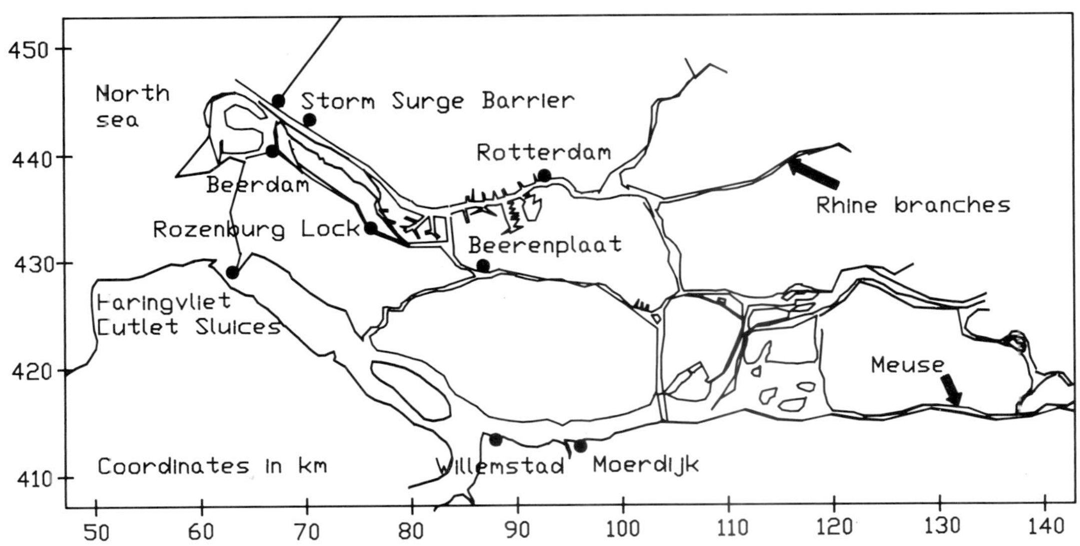

Fig. 2.36
Rhine Meuse Delta.

Hydraulic scale model shows the effect of embanking on flood levels in the Pampanga River Flood plain and delta in the Philippines. Scale models have been superseded by more convenient mathematical models.

Here, some examples will be given on the use of models in a river delta. This will be done for the Rhine-Meuse estuary near Rotterdam in the Netherlands (see Figure 2.36). In this estuary there have been many man-made changes in the past decades, which were accompanied with model investigations. Various types of mathematical models have been made for the area, making it a "showroom" of available techniques. In the past there was also a scale model of the whole area, mainly for the purpose of the investigation of sea water intrusion. Such a model is expensive to build and to operate and new techniques for mathematical modelling have made the scale model superfluous. This does not mean that a mathematical model is cheap and the costs of a model should always be balanced by the costs of the problem under consideration.

The Rhine-Meuse estuary is the main outlet of these rivers into the North Sea. The present geometry is, to a large extent, man-made. The river water flows freely past Rotterdam through an outlet at Hook of Holland which was dredged in the 19th century for the benefit of Rotterdam harbour. There is an outlet sluice in the southern branch of the delta, the Haringvliet estuary, which was built as part of the Rhine canalisation and the Delta works. These sluices are closed during low river discharges to minimise sea water intrusion and during very high sea levels to prevent the area from flooding.

Recent plans for the Rhine-Meuse estuary include the building of a storm surge barrier west of Rotterdam to safeguard the area further against flooding from the sea. At Beerdam and Rozenburg lock, new navigation channels between the harbour basins and the hinterland are planned, mainly to facilitate inland water container transport. In the southern branch there are plans to build an artificial island for the storage of polluted silt, which deposits in the delta and has a negative influence on the ecology of the whole area. For these and other interferences, model studies have been made, some of which will be presented here.

Fluid motion and the distribution of matter in water is always 3-dimensional in nature, but 3-dimensional mathematical models are still rather experimental. For practical applications in deltaic areas, simplifications have to be made. This is usually done by focussing on the dominant dimensions and phenomena. Three types of mathematical models are presented here; 1-dimensional, 2-dimensional horizontal and 2-dimensional vertical. Figure 2.37 shows the parts of the Rhine-Meuse delta where each type of model is used (see also Roelfzema et al. 1984, 1987).

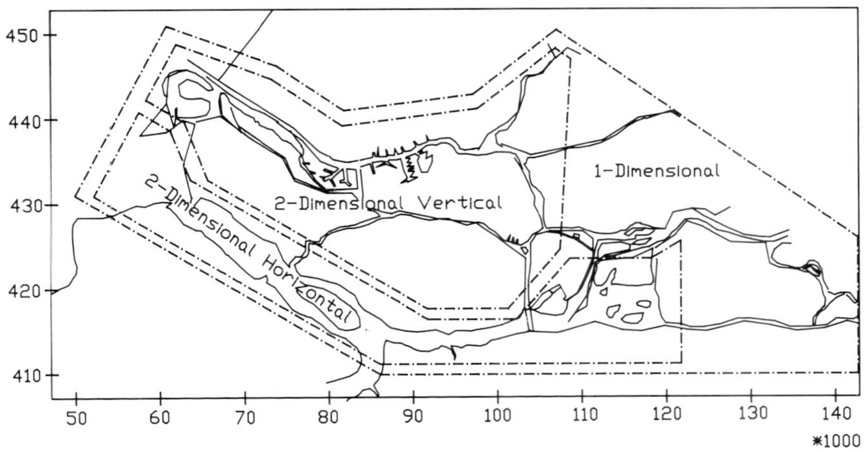

Fig. 2.37
Nesting of mathematical models in the Rhine-Meuse delta.

Each type of model has its own field of application. 1-dimensional models can be used for overall parameters, large areas and long time series or for a preliminary study. They, however, contain little physics and coefficients have to come from measurements or more detailed models. The three types of model will now be discussed more in detail.

1-dimensional model

When only information on water levels, discharges or overall concentrations of dissolved matter is wanted, a 1-dimensional model can be a useful tool. The basic hydrodynamic equations for surface water flow in the branches of a river delta for one dimension are:

Equation of continuity (conservation of mass):

$$\frac{\partial Q}{\partial x} + \frac{\partial A}{\partial t} = 0$$

Equation of motion (Newton's second law):

$$B\tau_w - P\tau_b - \rho g A I - \frac{1}{2} g h A \frac{\partial \rho}{\partial x} = \rho\left[\frac{\partial Q}{\partial t} + \frac{\partial}{\partial x}\left(\frac{Q|Q|}{A}\right)\right]$$

where:

A	= area of cross-section of river branch	(m^2)
B	= width of river branch	(m)
h	= waterdepth	(m)
I	= slope of the water surface	
P	= wetted perimeter of river branch = A/R	(m)
Q	= discharge in river branch	(m^3/s)
R	= hydraulic radius of river branch	(m)
τ_b	= bottom shear stress = $2gQ\,|Q|/C^2A^2$	(N/m^2)
C	= Chezy roughness coefficient	($m^{0.5}$/s)
τ_w	= wind shear stress = $C_w \rho W_x\,|W|$	(N/m^2)

C_w	= wind shear coefficient	
W	= wind velocity	(m/s)
W_x	= component of W in branch direction	(m/s)
ρ	= density of water	(kg/m^3)
t	= time	(s)
x	= main flow direction, branch direction	

The terms at the left hand side of the equation of motion are forces per unit length in the flow direction, due to wind, friction, waterlevel gradient and density gradient, respectively . At the right hand side are the acceleration terms; the local and convective inertia. These equations are valid for all kinds of non-stationary motion (e.g. tidal flow, varying river discharges, wind set-up and discharges or extractions by man-made structures). This makes a model, based on these equations, a valuable tool in hydraulic engineering in deltaic areas.

The transportation of chemicals, like salt, pollutant or heat can be described for one dimension by the dispersion relation, derived in the previous section:

$$\frac{\partial (Ac)}{\partial t} + \frac{\partial Q c}{\partial x} + \frac{\partial}{\partial x}\left[AD_x \frac{\partial c}{\partial x}\right] + \frac{Ac}{\gamma} = 0$$

where:

c = concentration (kg/m^3)
D_x = dispersion coefficient (m^2/s)
γ = dissipation coefficient for non-conserved properties (s)
and all parameters are averaged over the cross-section.

These equations can be solved analytically only for very simple situations. For practical applications in deltaic areas, approximate numerical solutions have to be obtained (e.g. by means of finite difference methods). DHL (1973) gives an introduction to these equations, their solutions and applications, while Abbott (1992) and Vreugdenhil (1989) give more details on computational schemes.

For numerical computations, a river delta has to be schematised with branches, the dimensions of which are derived from sounding or levelling. Boundary conditions are necessary for computations. These usually consist of water levels at the downstream end (e.g. a measured or predicted tidal fluctuation) and discharges at the upstream end (e.g. measured or predicted run-off). Calibration with field measurements is necessary, since the roughness and dispersion coefficients are hard to predict and all inaccuracies of the schematisation have to be included in those values. Moreover, as stated in the previous section, the dispersion coefficient is a collective parameter for all kinds of processes and the result of averaging concentrations. Since these processes are not incorporated in the 1-dimensional model, the dispersion coefficient can change considerably when the flow geometry changes. This will be demonstrated with an example of the 2-dimensional vertical model. Once a model is calibrated, it can be used to predict changes in the water movement due to interferences in the area, like closing off branches, building structures, extracting water, etc.

Figure 2.38 gives some examples of the 1-dimensional model for the Rhine-Meuse delta showing the schematisation, calibration with measurements and an application (in this case, the closing of the newly designed storm surge barrier near Rotterdam). The parameters in this example were calibrated from measurements.

Examples of the application of this model are the calculation of statistics of waterlevels in the area for dyke design (varying storm surges at the sea boundaries and river discharges at the upstream boundaries), improving the discharge programme of the Haringvliet outlet sluices and making time-series of salinity intrusion in the area for various channel geometries (using information on the dispersion coefficient from the 2-dimensional vertical model). The more detailed models are used to improve the coefficients in the 1-dimensional model, which is again used to give boundary conditions for these detailed models, for cases where no measurements are available. This shows the intense relationship between these models.

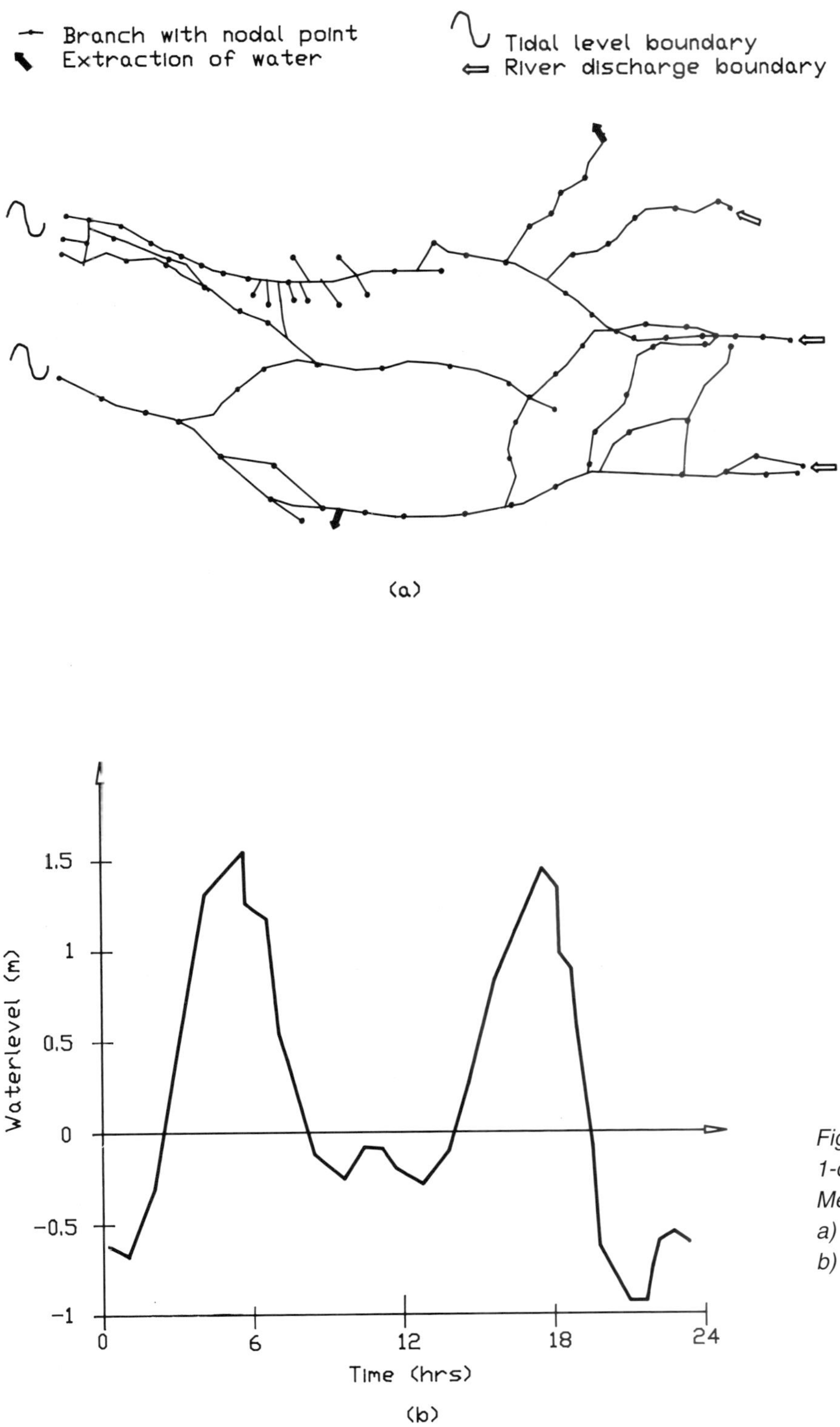

Fig. 2.38
1-dimensional model of the Rhine-Meuse estuary with
a) schematisation
b) boundary conditions.

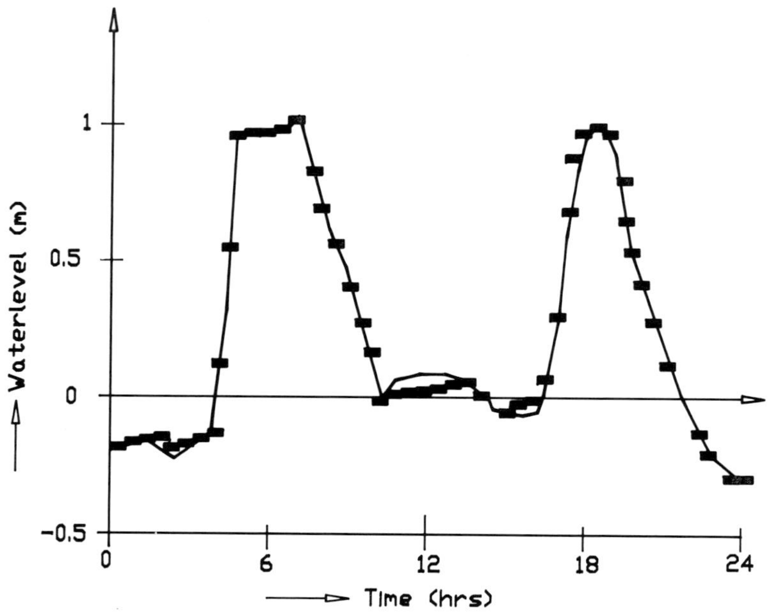

(c)

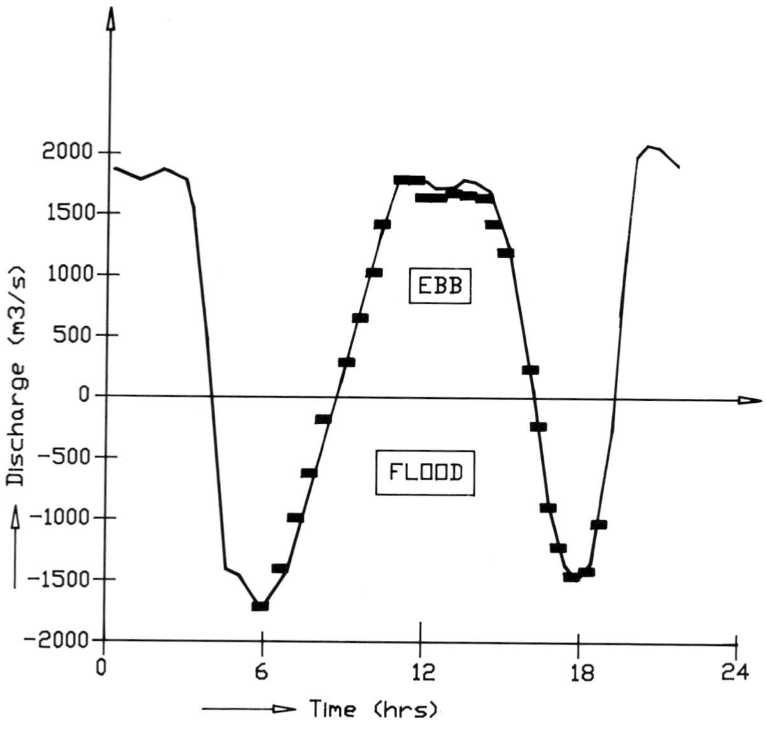

(d)

Fig. 2.38
1-dimensional model of the Rhine-Meuse estuary with ,
c) and d) calibration results.

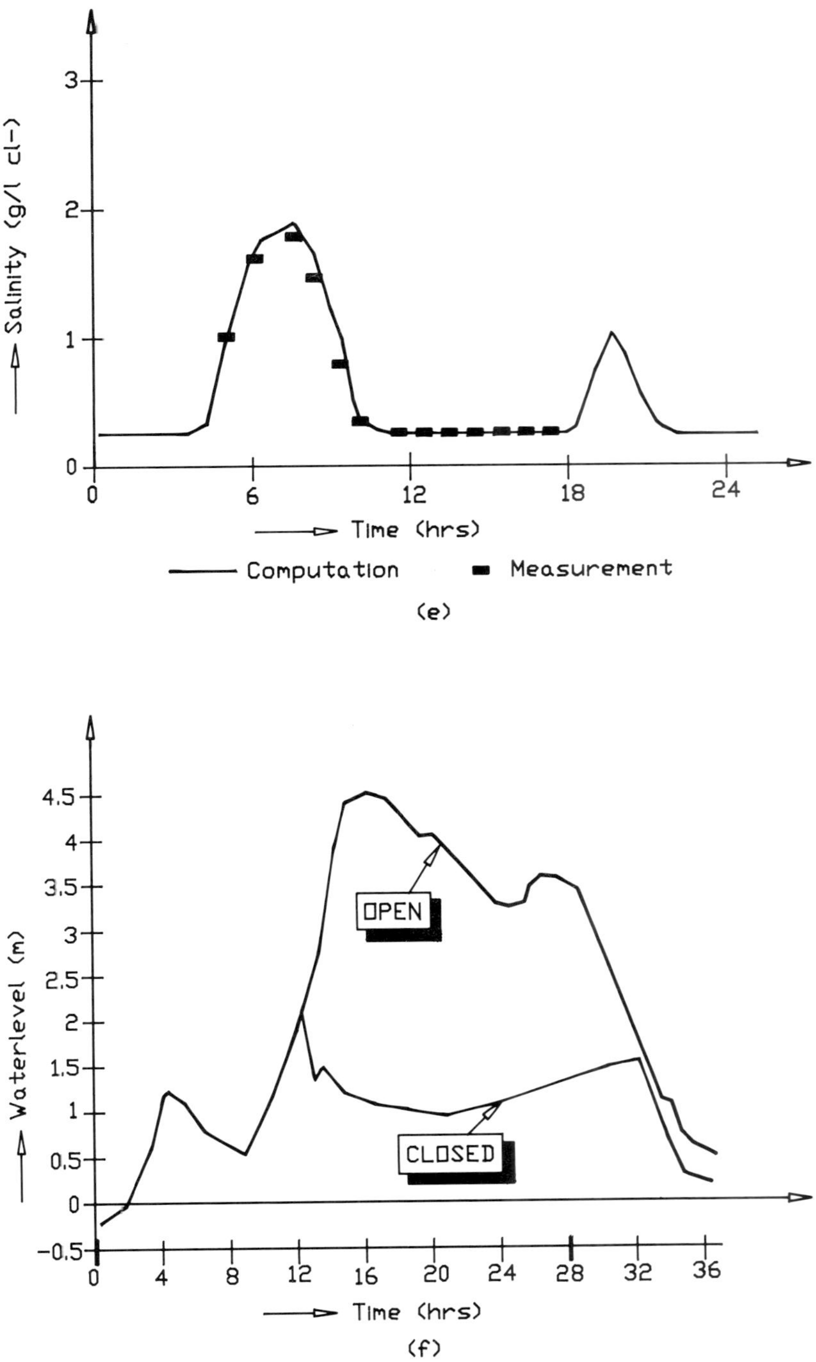

Fig. 2.38
1-dimensional model of the Rhine-Meuse estuary with e) calibration results and f) application.

2-dimensional horizontal model

When there is a less pronounced dominant flow direction, the 1-dimensional description becomes less accurate. This may be the case with a wide estuary or in seas. The choice for a 2-dimensional horizontal model also depends on the problem to be studied. When only water levels and discharges are important, a 1-dimensional model could even be used in a wide estuary, but when variations in the cross-section or more detail of the flow field is required, a 2-dimensional model is indispensable.

The basic equations are almost the same as for a 1-dimensional model, but extended into two dimensions (Leendertse, 1967). One major difference can be the inclusion of the Coriolis-acceleration in the equations. This acceleration is proportional to the flow velocity and acts perpendicular to it. It is important in seas and very wide estuaries, where the waterlevels and velocities can differ considerably at different shores, even when the main flow is in one direction.

Figure 2.39 gives the schematization for the southern branch of the Rhine-Meuse delta. It consists of a 100 m rectangular grid (for better alignment with complex geometries, a curvilinear grid is also possible nowadays). The black areas are shallow parts that can be dry or wet, depending on the waterlevel.

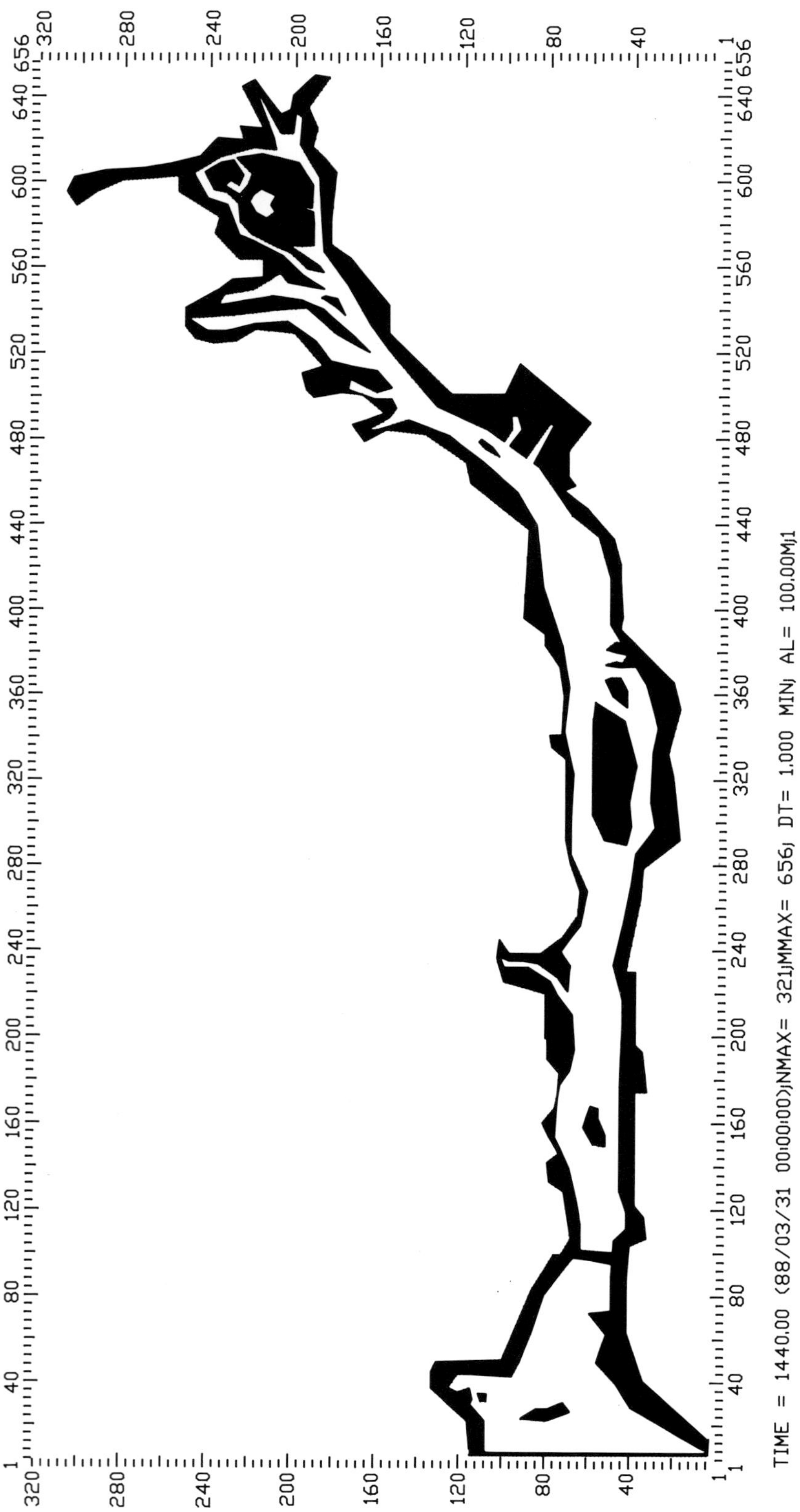

Fig. 2.39
2-dimensional horizontal schematisation.

Figure 2.40 shows the calibration of the model for the waterlevel at Moerdijk (Rijkswaterstaat, 1991). The results for the 1-dimensional model are also given, showing that for this purpose, a 2-dimensional model is not really necessary.

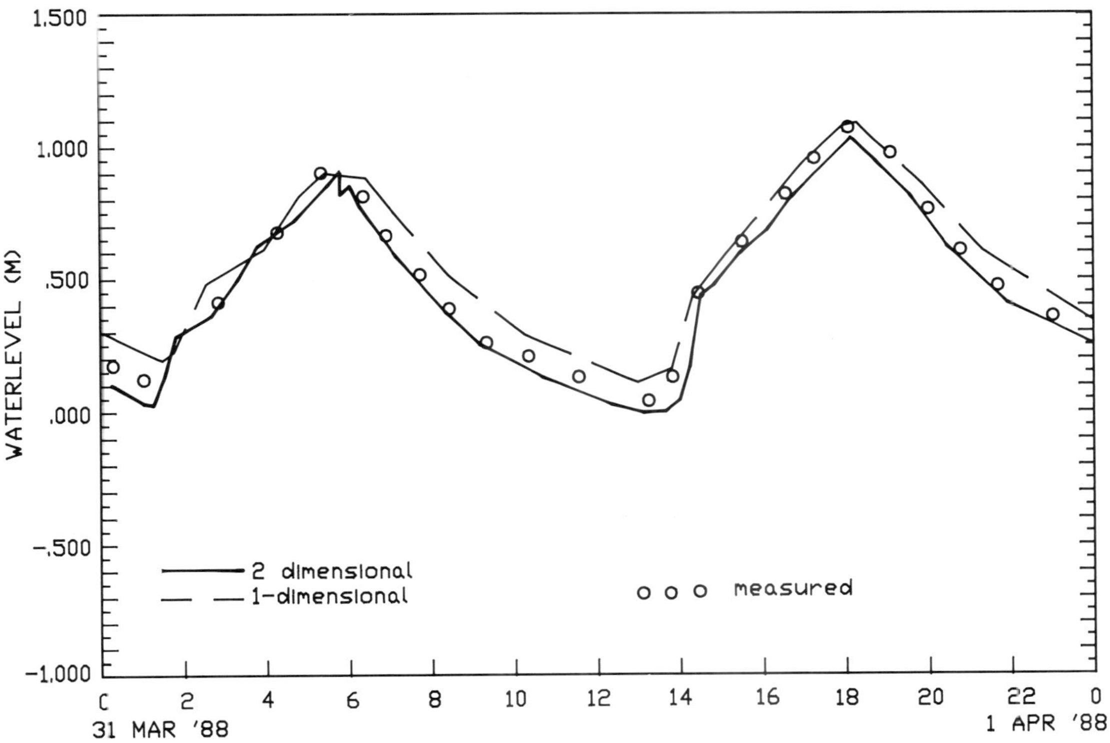

Fig. 2.40
Calibration of water levels at Moerdijk.

The true advantage of a 2-dimensional model is shown for the flow around a possible artificial island near Willemstad. With this model it is possible to study the influence of such an island on the distribution of the river discharge over the cross-section of the branch (Figure 2.41).

Another application of a two dimensional model is the spreading of pollutants. Figure 2.42 gives the diffusion of excess heat from a plant at Moerdijk (DHL, 1991), which is important for the wetlands on the other side of the estuary.

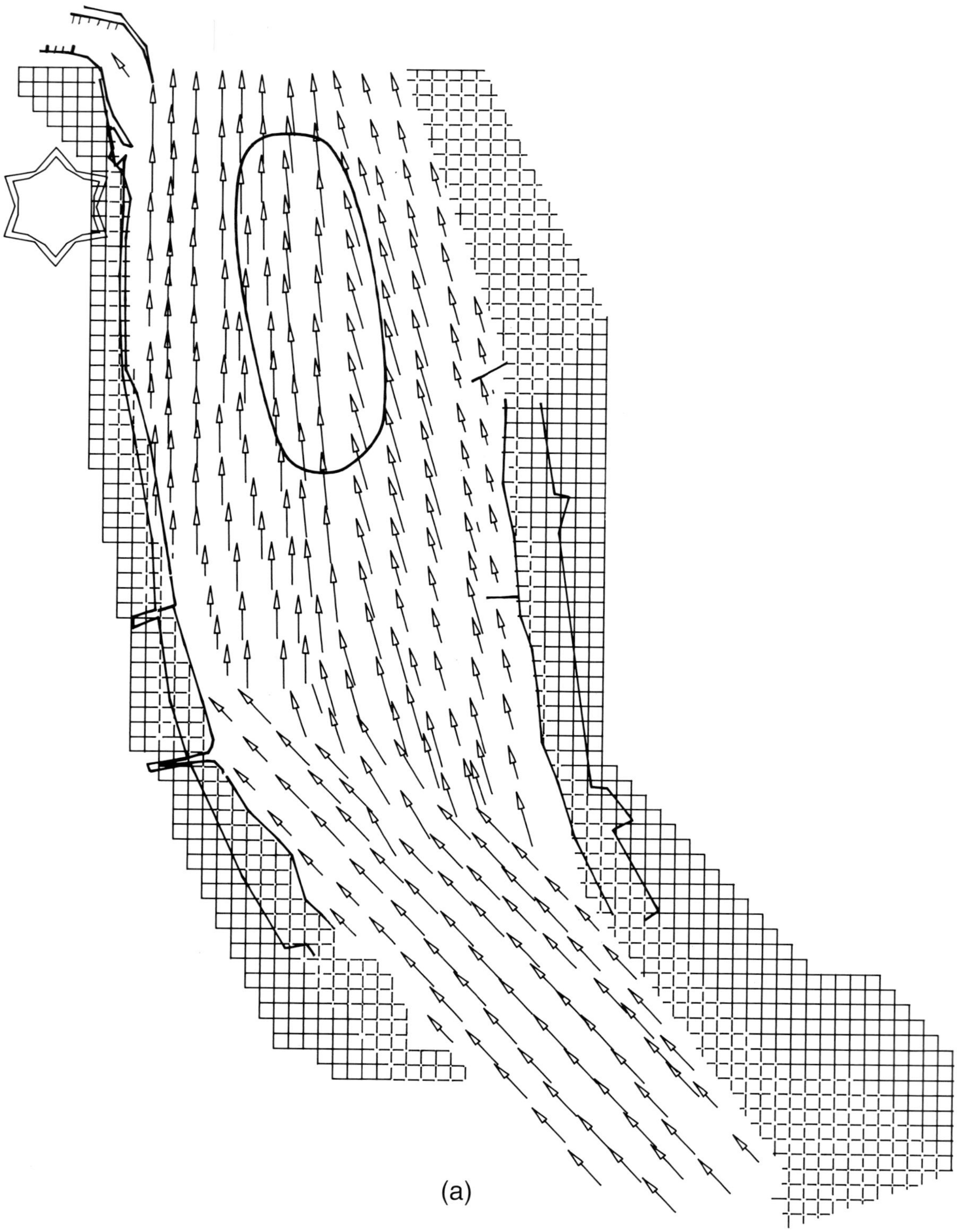

(a)

Fig. 2.41 (a and b)
Influence of artificial island on flow distribution.

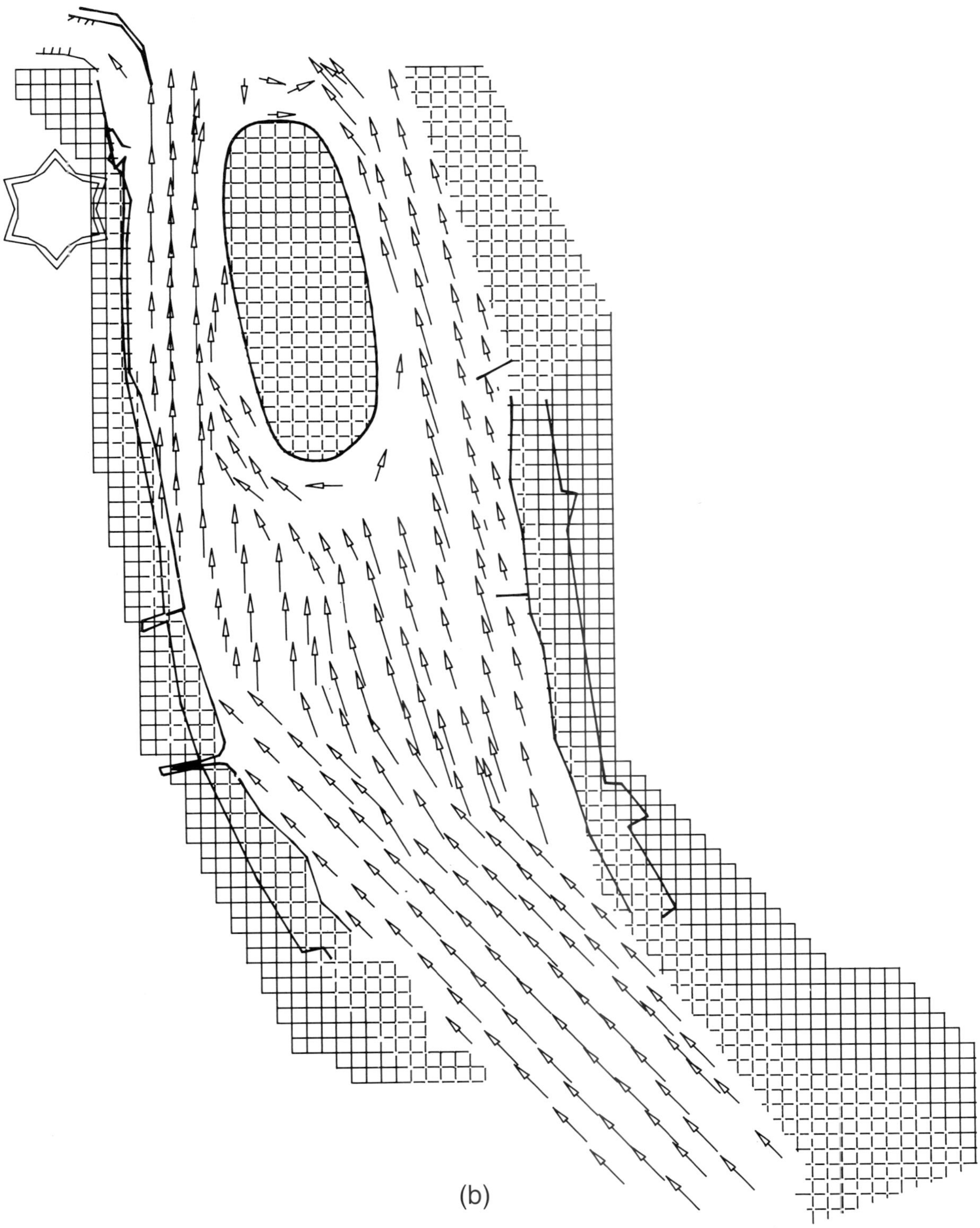

(b)

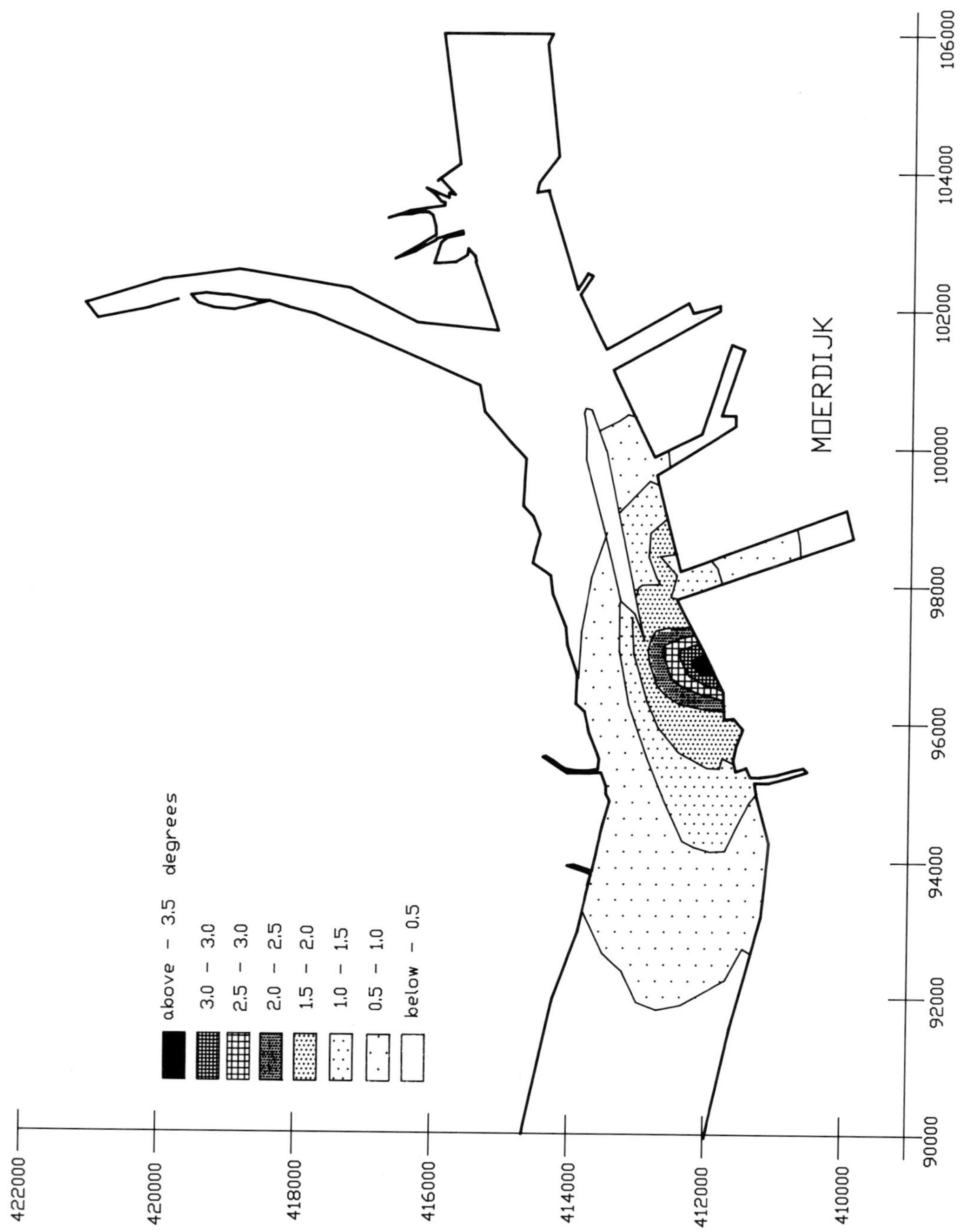

Fig. 2.42 Diffusion of excess heat.

2-dimensional vertical model

In the area of (partly) stratified sea water intrusion, a 2-dimensional vertical model can be very useful. The equations in the model include the vertical velocity and density distribution as well as the vertical mixing (see Perrels and Karelse, 1981). Figure 2.43 shows results of vertical velocity and salinity distributions in the Rhine-Meuse estuary at the location of the planned storm surge barrier (Verbeek, 1992).

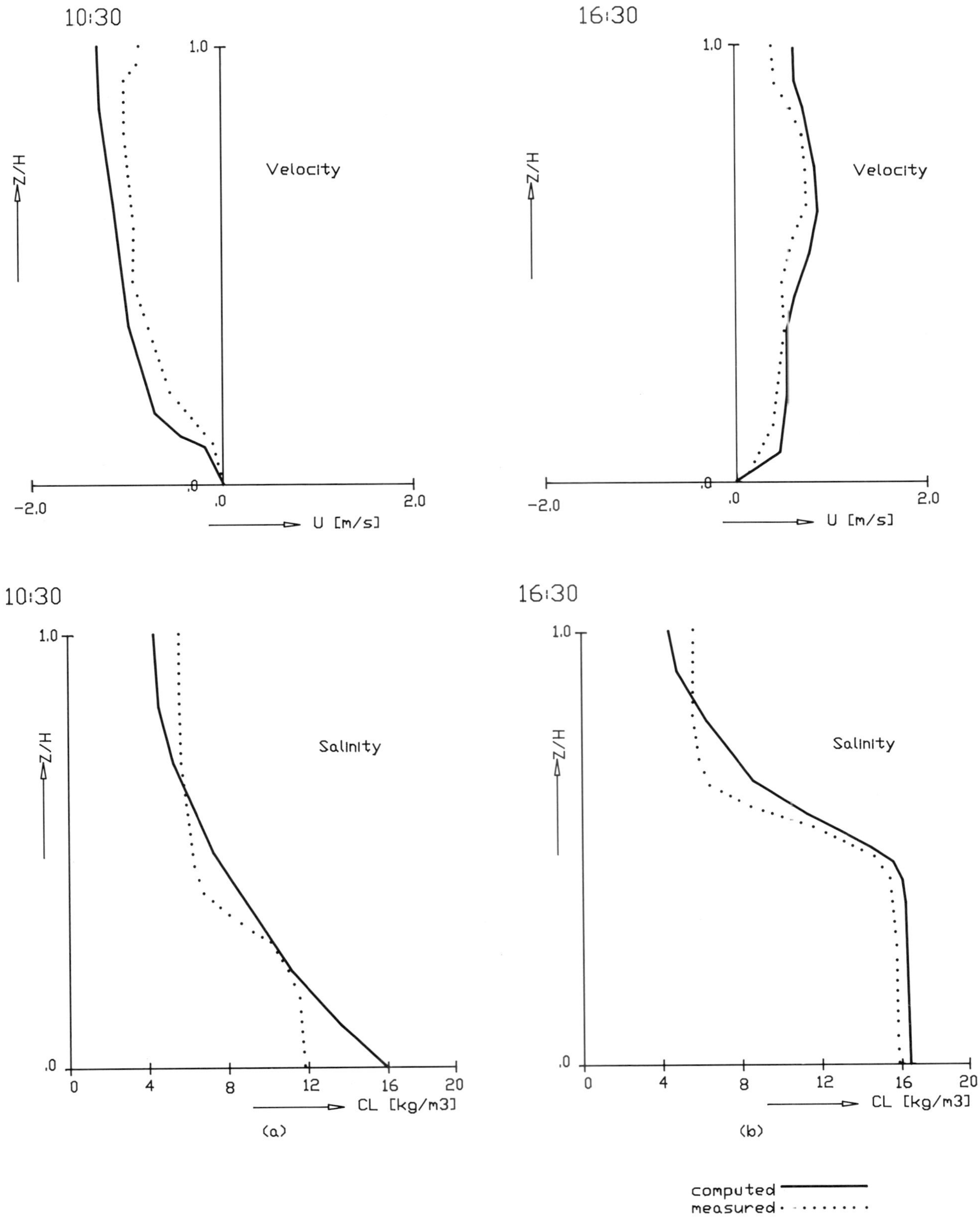

Fig. 2.43
Calibration velocity and salinity verticals at
a) LW-slack and
b) HW-slack.

The following application demonstrates the advantages of a 2-dimensional vertical model over a 1-dimensional model. The new navigation channels in the Rotterdam harbour area also offer new possibilities for the sea to intrude. The same amount of river water has to resist the sea water along two routes instead of only one (Figure 2.44).

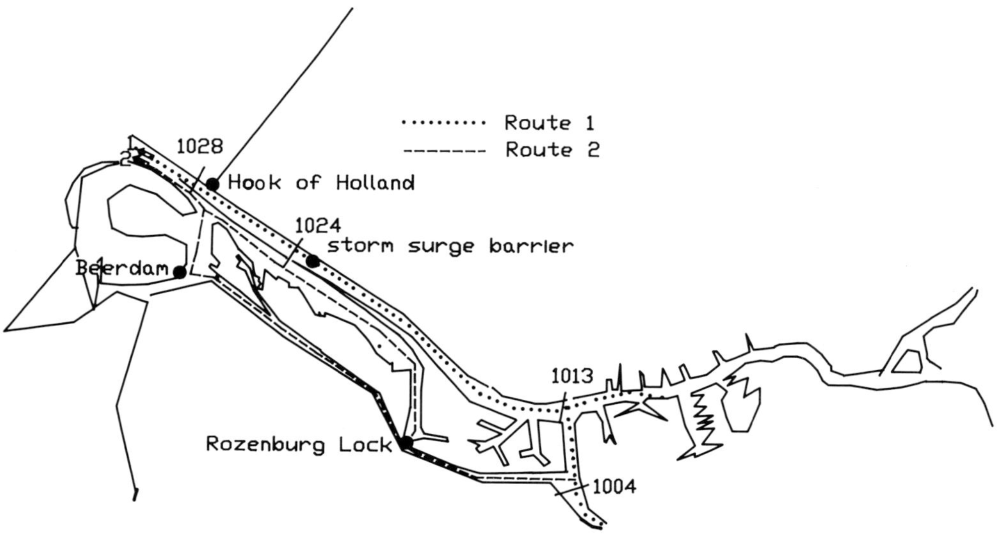

Fig. 2.44 Newly planned navigation channels.

One would expect more salinity intrusion into the hinterland and, indeed, that was the result of an analysis with a 1-dimensional model. 2-dimensional calculations however, gave a change in the vertical salt distribution in the western part, but no increase of the intrusion at all in the eastern part (see Figure 2.45). The reason for this is the phase-shift of the tide due to the new openings, causing extra large scale mixing at the nodal points of the river branches. For more information, see Rijkswaterstaat (1992) or Karelse et al. (1993). This shows that the dispersion coefficient in a 1-dimensional model still cannot do without measurements or a more sophisticated model.

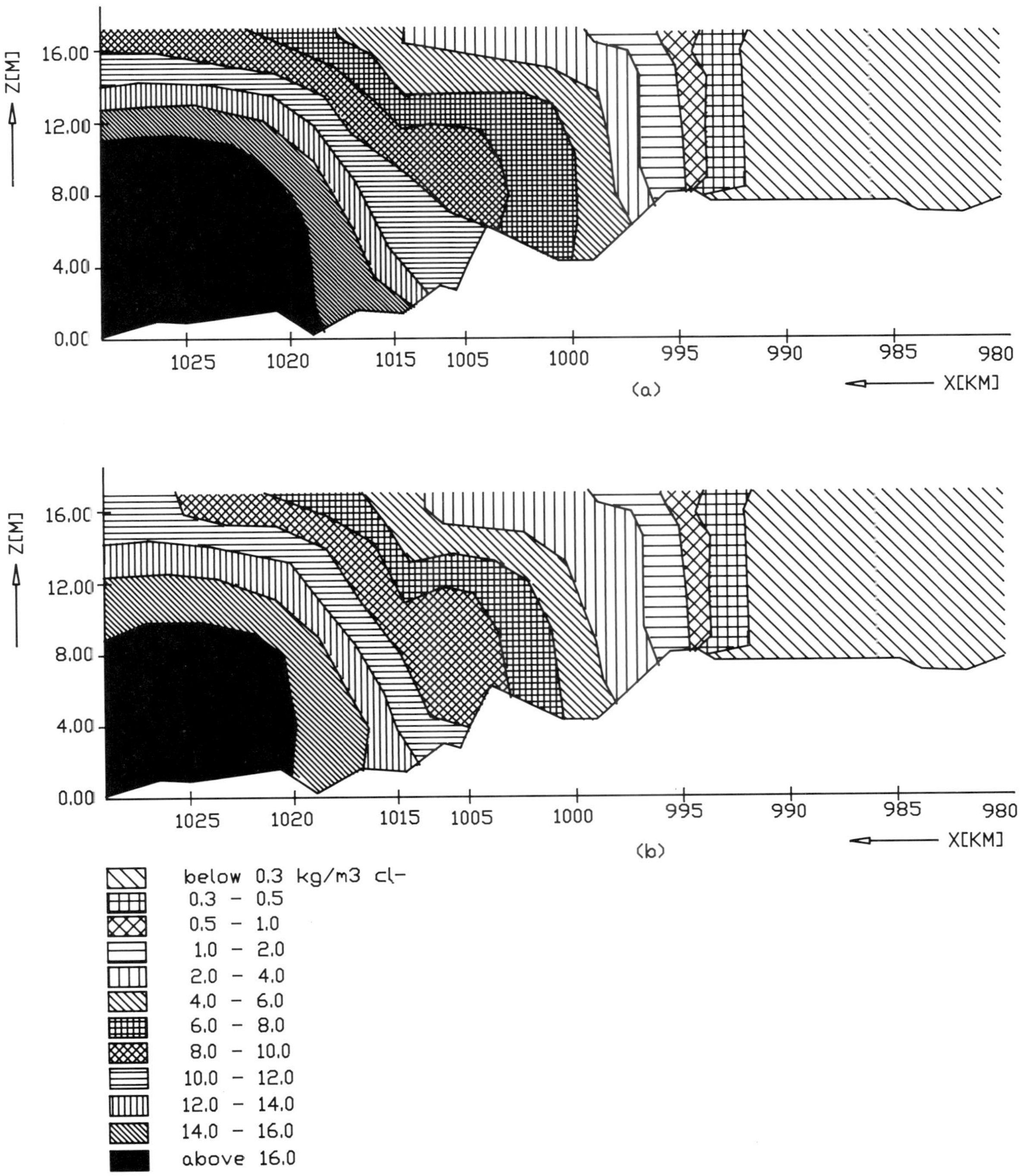

Fig. 2.45
Influence of new navigation channels upon sea water intrusion.

2.2 Groundwater

2.2.1 General

This section will deal with groundwater as a natural resource of deltaic areas. More specifically, it will deal with the occurrence, flow, and quality of groundwater in these areas. Groundwater is however, more than just a natural resource. It is an important feature of the ecology of deltas. While groundwater in deltaic areas may lead to environmental problems, it may also offer a medium for environmental solutions. Therefore, this section will also consider the influence of man on the groundwater conditions, and of these conditions on man.

2.2.2 Occurrence of groundwater

The occurrence of groundwater in deltas is controlled by the geological framework of the mass of deltaic sediments. This geological framework (i.e. the lithology, stratigraphy, and structure) is fairly well known for only a limited number of deltas (e.g. Mississippi, Rhine, and Nile). Deltas show a wide range of lithologies. Each type of sediment has its own environment or sub-environment of deposition, and its characteristic texture, composition, and set of hydraulic properties.

Geological formations made of permeable, or very permeable, sands and gravels are called aquifers. They are capable of transmitting large quantities of water under ordinary hydraulic gradients. Aquicludes are geological units that are incapable of transmitting significant quantities of water under ordinary hydraulic gradients. Very few geological units fit this definition. Even hard rocks are capable of transmitting water, especially when they are fractured, as many of them are. Nevertheless, the term aquicludes will be used to refer to all rocks or geological strata whose permeability is so low that it is, for all practical purposes, negligible. For the less permeable beds in a stratigraphic sequence, the term aquitard is used. An aquitard is a geological formation or bed, whose permeability is much lower than that of an aquifer, but high enough to transmit significant quantities of water. An aquitard's permeability is, however, not sufficient to sustain a successful pumped well. Hence, beds made of clay, silt, loam, silty-clay-loam, or any other fine-textured materials from the vocabulary of sedimentologists and soil scientists can be regarded as aquitards.

In deltas, several types of aquifers can be identified (Figure 2.46). The first is the confined aquifer, which is bounded above and below by aquicludes. This type of aquifer is fully saturated and contains water under pressure. The second is the unconfined aquifer, which is bounded below by a aquiclude and above by the water-table. An unconfined aquifer is, therefore, also called a water-table aquifer. It is partly saturated and occurs near the soil surface. The third type is the leaky aquifer, which is bounded above by an aquitard and below by an aquiclude or an aquitard. The term leaky aquifer is somewhat misleading because it is not the aquifer that leaks, but the confining aquitards. In literature, leaky aquifers are also called semi-confined aquitards. A fourth type is the multi-aquifer system, which in fact is a system common to most deltas. It consists of two or more aquifers separated by aquitards.

In deltaic areas, the first type of aquifer, a truly confined aquifer, does not occur near the soil surface. It may occur at a great depth, where the load of overburden has compressed the fine-textured sediments to such an extent that their permeability has become so low that they can be regarded as aquicludes. Since the geological framework of deltas throughout the world is not identical, a universal hydro-geological model of such areas cannot be presented. Even within a single arbitrary delta there are many facies sequences. The complexity of some sequences makes it difficult to define the type of aquifer and its boundary conditions. The lower boundary of a sand bed is usually well defined, but owing to an upward fining of the channel deposits, there is no sharp upper boundary between channel sand (aquifer), and flood-plain deposits (aquitard).

In spite of their variability in hydro-geological conditions, deltaic plains have a lot in common. For instance, they are flat and slope very gently toward the sea. The land surface of most of them lies only a few metres above sea-level, but may rise to several tens of metres in the upstream areas. The position and hydraulic gradient of the regional water-table is a function of these topographical conditions. The water-table in deltas is shallow. In the coastal tracts it is at, or close to, the land

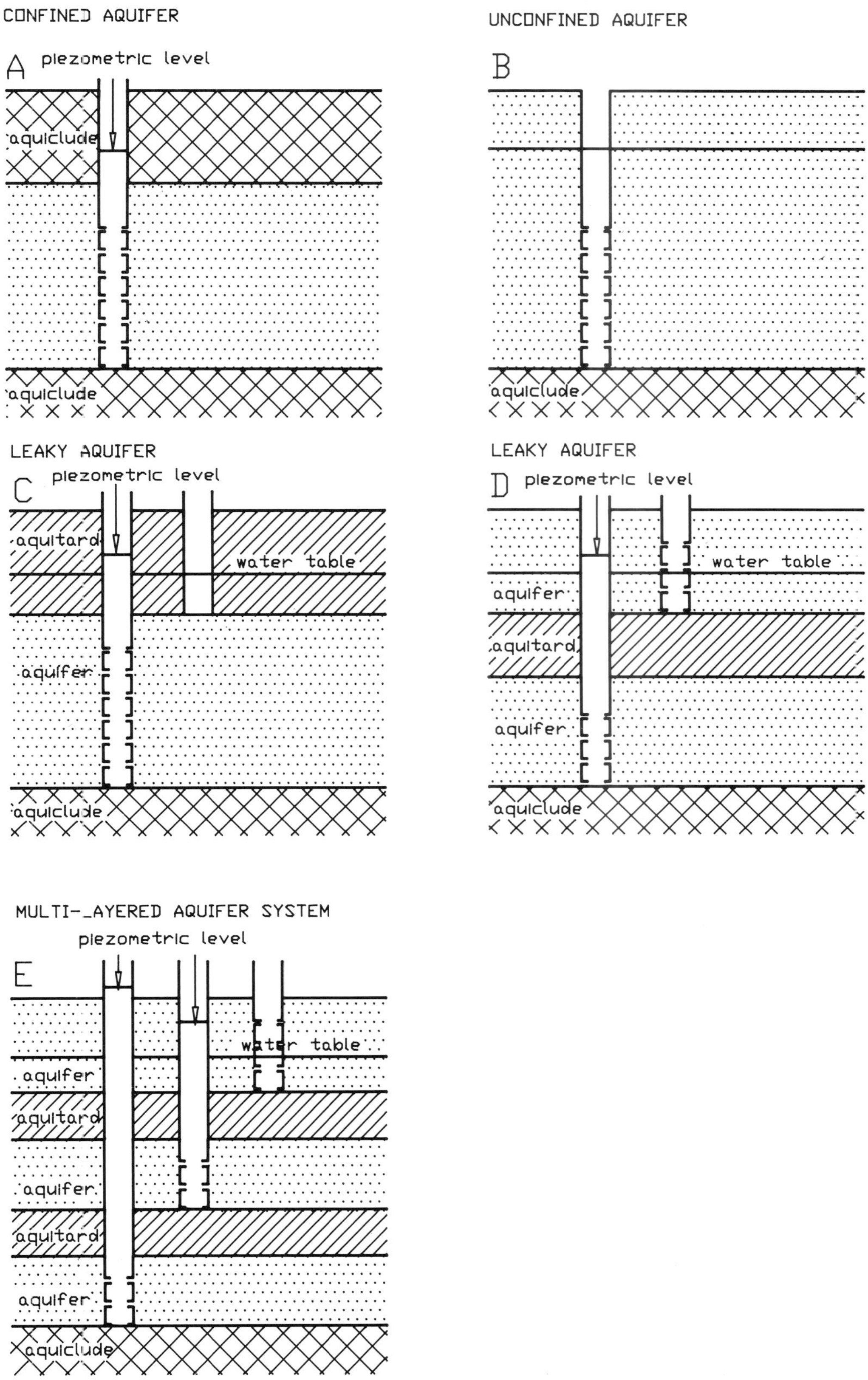

Fig. 2.46
Different types of aquifers.

surface, but in the higher, upper reaches of the delta it may be several metres or even several tens of metres deep. For example, Anderson (1978) reports that the regional water-table in the Holocene area of the Mekong delta is less than 5 metres above sea-level. In the topographically higher tracts, which border the modern deltaic plain, it is 30 metres or less above sea-level. Owing to the relatively flat nature of deltaic plains, water-table gradients are low.

The water-table in a delta is, however, not only a function of the topographical conditions, but also of tidal, climatic, flood and the drainage conditions. Rivers such as the Volga, the Danube, the Po and the Nile discharge into inland seas, where the tidal range is very small. Other delta-building rivers outfall into seas or oceans where the tidal range is several metres (e.g. the Rhine-Meuse, the Mekong and the Ganges-Brahmaputra). The tide influences the water level in the tributaries, which in turn, influences the water-table in the adjacent lands. The climate has a marked influence on the position and variation of the water-table in deltas. Figure 2.47 shows the effect of rainfall on the water-table in the Mekong Delta north of Ho Chi Minh city (Anderson, 1978). In most deltaic plains, high water-tables prevail, so that the annual water-table fluctuations are much lower than 6 to 7 metres. They may vary from 1 or 2 metres to less than 0.5 metres. In these high water-table areas, high rainfall or floods cause the water-table to rise to the surface. After the rain ceases and the floods recede, natural drainage and evapotranspiration cause the water-table to drop. In most cases, the permeable deltaic soils and highly permeable underground are responsible for the good natural drainage conditions in deltas. A classic example is the Nile Delta where, until recently, the natural drainage could easily cope with the annual floods that soaked the fertile soils. This prevented the permanent waterlogging and subsequent salinisation of the deltaic soils.

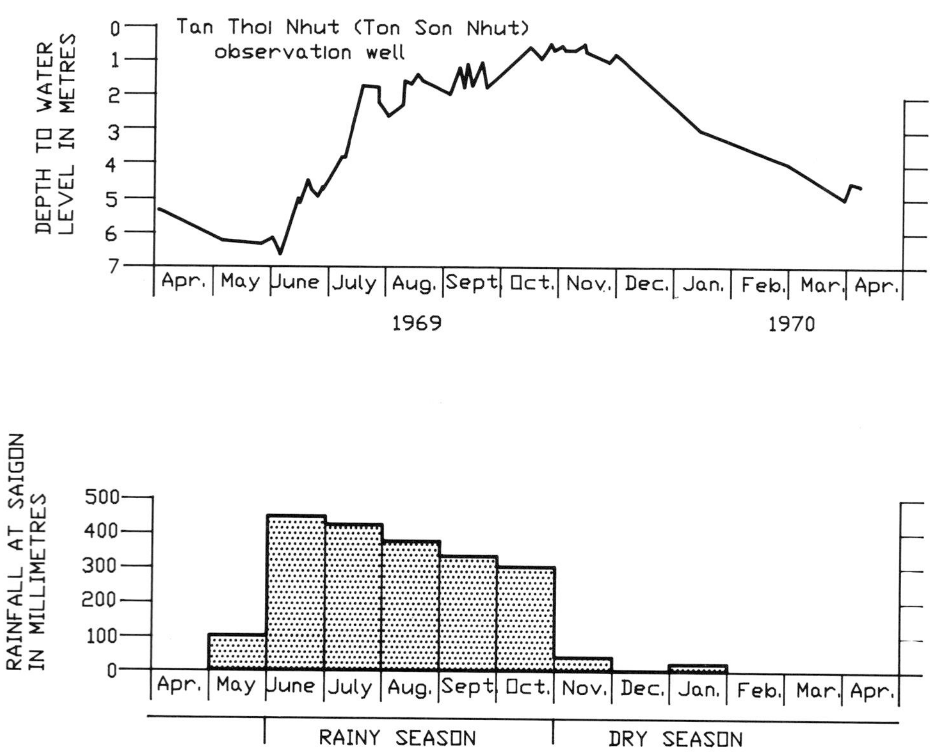

Fig. 2.47 Fluctuations of the water level in a well in the Mekong Delta.

As a result of variability of the soil conditions in deltas, there are of course, major areas in deltaic plains which suffer from inadequate drainage and excessively high water-tables. The areas between the tributaries are often concave because of the natural levees built up on either side. These interfluvial areas, made up of very heavy clay with poor internal drainage, are flat and usually flooded for long periods. Here, artificial drainage is needed to control the water-table and to transform the areas into safe dwelling places and productive agricultural zones. When viewed in more detail, deltaic plains are not just featureless flat areas. As already mentioned, tributaries diverge out over the plain with water levels that rise and fall during the course of the year. Depending on whether these levels are above or below the water-table of the adjacent land, recharge or discharge of the groundwater will occur. Other typical features of deltaic plains are partly or fully silted-up former tributaries, coastal sand ridges and dunes, coastal lakes and ponded areas. All of these morphological features affect the position of the water-table. Fully silted-up channels usually rise a few metres above the surface of the plain, as do the barrier sand ridges and dune formations near the coast. In humid climates, these ridges are the areas where rain water infiltrates causing the water-table to rise, while a fresh water lens may develop. An example of this, from the Mekong delta, is shown in Figure 2.48 (Anderson, 1978).

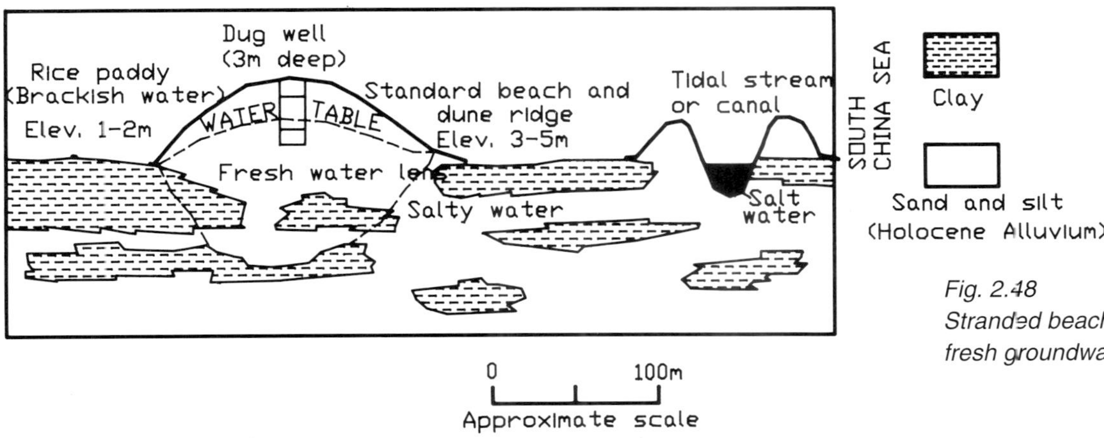

Fig. 2.48
Stranded beach ridge with a fresh groundwater lens.

Groundwater confined in deltaic aquifers is under pressure. The water-level of a well penetrating such aquifers lies above the boundary of the aquifer and the overlying aquitard. The pressure in deep aquifers may be so high that the water level of the well rises above the soil surface, causing it to flow freely. Jones (1970) reports that the groundwater in the Holocene entrenched valley aquifer is under pressure throughout most of the Mississippi Delta region. Wells that tap the aquifer in the coastal marsh and in the Balize Delta contain salty water. When piezometric heads are used for mapping or calculating, it must be remembered that the salinity of the groundwater in deltaic areas often varies significantly, both in the vertical and horizontal directions. This will affect the measurement of these heads which have to be corrected in two ways. The first correction is to convert the measured head, a column of water with a certain density, into a column of water with a reference density, usually that of fresh water. Secondly, it must be determined whether or not the pressure distribution is hydrostatic.

2.2.3 Groundwater flow

The local occurrence of groundwater in deltas is not merely a product of chance, but the consequence of a combination of climatological, hydrological, geological, topographical, ecological, and soil-forming factors, which together, form an integrated dynamic system. These factors are inter-related in such a way that each of them provides some insight into the functioning of the total system. The most important relationships are the physical characteristics of the framework in which the groundwater system occurs, the balance between groundwater recharge and discharge, and the relationship between factors affecting the movement of groundwater from the areas of recharge to the areas of discharge. The groundwater regime is defined in terms of the nature of the basic conditions, and the changes in the quality, quantity, and physical properties of the groundwater in space and time. In a typical delta, precipitation and flooding are the principal sources of groundwater recharge. However, in the semi-arid and arid zones deltas receive little precipitation and their groundwater is recharged annually by flooding. A classic example is the Nile, which flows through a true desert and receives its water from the rain that falls on the Ethiopian Plateau and the Equatorial Plateau of Central Africa. It is the rainfall on the Ethiopian Plateau which produces, via the Blue Nile, the regular floods. Precipitation and flooding are common sources of groundwater recharge in deltas in the humid zones. The topography of the delta is a governing factor with respect to flooding.

Obviously, the low-lying downstream parts of the delta and the interfluvials are prone to flooding. The higher parts are inundated only in event of high floods, whereas the uplands near the apex of the delta are never flooded. The uplands are the areas where the deep aquifers in the delta reach the surface. Precipitation on these outcrops percolates and recharges the groundwater confined in the aquifers. An example is the Quaternary Old Alluvium of the Mekong Delta, which is hundreds of metres thick in most places, but which thins out towards the margins. It contains several aquifers, which are the principal sources of fresh water supply in the outer (seaward) part of the delta, and to a large extent, in the inner (inland) part. The intake area of the Long, an aquifer 200 metres below the soil surface, is believed to correspond to the outcrop area of the Old Alluvium to the north of the delta (Anderson 1978). In the Ho Chi Minh area, aquifers in the Old Alluvium are known as the Saigon Tri-Partite sands.

Depending on the seasonal rainfall characteristics, the water-level of the deltaic rivers may be higher or lower than that in the adjacent land. If it is higher, the rivers will recharge the groundwater in the shallow aquifers. If the water level falls below the water-table, groundwater will discharge into the river channels. In the course of the year, delta channels experience alternating groundwater recharge and discharge.

In the Rhine-Meuse delta in The Netherlands, the river level is about 3 metres higher than the average daily water-level. The aquifer is leaky and the overlying Holocene aquitard is some 3 to 8 metres thick. The river channel is deeply incised and in hydraulic contact with the aquifer. The distribution of the head in the aquifer at high river stages is shown in Figure 2.49 (Colenbrander, 1962a). The water levels indicated in this figure refer to piezometers which were placed along a line perpendicular to the two rivers. The artesian conditions of the aquifer at high river stages is clear. The head varies with the water level in the river and decreases with the distance from the river. The concave shape of the land surface in the interfluvial area should be noted.

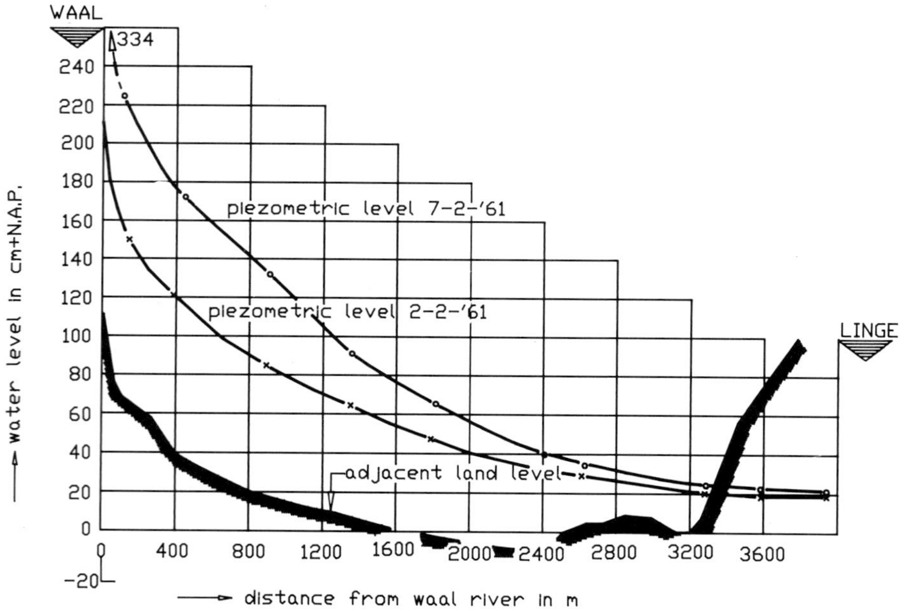

Fig. 2.49
The distribution of piezometric heads perpendicular to the river.

Figure 2.50 shows how leaky aquifers in contact with river channels react quickly to changes in the water-level in the channels (Colenbrander, 1962a). As far away as a few kilometres from the river, the hydraulic head in the aquifer still reacts to changes in the river stage. However, the fluctuations of the head are damped as the distance from the river increases. Jones (1970) reports similar conditions for the Holocene entrenched valley aquifer in the Mississippi Delta where, a few miles away from the river, the head in the aquifer rises and falls with the river as a result of direct hydraulic connection.

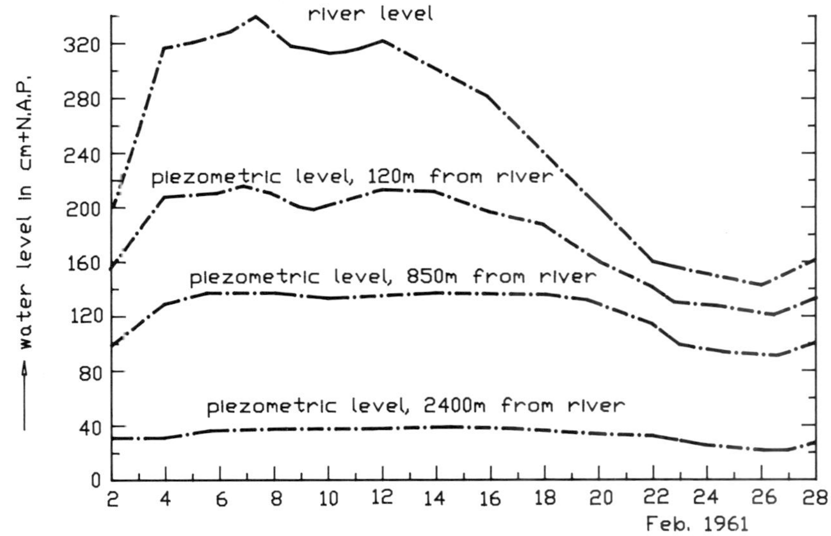

Fig. 2.50
Changes in river water levels and piezometric levels away from the river.

As stated previously, multi-aquifer systems are common in the mass of sediments formed during delta building. In such a system, there may be a significant vertical exchange of water between the individual aquifers, depending on the head in the aquifers and the hydraulic resistance of the separating aquitards. A particular aquifer is recharged if its head is lower than that in an underlying or overlying aquifer. When precipitation ceases and the floods recede, the groundwater in the delta starts discharging into the network of major and minor natural drainage channels in the flood plain and higher grounds of the delta. Shallow water-tables are characteristic of major parts of the delta, and evapotranspiration must therefore be regarded as an important discharge factor.

As deltaic areas are usually densely populated, many of man's activities can strongly affect the groundwater regime. Flood protection, land reclamation, irrigation, and groundwater recovery have, in varying degrees, an impact on the groundwater regime of the delta. For a proper understanding of this regime, all the components of recharge and discharge of the system must be considered and quantified. This is usually done by assessing a water balance of the delta or a part thereof.

Groundwater balance

In hydrology, the statement of the law of conservation of matter is called the water balance. It states that, in a specified period of time, all water entering a specified area must either go into storage within its boundaries, be consumed therein, be exported from, or flow out, either on the surface or underground. Although groundwater is closely related to surface water and soil water, and is thus not a separate source of water, it may be desirable to assess the water balance for the groundwater regime only. For instance, the problem of the maximum quantity of groundwater that can safely be withdrawn from an area must be determined from the annual groundwater balance of that area. Moreover, problems of water-table control in deltas also require an assessment of the groundwater balance, be it a problem of excess precipitation, seepage, or irrigation percolation losses to the underground.

The groundwater balance equation includes the following items (Figure 2.51):

Inflow
- precipitation that percolates to the water-table, P_e,
- surface water (rivers, lakes, ponds) that percolates to the water-table, Q_{in},
- lateral groundwater inflow through the aquifer's boundaries, Q_{si},
- vertical groundwater inflow from a deep aquifer through an aquitard, Q_{up},
- artificial recharge from reservoir irrigation, and injection well, Q_{iw}.

Outflow
- evaporation from the capillary fringe and shallow water-table, E,
- transpiration by phreatophytes,
- groundwater outflow into river, lakes and ponds, and spring flow, Q_{dr},
- lateral groundwater outflow through the aquifer's boundaries, Q_{so},
- vertical groundwater outflow to a deep aquifer through an aquitard, Q_{do},
- artificial discharge by drains, and pumped or free flowing wells, Q_{dw}.

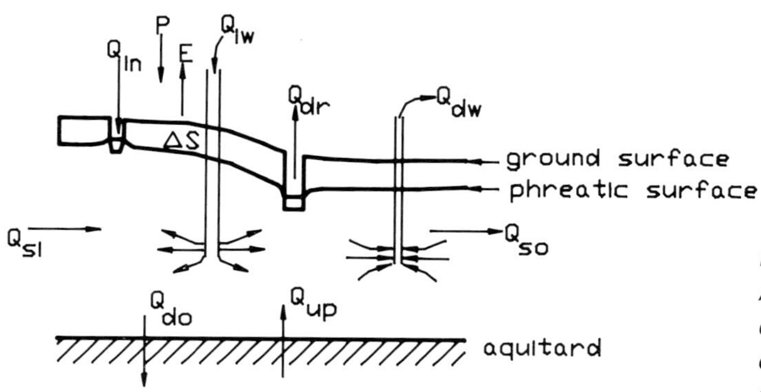

Fig. 2.51 Aquifer with different inflow and outflow components.

A water balance requires that the total inflow into the area of study balances the total outflow from the area. A marked difference usually exists between the natural inflow and natural outflow over the period of time considered, which can be a week, month, season, year, or even several years. For groundwater, this difference is accounted for by changes in storage in the saturated zone, the groundwater domain. If only short periods are considered, the unsteady state of the system is apparently due to the climatic fluctuations. The longer the period over which the water balance is assessed, the closer the various items approach the steady state. Longer periods approach average climatic conditions.

Assessing the groundwater balance of an area is more complex than assessing, for example, the surface water balance, because few of the above items are directly measurable. Consider for example, the groundwater inflow through the boundaries of an aquifer. Calculation of the horizontal inflow Q_{si}, and outflow Q_{so} through the vertical boundaries requires data on the aquifer's transmissivity and water-table gradient at these locations. The problem can be greatly simplified if these boundaries are no-flow boundaries. For instance, if the boundary on the left-hand side of Figure 2.51 coincides with a groundwater divide or an impermeable valley wall, it can be selected to represent a no-flow boundary. A deeply-incised draining tributary channel may serve this purpose. An aquifer is assumed to rest upon an aquitard, which in turn, rests upon an aquifer containing groundwater under pressure. Where this happens, the shallow aquifer receives water from the deep one by vertical flow through the aquitard (Q_{up}). Although this rate of flow may be small, it can be quite appreciable when taken over a large area. Obviously, we may think of the reverse situation if the shallow aquifer loses water to the deep aquifer by vertical downward flow through the aquitard (Q_{do}). This happens when the head in the deep aquifer is lower than that in the shallow one (e.g. in deltas in the arid zones, where irrigation creates high water-tables in the shallow aquifer). Determination of the rate of upward or downward flow through the aquitard requires a lot of field work, including exploratory drilling, placing double piezometers (one each in the upper and lower aquifer), measuring water-levels in these piezometers regularly during the time for which the groundwater balance is to be assessed, and conducting pumping tests to determine the hydraulic resistance of the aquitard.

This type of detailed investigation may be justified in experimental fields of relatively small size. For obvious reasons, it is impossible for regional studies, in which case a less expensive approach can be adopted by lumping the unknown terms of groundwater inflow and outflow through the boundaries of the groundwater balance area, into a single term called the net sub-surface inflow, I_{ss} (Ernst et al. 1970). This term is the algebraic sum of the components Q_s, Q_{so}, Q_{up}, and Q_{do}. This approach simplifies the groundwater balance equation, which can now be written as:

$$P_e + I_{SS} + \frac{Q_{in}}{A} - E = \frac{Q_{dr} + Q_{dw}}{A} + \frac{\Delta S}{\Delta t}$$

where:

P_e = effective precipitation that reaches the water-table,
E = evaporation,
A = horizontal surface area,
I_{ss} = net sub-surface inflow per unit of horizontal surface area,
Q_{in} = seepage losses from open channels or lakes with high water-levels,
Q_{dr} = groundwater outflow into open conduits with low water- levels,
Q_{dw} = discharge by deep wells,
ΔS = change in storage of groundwater per unit of horizontal surface area.

If an area where pumped wells or injecting wells and leaking channels do not occur, is selected, and if only periods of long duration are considered, so that the change in groundwater storage is so small as to be negligible, then the above equation reduces to:

$$I_{ss} = \frac{Q_{dr}}{A} - (P_e - E)$$

This equation shows clearly the practical significance of knowing the value of the net sub-surface inflow. I_{ss} represents the difference between the quantity of water that must be drained from the area, and the average supply. Obviously, I_{ss} can be positive or negative, depending on the difference between the groundwater levels in the area under consideration and those of the surrounding areas. Although relatively flat, most deltas show minor topographical features and thus minor differences in water-table height. In undulating terrain, topographical depressions and river valley bottoms receive a positive net sub-surface inflow from the neighbouring ground, which has a high water-table. In topographically high terrain, the net sub-surface inflow is negative. The large distance over which the major differences in water-level usually occur, and the time independence of these differences, explain why the value of I_{ss} hardly changes with time when compared with the large and frequent variations in the value of Q_{dr} caused by rainfall.

Seepage

Water management schemes have significant practical consequences for the hydrology of deltas and induce a wide range of seepage and groundwater flow problems. Many of these problems and their solutions have already been discussed elsewhere, e.g. Verruijt (1970), Bear (1972), Huisman (1972), Rushton and Redshaw (1979), Freeze and Cherry (1979), and De Marsily (1986). The first problem to be considered here will be that of groundwater flow in deltas with embanked channels. If the river discharge is high, the water-level in the channels can rise far above the level of the adjacent land. When this happens, a hydraulic gradient is formed in the underlying aquifer, which is assumed to be in hydraulic contact with the channel. Consequently, infiltrating river water starts moving inland through the aquifer, where it may be discharged as seepage through an overlying aquitard. Two significant quantities to be calculated for given conditions are the rate of groundwater inflow at the river, and the seepage rate at different distances from the river channel. The general case will be considered first (Figure 2.52).

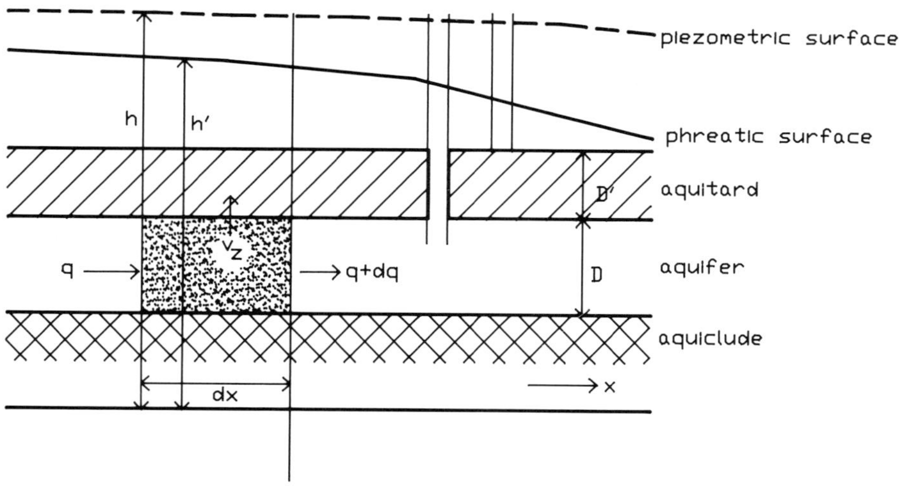

Fig. 2.52 One-dimensional flow through a section of a leaky aquifer.

This problem of one-dimensional flow is solved on the basis of the following assumptions: the aquifer is leaky, horizontal, homogeneous, isotropic and of an equal thickness; the flow through the aquifer is steady and one-dimensional; the overlying aquitard is homogeneous, isotropic and of an equal thickness, and the level of the river and the hydraulic head in the aquitard is constant. The flow through the aquifer can be expressed by:

$$q = -kD\frac{dh}{dx}$$

where:
q = the discharge per unit width of the aquifer ($m^3/s/m$)
k = the hydraulic conductivity of the aquifer (m/s)

Small quantities of water leave the aquifer through the overlying aquitard. The principle of continuity requires that the change in flow in the aquifer be equated to the rate of upward flow through the aquitard, v_z, which can be expressed as follows:

$$v_z = -\frac{dq}{dx} = -k_z\frac{h' - h}{D'} = \frac{h - h'}{c}$$

where:
kz = hydraulic conductivity of aquitard for vertical flow (m/s)
c = hydraulic resistance of aquitard = D'/k_z (s)

Differentiating the aquifer flow equation and substituting the expression for v_z gives the general differential equation for steady one-dimensional flow in a leaky aquifer:

$$\frac{d^2h}{dx^2} = \frac{h - h'}{kDc} = \frac{h - h'}{\lambda^2}$$

where:
λ = the leakage factor of the aquifer (m)

This equation is based on the assumption that the flow through the aquifer is horizontal and that through the aquitard is vertical. This assumption is justified if kc >> D or if λ >> D. The general solution of this equation, as given in handbooks on integral calculus, is:

$$h - h' = C_1 e^{\frac{x}{\lambda}} + C_2 e^{-\frac{x}{\lambda}}$$

where: C_1, C_2 = unknown constants determined from boundary conditions.

For an infinite aquifer, and a phreatic surface in the upper aquifer that is horizontal and constant (i.e. not affected by changes in the hydraulic head in the leaky aquifer) the following boundary conditions of $h = h_0$ for $x = 0$, and $h = h'$ for $x = \infty$ can be introduced (see Figure 2.53), which results in the following solutions:

$$h - h' = (h_0 - h')\ e^{-\frac{x}{\lambda}}$$

which gives the relation between the head in the aquifer, h and the distance, x, and:

$$q_x = \frac{kD}{\lambda}(h_0 - h')\ e^{-\frac{x}{\lambda}} = q_0\ e^{-\frac{x}{\lambda}}$$

which is a relation between the flow per unit width of the aquifer, q and the distance, x.

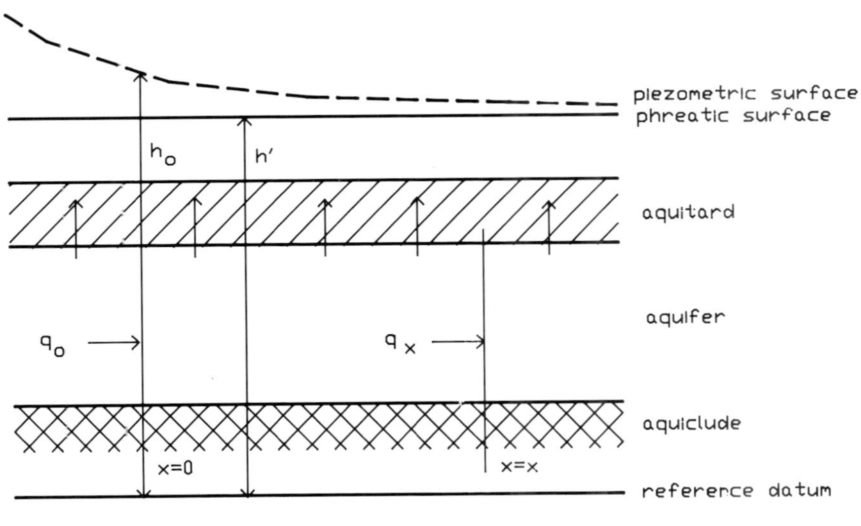

Fig. 2.53
Leaky aquifer with a constant phreatic surface in the upper, unconfined aquifer.

This example shows that the spatial distribution of the seepage from the aquifer into the over-lying aquitard depends only on the leakage factor, λ. For some values of x, the corresponding ratios, q_x/q_o, and the seepage as a percentage of the flow entering the aquifer at x = 0 (river bank) are given in Table 2.4 (Santing, 1963).

Distance from the river	q_x/q_o	Seepage over distance x, as percentage of q_o
x = 0.5 λ	0.607	39.3
x = 1.0 λ	0.368	63.2
x = 2.0 λ	0.135	86.5
x = 3.0 λ	0.050	95.0

Table 2.4
Seepage into aquitard as function of leakage factor.

These figures indicate that about 95% of the water entering the aquifer at x = 0 is lost to seepage over a zone extending from x = 0 to x = 3λ. It should be noted that the above equations contain a damping exponential function which is governed by the leakage factor, λ. λ is thus the only physical characteristic which must be known in order to calculate the seepage rate at various distances from a river channel. One method to determine λ is to conduct one or more pumping tests to determine the values of the transmissivity of the aquifer, kD and the hydraulic resistance of the aquitard, c, and substituting into the expression λ^2 = kDc (Kruseman & De Ridder, 1983). Another, less costly method is to use water level data from piezometers sunk in rows perpendicular to the river channel. Taking the logarithm of the above solution and rewriting it yields:

$$\lambda = \frac{x}{2.30\ \{\log(h_0 - h') - \log(h - h')\}}$$

Hence, by plotting (h-h′) against the distance x, on semi-log paper, a straight line is obtained with a slope of 2.30 λ. It should be noted that, at low river stages, the flow through the aquifer may reverse, causing the flow in the overlying aquitard to become vertical and downward. The same formulae apply, but (h_0-h') becomes negative, provided that all other conditions remain the same.

Another case of practical interest, where the flow through the aquifer is in one direction only, is that of the flow from a lake into the adjacent land, separated from the lake by a straight impermeable dyke (Figure 2.54). It is assumed that the dyke rests on an aquitard that is overlying an aquifer. To the left of the dyke, water from the lake will percolate through the aquitard into the aquifer. To the right, water from the aquifer will seep up vertically into the aquitard. To solve this problem, the aquifer has to be divided into three parts from $-\infty < x < -L$, $-L < x < +L$, and $+L < x < +\infty$ (Verruijt, 1970). At x = - L and x = + L the heads are h_2 and h_3 respectively. These heads are the unknowns which will be determined from the boundary conditions.

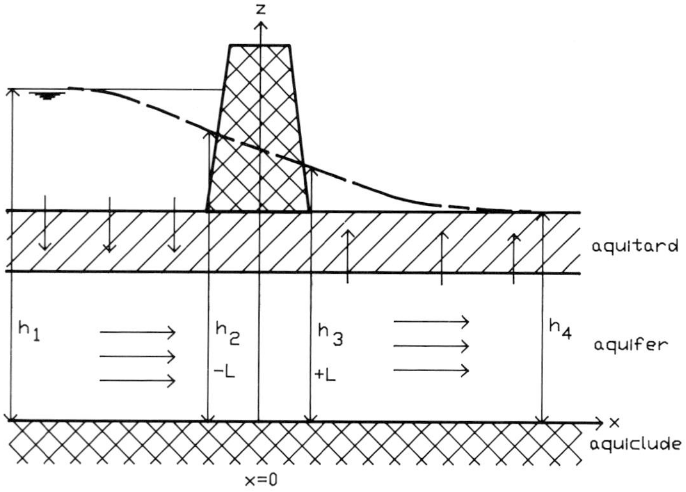

Fig. 2.54
Flow beneath an impermeable dyke.

The solutions of the general flow equation are:

$$h = h_1 - (h_1 - h_2)\, e^{\frac{(x+L)}{\lambda}}$$

for the region $-\infty < x < -L$, and:

$$h = h_4 - (h_4 - h_3)\, e^{\frac{-(x+L)}{\lambda}}$$

for the region $+L < x < +\infty$.

The total discharge per metre length of the dyke is:

$$q = kD \frac{(h_1 - h_2)}{\lambda}$$

at $x = -L$, and:

$$q = kD \frac{(h_3 - h_4)}{\lambda}$$

at $x = +L$.

The head under the impermeable dyke is given by:

$$h = 0.5\,(h_2 + h_3) - 0.5\,(h_2 - h_3)\frac{x}{L}$$

and the discharge is:

$$q = kD \frac{(h_2 - h_3)}{2L}$$

Continuity of flow at the common boundaries $x = -L$ and $x = +L$ requires that the discharge, q be the same in all cases. These three equations contain three unknown quantities; h_2, h_3, and q, which can be solved to give:

$$h_2 = h_1 - (h_1 - h_4)\frac{\lambda}{2L + 2\lambda}$$

$$h_3 = h_4 + (h_1 - h_4)\frac{\lambda}{2L + 2\lambda}$$

$$q = kD \frac{(h_1 - h_4)}{2L + 2\lambda}$$

The equations are rather simple and can be used, as a first approximation, to calculate the flow rate beneath a dyke from a river that is at high stage. However, most of the assumptions and conditions underlying these equations, are seldom satisfied in the flood plains of modern deltas. For instance, aquitards that extend from the surface down to the aquifer are rarely homogeneous, continuous or of constant thickness. These features are clearly illustrated in Figures 2.55 and 2.56, which show a number of abandoned and silted-up Holocene river channels in the central part of the Rhine-Meuse delta (Verbraeck & De Ridder, 1962). The former channels are often incised into the Pleistocene aquifer and filled with sands and finally covered with a thin layer of peat and clay. It is obvious that the aquitard is discontinuous at these places and that the silted-up former river channel is a major leak.

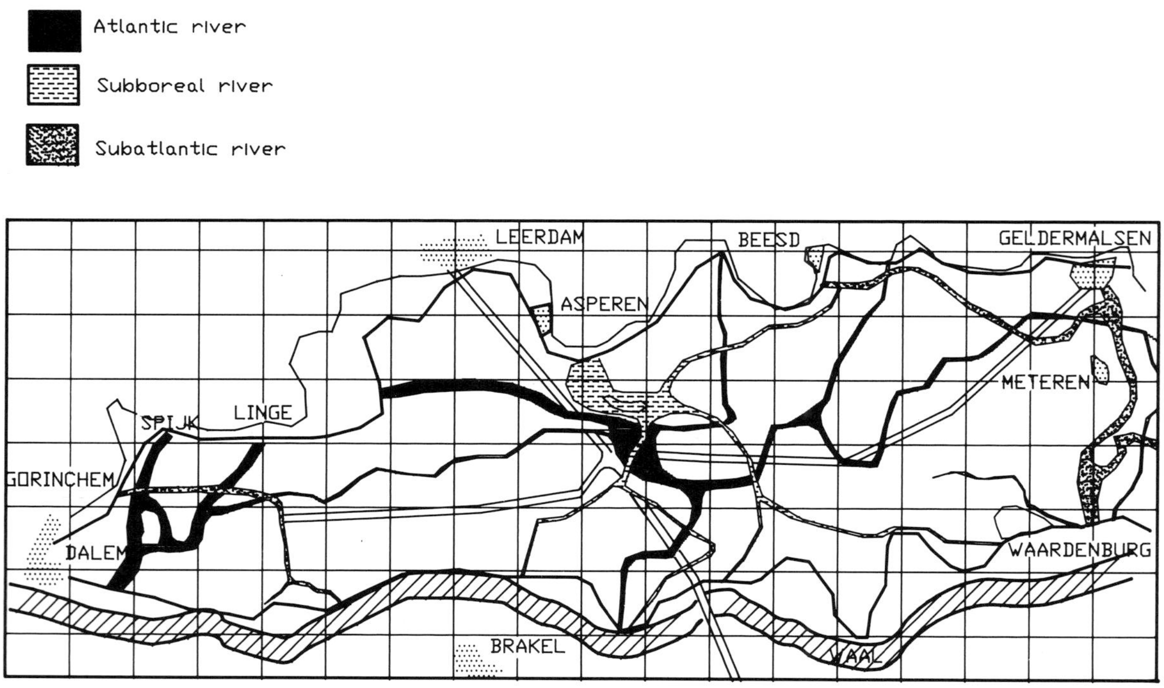

Fig. 2.55
Abandoned and silted-up river channels.

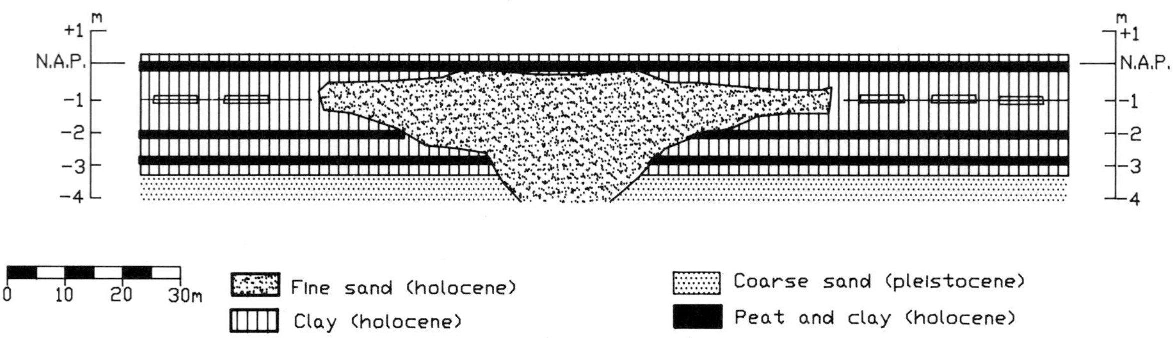

Fig. 2.56
Cross-section of a silted-up river channel.

It can therefore be concluded that the flood plains of modern deltas contain numerous leaks through which significant quantities of water can flow upward or downward, depending on the river stage, and on the head distribution in the aquifer. To solve problems of this type, it is advisable not to rely on one single method, but also to apply other approaches. The results can then be compared and the order of magnitude of the desired quantity can be determined.

Tidal transmission

Most deltas, except for those in land-locked seas, are affected by the tide which is felt far upstream in the river channels. In deeply incised channels, that are in direct hydraulic contact with the aquifer, the tidal movement induces corresponding variations in the groundwater pressure of confined and leaky aquifers, More specifically, if the level of the sea or ocean varies with a simple harmonic motion, a sequence of sinusoidal waves is propagated inland from the underwater outcrop of the aquifer. Figure 2.57 shows how water levels in piezometers placed in the aquifer at different distances from the river demonstrate a similar sinusoidal motion (Wesseling & Colenbrander,1962).

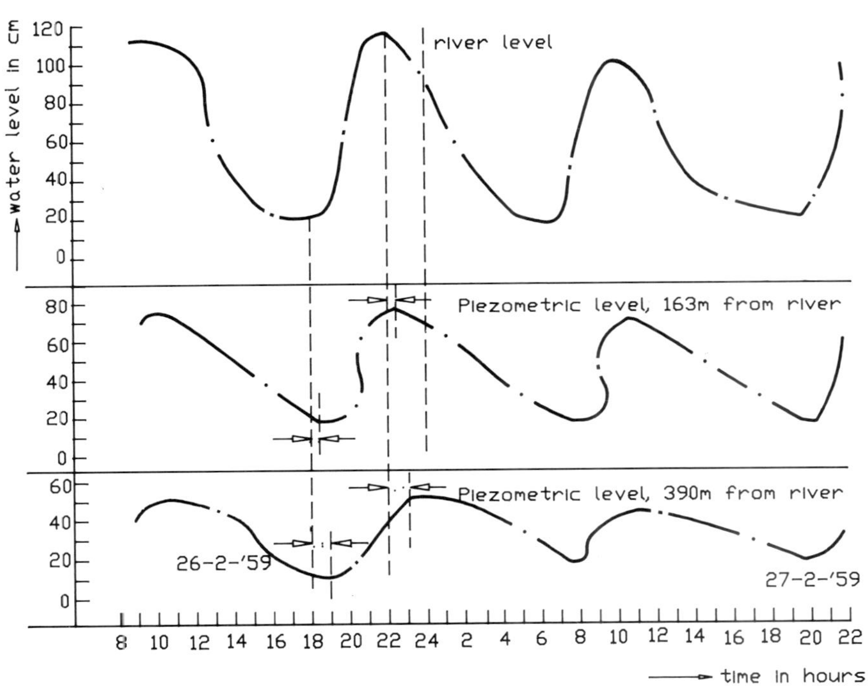

Fig. 2.57
Periodic water-level fluctuations in one of the major tributaries of the Rhine.

However, the amplitude of the sinusoids decreases with distance from the river. In other words, the waves attenuate inland. Also, the time lag, or phase shift, of a given maximum or minimum water-level increases inland. It is clear that a relationship must exist between the attenuation and the phase shift on the one hand, and the aquifer characteristics on the other. The harmonic motion of the sea-level can be described by:

$$y_o = y_m + A \sin \omega t$$

where:

y_o	= water-level elevation of the channel	(m)
y_m	= mean water-level elevation in the aquifer	(m)
A	= amplitude of the tidal wave	(m)
ω	= $2\pi/T$ is the wave frequency	(rad/day)
T	= period required for a full cycle	(days)
t	= time elapsed within any cycle	(days)

Steggewentz (1933) was probably one of the first hydrologists to study the propagation of tidal waves in deltaic aquifers. He assumed that the aquifer and the water in it were incompressible, and so derived the following equation for a leaky aquifer:

$$h(x,t) = h_m + A e^{-ax} \sin(\omega t - bx)$$

where:

$h(x,t)$	= hydraulic head	(m)
h_m	= mean hydraulic head	(m)
bx	= phase shift	(rad)
e^{-ax}	= amplitude reduction factor.	

For an unconfined aquifer, Steggewentz found the following relation between a and b and the aquifer characteristics:

$$a = b = \sqrt{\frac{\mu\omega}{2kD}}$$

where:
μ = (dimensionless) storage coefficient of the aquifer.

The parameters a and b, describing the attenuation and phase shift, are equal. If they are not, the aquifer will be confined or leaky. In the flood plains of modern deltas, leaky aquifers are common, which means that the compressibility of the water, in the aquifer and the aquitards must be taken into account. Jacob (1940, 1950) derived expressions for the propagation of tidal fluctuations through truly confined aquifers. Bosch (1951) extended Jacob's theory to leaky aquifers by including the effect of leakage through the overlying aquitard. The differential equation describing this one-dimensional, unsteady flow problem is:

$$\frac{\partial^2 h}{\partial x^2} - \frac{h - h'}{kDc} - \frac{S}{Kd}\frac{\partial h}{\partial t} = 0$$

In this equation, S is the (dimensionless) storage coefficient of the aquifer, which is defined as the volume of water that an aquifer of thickness, D, releases from or takes into storage, per unit surface area of aquifer, per unit change in hydraulic head. The storage coefficient is generally described as:

$$S = \rho g(\alpha + n\beta)D$$

where:

ρ	= fluid density	(kg/m^3)
g	= acceleration due to gravity	(m/s^2)
α	= vertical compressibility of the aquifer	(m^2/N)
β	= compressibility of the fluid	(m^2/N)
n	= porosity of the sediment in the aquifer	(%)

Meinzer (1928) emphasises the importance of compressibility and elasticity of low-permeable formations. For further information on this subject, see the review article by Streltsova-Adams (1978), and publications by van der Kamp (1973), De Lange and Maas (1986), and Stroet (1987). For a thorough treatment of the concept of compressibility of porous media and water, refer to Verruijt (1970), and Freeze and Cherry (1979).

2.2.4 Groundwater abstraction

Today most groundwater supplies developed in coastal areas come from deltaic formations. This is not surprising as the large mass of permeable sediments, the generally shallow water-tables, and the readily available recharge from river channels are ideal conditions for groundwater supply. The study of the hydraulics of pumped wells has developed into an important branch of geo-hydrology (Hantush (1964), Walton (1970), Huisman (1972), Kruseman & De Ridder (1986)). Groundwater exploitation, however, can bring about undesired environmental side effects. For examples, damage to ecosystems and agriculture from water-table drawdown, deterioration of groundwater quality due to saltwater intrusion near the coast, leakage from an adjacent aquifer, upconing of the freshwater-saltwater interface, induced recharge from polluted streams, and land subsidence from the dewatering of a thick sequence of poorly consolidated sediments. Some of these problems will be discussed in the following sections.

Induced recharge

To ensure a sustained water supply and avoid excessive water-table drawdowns, wells are often sunk near perennial streams (Figure 2.58). The stream will represent a recharge boundary, if the aquifer is in hydraulic connection with it.

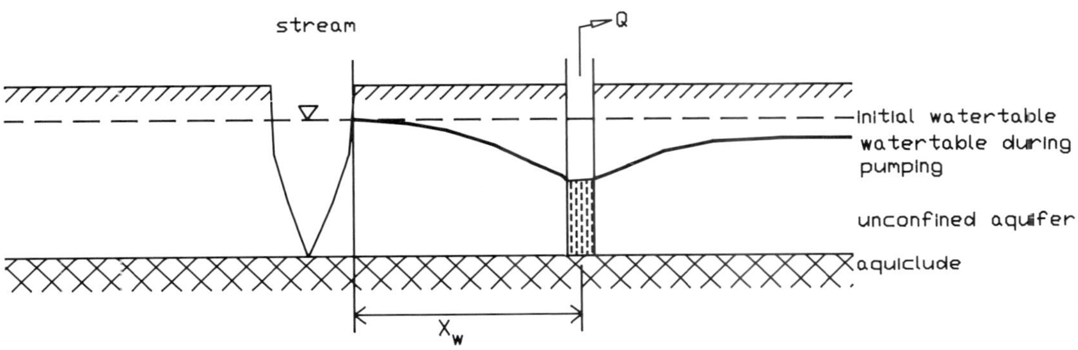

Fig. 2.58 Pumped well near a perennial stream.

When the well is pumped, its discharge is extracted initially from the groundwater. As pumping continues, the cone of depression reaches the stream. As a result of the hydraulic gradient thus developed, water from the stream starts flowing toward the well, thus recharging the aquifer (induced recharge), and stabilizing the cone of depression. In other words, a steady-state flow system will have developed, and all the pumped water would come from the river. If the drawdown at any point (x,y) of the flow domain induced by the pumping is required, the theory of image wells and the principle of superposition (Ferris et al. 1962) can be applied. The drawdown equation reads:

$$s = \frac{Q}{4\pi kD} \ln \frac{(x + x_w)^2 + (y - y_w)^2}{(x - x_w)^2 + (y - y_w)^2}$$

where:

s	= drawdown	(m)
Q	= (constant) well discharge	(m^3/day)
kD	= transmissivity of the aquifer	(m^2/day)
x_w, y_w	= the coordinates of the pumped well.	

An advantage of obtaining water from the induced recharge technique rather than directly from a river, lies in the chemical and biological purification it undergoes on its way through the alluvial sediments to the well. Some streams, however, are so polluted with toxic chemical waste that the aquifer becomes contaminated.

Upconing of saline water

Fresh groundwater may be underlain by saline groundwater. When a well in the freshwater zone is pumped, the freshwater-saltwater interface beneath the well will rise (Figure 2.59). This phenomenon, known as upconing, is caused by the reduction of pressure on the interface due to the drawdown of the water-table around the well. If the bottom of the well is close to the saline water, or if the well discharge is relatively high, the saltwater cone may reach into the well, causing the pumped water to be a mixture of fresh and saline.

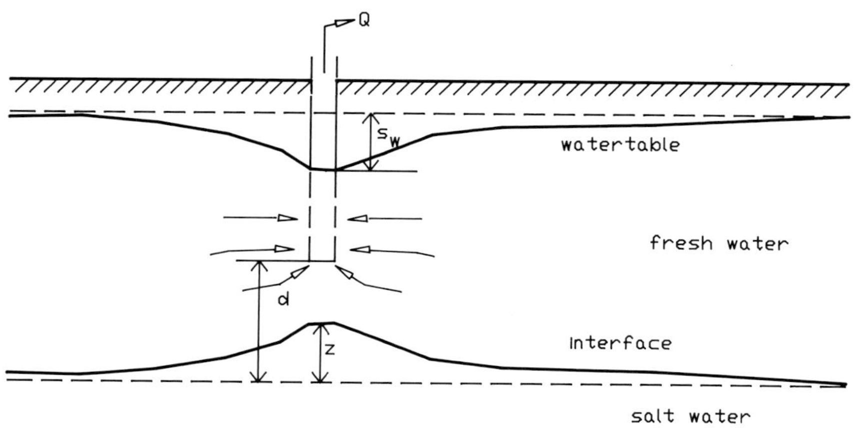

Fig. 2.59. Upconing of the freshwater-saltwater interface.

Land subsidence

A harmful environmental side effect of extensive groundwater exploitation (and oil or gas production), is deformation of the mass of sediments, resulting in movement of the land surface. Downward movements of as much as 10 metres and lateral movements of several metres have been measured (Bouwer 1978; and Freeze and Cherry 1979). Downward displacement of the land surface is called subsidence which may cause damage to buildings, road systems, pipelines and powerlines. The flood hazard of already low-lying areas will also be increased and the natural and artificial drainage system, whose gradients become reversed, will be disrupted. Subsidence is largest where the decline of groundwater levels is greatest and the aquifers and aquitards are thickest and most compressible.

There is one feature that is common to all groundwater conditions in deltas. That is a thick sequence of unconsolidated sediments forming an inter-bedded aquifer-aquitard system. Pumping, of course, is from the sand and gravel aquifers, but a large percentage of the pierced geological formations consist of highly compressible clays and fined-grained sediments. Dewatering of these fine-grained materials leads to compaction, just as drainage of the aquifer by pumping leads to compaction of the aquifer. However, the compressibility of clays is one or two orders of magnitude greater than that of sand and gravel and thus the total potential compaction of an aquitard is much greater than that of an aquifer. What is perhaps more important is the hydraulic conductivity of the different geological materials; for clays this is usually several orders of magnitude less than it is for sand and gravel. Consequently, the process of dewatering and hence the process of compaction of aquitards is much slower than it is for aquifers.

With regard to the compressibility of a porous medium, a reduction in volume can be achieved by compression of the water in the pores, compression of the individual particles of the medium, and re-arrangement of the particles into a more closely packed configuration. The first of these mechanisms is controlled by the fluid compressibility, β. The compressibility of the individual soil particles is small, so that it is reasonable to assume that the second mechanism can be neglected. Following Terzaghi (1925), the third mechanism requires the introduction of the concept of effective stress, also called inter-granular or effective pressure (σ_e). Re-arrangement of the soil particles and the resulting compression of the granular skeleton is caused by changes in the effective stress and not by changes in the total stress σ_T, which is exerted on the system. For the stress equilibrium on an imaginary plane through a saturated medium at a certain depth, the following simple relation holds:

$$\sigma_T = \sigma_e + p$$

where:
p = fluid pressure of the water in the pores at that depth.

The total stress, σ_T, acting downward on the plane is due to the weight of the overlying soil and water masses. If the weight of soil and water masses overlying each point in the system does not change with time, then, if the fluid pressure increases, the effective stress decreases by the same amount, and vice versa. The effective stress at any point in the medium and the resulting volumetric deformation at that point, is controlled by the fluid pressure at that point. Fluid pressure is defined as p= $\rho g \varphi$, where ρ is the density of the fluid, g is acceleration due to gravity, and φ is the pressure head. Changes in the effective stress at a point in the medium are therefore controlled by changes in the hydraulic head, h, at that point or:

$$\partial\sigma_e = -\rho g \partial\varphi = -\rho g \partial h$$

If the hydraulic conductivity of the aquitards is at least two orders of magnitude less than the hydraulic conductivity of the aquifers, it is generally assumed that the flow in the aquifers is horizontal and that the flow through the aquitards is vertical. If the aquifer is pumped and if data on the hydraulic head at some point in the aquifer are available, the hydraulic head h(z,t) at an overlying point in the aquitard can be predicted using Terzaghi's one-dimensional flow theory. The diffusion equation in one-dimensional form is:

$$\frac{\partial^2 h}{\partial z^2} = \frac{\rho g\,(\alpha' + n'\beta)}{k'}\frac{\partial h}{\partial t}$$

where:
α' = compressibility of the aquitard (m^2/N)
n' = porosity of the aquitard (%)
k' = vertical hydraulic conductivity of aquitard (m/s)

The initial condition is given by $h(z,0) = h_0$, and the boundary conditions are $h(0,t) = h_0 - \Delta h$ and $h(D',t) = h_0 - \Delta h$. Terzaghi (1925) derived an analytical solution to this boundary-value problem. He noted, however, that for aquitards (clays), $n'\beta \ll \alpha'$. He grouped the remaining aquitard parameters into a single parameter c_v, which is called the coefficient of consolidation defined as:

$$c_v = \frac{k'}{\rho g \alpha'}$$

Terzaghi further introduced the dimensionless time factor T_f, defined as:

$$T_f = \frac{4c_v t}{D'^2}$$

where:
D' = thickness of the saturated aquitard (m)

If the consolidation coefficient and the aquitard thickness are known, T_f can be calculated for any time t, after pumping began. Terzaghi presented graphs for his solution, which relate $h(z,T_f)$ and D' (Figure 2.37). These graphs allow the hydraulic head to be predicted at any elevation, z, and at any time, t, in an aquitard which separates two producing aquifers, provided that the drawdown, in the aquifers is known. Terzaghi's solution can also be used for a situation where the aquitard drains to only a single aquifer. If the lower aquitard is impermeable, only the upper half of the curves in Figure 2.60 are used for the prediction of h(z,t).

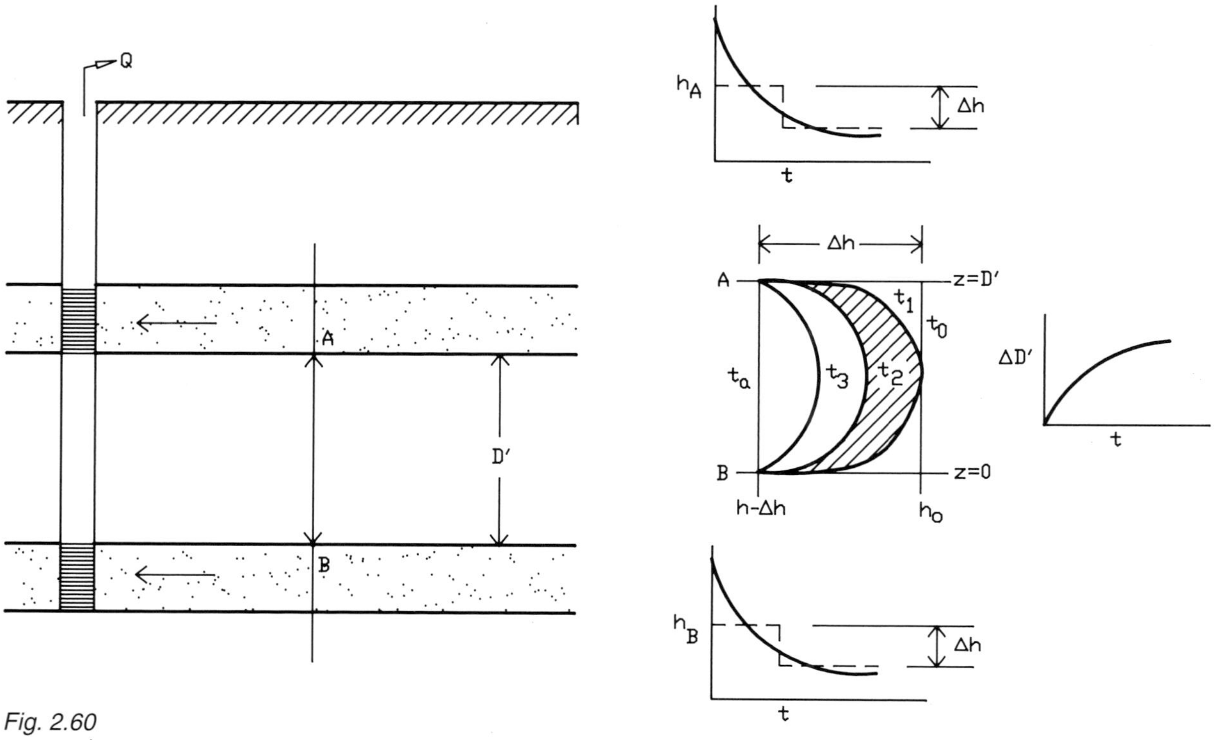

Fig. 2.60
One-dimensional consolidation of an aquitard.

To calculate the compaction of the aquitard, one must know the drop in hydraulic head at any point in the aquitard as a function of time. In the aquitard, the drop in head, δh, at any point z (n.b. $z = 0$ at the bottom of the aquitard and $z = D'$ at the top of the aquitard), between the times t_1 and t_2 induces a drop in fluid pressure given by:

$$\partial p = \rho g \partial \varphi (h - z) = \rho g \partial h$$

which is reflected by an equivalent increase in the effective stress. The increase in σ_e, acting through the aquitard's compressibility, α', causes the compaction of the aquitard $\partial D'$. To calculate $\partial D'$ between the times t_1 and t_2, the aquitard must be divided into m slices and the following equation can be applied:

$$\partial D'_{t_1 - t_2} = D' \sum_{i=1}^{m} \rho g \alpha' \partial h_i$$

where:
∂h_i = average head drop in the i-th slice.

For the compaction of aquifers we arrive at a similar expression:

$$\partial D = -\alpha D \partial \sigma_e = -\alpha D \rho g \partial h$$

For a multi-aquifer system with several pumped wells, the land subsidence as a function of time is calculated as the summation of all the aquitard and aquifer compactions (Domenico & Mifflin (1965) and Terzaghi & Peck (1967)).

Finally, the question arises whether land subsidence can be stopped or even restored by injecting water back into the system. In principle this should increase the head in the aquifers, drive water back into the aquitards, and cause an expansion of both the aquifer and aquitards. Unfortunately, this technique does not lead to the desired result because the compressibilities of aquitards in expansion are only about one-tenth the value in compression (Freeze and Cherry, 1979). Environmental damage is difficult to mend and, if it can be cured, the costs are often significant. With proper planning and investigation the problem of land subsidence due to overpumping can be minimised. Reclamation of low-lying, waterlogged deltaic areas generally entails lowering of the water-table by artificial drainage of the land, which also gives rise to land subsidence. However, there is evidence that for this situation, the linear relationship between effective stress and pressure head in the saturated porous media does not hold. For unsaturated flow, Bishop and Blight (1963), suggest the following equation:

$$\partial \sigma_e = \rho g X \partial \varphi$$

where:
X is a parameter depending upon the degree of saturation, the soil structure, and the wetting-drying history of the soil.

2.2.5 Groundwater quality

In a typical delta, the quality of groundwater, as determined by its chemical and biological constituents, shows large variations, both in the horizontal and vertical directions. This should not come as a surprise in view of the complex geological history of deltas. The chemical composition of groundwater in deltas can be indicative of its origin and history, and of the sediments the water has been in contact with. The different groundwater qualities and their spatial distribution in deltas are only now beginning to be understood. Several natural and human factors contribute towards the groundwater quality. The natural factors include the mixing of fresh and saline water, dispersion and diffusion, geochemical processes such as precipitation, solution, oxidation, reduction and adsorption, the exchange of gases and ions, and soil and biological conditions. The human factors include land reclamation, intensive farming, excessive pumping of groundwater, and disposing of waste in the soil. The following sections, will deal with some of the natural factors governing the quality of groundwater in deltas.

Occurrence of fresh groundwater

Atmospheric precipitation is the main source of fresh groundwater in deltas. This groundwater forms part of the hydrological cycle and is known as meteoric water. The occurrence of fresh meteoric groundwater is widespread in the upper and middle tracts of deltas and other topographically high ground like coastal dunes, abandoned sand-filled former river channels, and stranded beach ridges, etc. Badon Ghijben (1888) and Herzberg (1901) were probably the first to give an explanation for the occurrence of fresh meteoric groundwater at great depths near the sea. They explained this phenomenon by assuming that the lighter, fresh groundwater floats on the heavier, saline groundwater. Meteoric groundwater can have different ages, ranging from one day to more than 50,000 years. Its quality is usually good, except where it has been stagnant or isolated from the hydrological cycle in mineral-rich or clay-rich aquifers, or where it has been in contact with saline groundwater or evaporites. It is distinct from connate groundwater, which is water that was entrapped in the pores of the sediments at the time of their deposition. Connate groundwater can still be of atmospheric origin, but has been isolated from the hydrological cycle for up to millions of years. It is usually found in the deeper parts of deltas and is often of poor quality, especially when it has been in contact with evaporites.

Occurrence of brackish and saline groundwater

Brackish and saline groundwater not only occurs in the deeper parts of deltas, but also at shallow depths in a wide zone along the coast. However, small lenses of fresh groundwater have been found in some deltas near the coast. The water quality profile in this part of a delta is often very irregular, as shown in Figure 2.61.

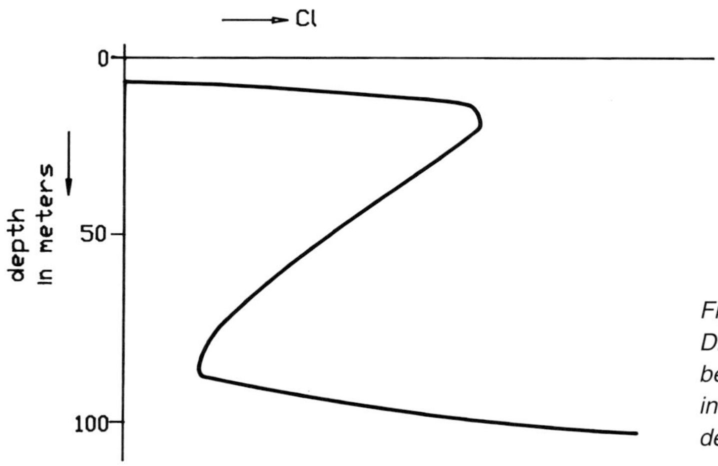

Fig. 2.61
Distribution of chloride beneath the IJssel Lake in the Rhine-Meuse delta.

The groundwater in the upper few metres of the lake bottom is fresh. Underneath this layer, the chloride content of the groundwater increases, and then gradually decreases from 20 metres to about 80 metres. From 80 metres down to about 100 metres below the lake bottom, the chloride content increases again. Volker (1961), demonstrated that the distribution process of the chloride ions can be explained exclusively by diffusion. This does not mean that other distribution mechanisms are rejected, but, in contrast with diffusion, they are not always necessarily present.

The climatic changes and their associated sea-level changes during the Quaternary must also be considered when trying to explain spatial variations in groundwater qualities. During the Pleistocene glacial periods, the sea retreated, and the aquifers which were formed were filled with fresh-water. The sea subsequently advanced in an interglacial period, flooding the lower part of the delta. Geirnaert (1973) reported that the saline groundwater in the 30 to 50 kilometre wide coastal zone of the Rhine-Meuse delta originated from seawater that infiltrated in the fluviatile Pleistocene formations during transgressions. Part of the saline water should then have been replaced by fresh groundwater flowing seaward from the higher lands in the middle of the delta. Meinardi (1973) rejected Geirnaert's hypothesis of the replacement of saline groundwater by fresh groundwater flowing from the higher parts of the delta towards the sea, as being too slow, and mentioned three causes of the present chloride distribution in the groundwater underneath the low-lying coastal strip of the Rhine-Meuse delta as being dispersion, diffusion and reclamation.

Land reclamation in waterlogged areas in the lower parts of deltas (e.g. impoldering and artificially controlling the water-table by drainage) can have a significant effect on the flow and quality of the groundwater. The lowering of the water-table in polders has the same effect as the pumping of a well in a fresh groundwater zone underlain by a saline groundwater zone. There will be a local upconing of the freshwater-saltwater interface which may reach the drains and introduce substantial amounts of salts into the water courses and canals (Wesseling, 1979).

The effect of overburden

Compaction of sediments that have little load-bearing strength (e.g. montmorillonite clays) can be an important natural mechanism for the expulsion and circulation of connate water in deltas

(Jones, 1970). Montmorillonite clays in a marine environment absorb large amounts of water. Under the increasing load of growing deltas, these muds are compressed and their water expelled into adjacent sandy formations. If these formations are closed seaward by marine clays, the only escape route is landward as shown in Figure 2.62 (Jones, 1970). This may explain why the groundwater and marshy areas behind the coastline of many deltas are so salty.

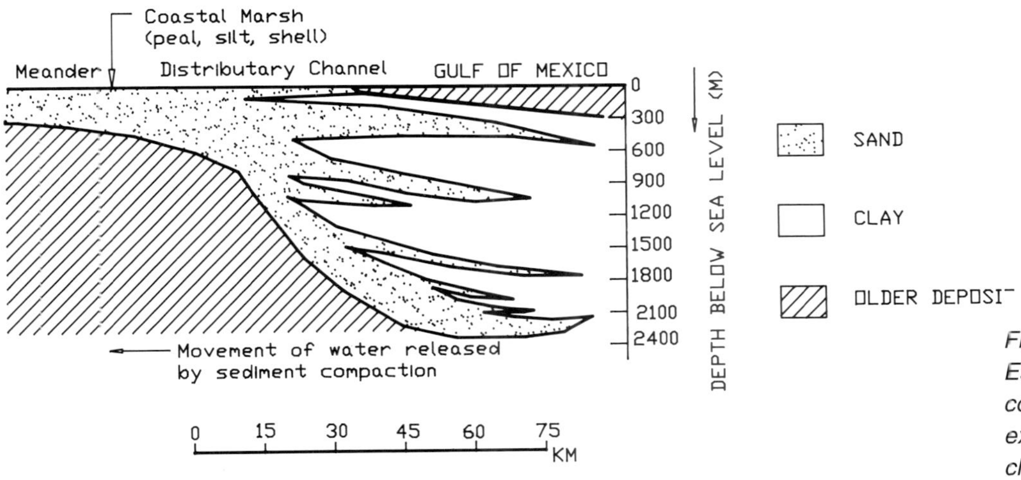

Fig. 2.62 Escape route of connate water expelled from clay beds.

Spatial variations in groundwater quality

Figure 2.63 (Anderson, 1978) shows spatial variations in the groundwater quality of a section along the coast of the Mekong delta. Fresh groundwater occurs in the foothills bordering the delta on the north. These foothills are recharge areas where rain infiltrates and recharges the shallow groundwater. The chloride content of the groundwater increases rapidly from the foothills towards the delta, reaching values as high as 10,000 to 15,000 milligrams per litre beneath the Mekong and its tributaries. Even in the inland sector of the delta, where the water in the tributary channels is fresh, the shallow Holocene aquifers contain brackish and saline, probably connate, water. Seepage from these channels at high stages is so small that it can be neglected. The water in the coastal fringe of the 100 metre and 200 metre aquifers is also brackish or salty. Further inland, these aquifers contain only slightly mineralised water. Surprisingly, a well at Bai Xao shows chloride contents increasing with depth, up to more than 6,000 milligrams per litre. However, at 450 metres, a freshwater aquifer occurs with a chloride content of only 130 milligrams per litre, contradicting earlier statements that connate water in deep deltaic aquifers is usually of poor quality because of mixing, geochemical processes, and long residence times.

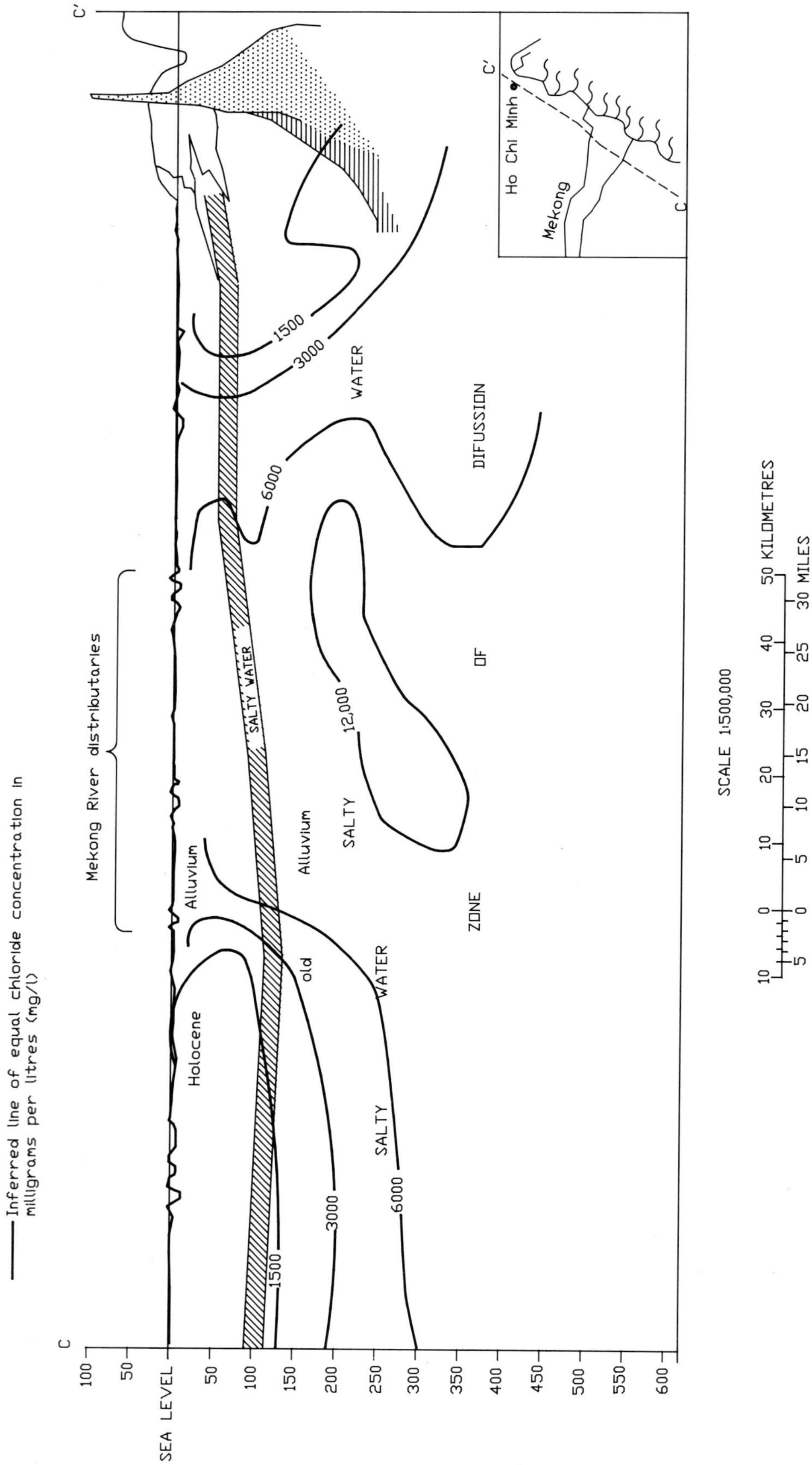

Fig. 2.63
Section along the outer fringe of the Mekong Delta.

2.2.6 Salt water intrusion

Fresh groundwater underlain by brackish and saline groundwater is a common phenomenon in most deltas. The present distribution in the subsoil of fresh, brackish and saline water developed during the geological past and is affected by several natural processes. More recently, human intervention has also had a significant effect.

Sedimentation of rock material in coastal areas generally took place in a marine environment and subsequently lead to saline water in these formations. The actual distribution of fresh and saline water in the subsoil is affected by:

- natural recharge of the groundwater system,
- transport processes and
- density differences between fresh and saline water.

Natural recharge and rainfall infiltration are the main sources of fresh water in these formations. This water travels from the recharge area through permeable formations (aquifers) to the sea or ocean. On its way it may encounter saline aquifers. The distribution of fresh and saline groundwater depends upon geohydrological parameters and the density differences between the liquids. Climatic changes with different rainfall regimes have also influenced the salinisation process. Seas, oceans, lakes and rivers act mostly as outflow boundaries for groundwater systems. Long term differences in their levels (e.g. transgressions, regressions) will therefore interfere in the groundwater regime.

The main transport mechanism is convection (i.e. flow of water due to gravity). If density differences occur and the water is stagnant then the least dense water will float on top of the more dense water and a horizontal layering will occur. Flowing water will create sloping interfaces. Apart from this, a movement of ions occurs (i.e. molecular diffusion). This process is caused by differences in concentrations, is independent of the flow, and develops proportional to the square root of time. Diffusion will smooth the concentration gradients. In the case of an originally sharp interface between fresh and saline water, a transition zone of brackish water will develop. Flowing water will lead to dispersion - a mixing process caused by the different magnitudes and orientations of the velocity vectors of the groundwater in the pores. On a large scale, irregularities in an aquifer have a similar effect. The displacement of ions by dispersion is proportional to the velocity of the groundwater. Both dispersion and diffusion have a smoothing influence on differences in concentration. Dispersion and diffusion are difficult to distinguish, therefore their joint effect in groundwater is taken into account as hydrodynamic dispersion (Bear, 1979). Density differences and their effect on the distribution of fresh and saline water in the subsoil will be treated later in this section.

Reclamation of coastal areas, impoldering, extraction of groundwater, artificial recharge of groundwater, lowering of groundwater tables, irrigation and drainage, construction of canals etc. have all led to changes in the geo-hydrological regime. The response of the groundwater system is to reach a new state of equilibrium, with a different distribution of fresh and saline groundwater. This may last for decades or even centuries, as groundwater flow is generally a very slow process. This also means that once an unwanted effect is observed, it will take a long time before countermeasures become effective.

In this section, some properties of groundwater and the groundwater system related to salinisation will be treated and a few cases of salinisation in aquifers will be given (see also van Dam, 1992). The effects of human intervention and sea level rise will then be discussed. A few case studies on salinisation in deltaic areas complete this section.

The groundwater system and salinisation

The groundwater system is basically a combination of rock/sediment matrix and water. The size and shape of the voids of the matrix and viscosity and density of the pore water, are determining factors for the permeability (k):

$$k = \frac{\kappa \rho g}{\mu} \text{ (m/s)}$$

where:
κ = intrinsic permeability (m^2)
μ = dynamic viscosity (kg/m/s)

The influence of density differences on the permeability is of little practical importance with regard to the accuracy of the determination of the permeability itself. More important may be the phenomena of changes in the permeability in the case where fresh groundwater is replaced by saline water. Irreversible changes in permeability may occur, as have been reported in several publications (see Goldenberg et al., 1986).

Flow is directed by differences in hydraulic head, which is the sum of the elevation and the pressure heads. Density is included in the latter. Density depends on salinity, temperature and, to a lesser extent, pressure. Salinity is the concentration of dissolved solids in water given in mass per volume or in ppt (parts per thousand) or ppm (parts per million). Ocean water may contain approximately 34,500 mg/l of dissolved solids. The main contributor is the Chloride (Cl^-) ion at about 19,000 mg/l.

In coastal aquifers the Cl^- content varies between 5 mg/l Cl^- (infiltrated precipitation) to 21,000 mg/l Cl^-. The following equation (van Dam, 1992) gives a rather good approximation for the relation between density, salinity and temperature, for temperatures less than 15 °C and salinity values less than 20,000 ppm (≈20,000 mg/l).

$$\rho_{(S,T)} = 1000 + 0.8054S - 0.0065\,(T - 4 + 0.2214S)^2$$

where:

$\rho_{(S,T)}$	= density	(kg/m^3)
S	= salinity in p.p.t.	(≈ g/l)
T	= temperature	(°C.)

Densities may occur ranging from 1000 kg/m^3 for fresh water to 1025 kg/m^3 for ocean water. Although the differences appear to be small, they have a significant effect on hydraulic heads and thus on the groundwater regime. The following simple equation can be derived from an equilibrium state of a sharp interface between fresh and saline water (see Figure 2.64).

$$h = \left[\frac{\rho_s - \rho_f}{\rho_f}\right] H = \Delta H$$

where:

ρ_s	= density of salt water	(kg/m^3)
ρ_f	= density of fresh water	(kg/m^3)
Δ	= relative density difference	(-)

This equation is called the Badon Ghijben-Herzberg (BG-H) relation (Badon Ghijben (1888) and Herzberg (1901)). The equation is correct under hydrostatic conditions and if there is only horizontal flow in the fresh water zone and if the saline water is stagnant. Although the position of the interface is not correct in the outflow zone, the use of the equation still gives a rather good approximation to the real situation. The use of hydraulic heads in the situation where density differences occur in groundwater leads to some difficulties. This is illustrated in Figure 2.65. Water levels in an observation well are given at one location, at different depths with different densities. Although the situation is assumed to be hydrostatic, the levels are not the same, due to the density differences.

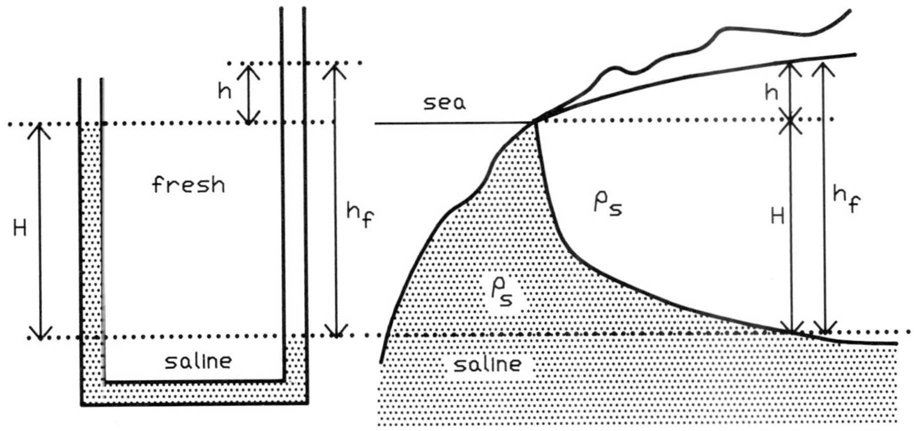

Fig. 2.64
Fresh water / saline water interface.

For horizontal flow, the pressures (or hydraulic heads) in the horizontal plane, have to be determined at the different locations of the observation wells. They can be derived from the hydraulic heads and the vertical density profiles, which have to be known beforehand. Firstly, the heads are expressed in terms of a reference density, usually that of fresh water. The length of the column of water in the observation well is converted into an equivalent column of fresh water.

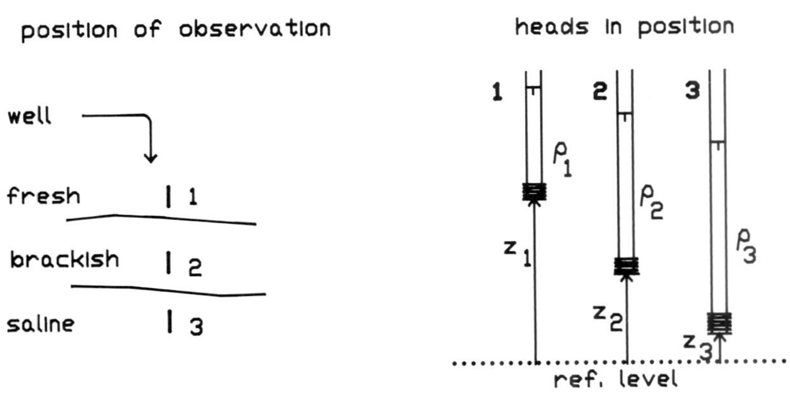

Fig. 2.65
Hydraulic heads at one location but at different depths with differing densities.

$$h_f = h_i \frac{\rho_i}{\rho_f} = (1 + \Delta_i)\, h_i \quad (i = 1,2,3)$$

In this case, the expression for the hydraulic head (ϕ_f) becomes:

$$\phi_f = z_i + (1 + \Delta_i)\, h_i \quad (i = 1,2,3)$$

where:
z_i = elevation head of the screen relative to a reference level.

From the pressures at the position of the screen, the pressures at the adopted plane are computed and converted into fresh water heads (see Figure 2.66):

$$\phi_A = z_2 + \frac{\rho_1}{\rho_f} h_1 + \frac{1}{\rho_f g} \int_{z_2}^{z_1} \rho(z) g dz$$

$$\phi_B = z_2 + \frac{\rho_2}{\rho_f} h_2$$

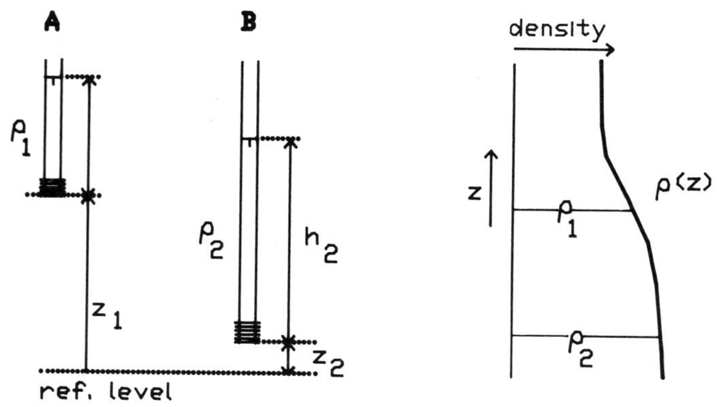

Fig. 2.66
Determination of hydraulic head.

For vertical flow through an aquitard, the hydraulic heads in the bounding aquifers should be known. From these data and the vertical density profile it is possible to determine whether the vertical pressure distribution is hydrostatic or not. A variation from hydrostatic conditions will cause vertical flows. Considering the two observation wells in Figure 2.66 to be at the same location but at different depths with different densities, the hydrostatic hydraulic head can be calculated for level z_1 by measuring the pressure at level z_2 and integrating the specific weight (ρg) from z_2 to z_1. If there is a difference between the result of this and the actual head at z_1 (obtained by direct measurement), there will be vertical flow. The magnitude and direction follows from Darcy's Law. A comprehensive description of the method can be found in van Dam (1977).

For a correct description of salt water intrusion into groundwater systems the mass balance equations for fresh and saline groundwater have to be taken into account. Even then, the actual situation can only be approximated as hydrodynamic dispersion is ignored. To achieve a more realistic image of salinisation, computer codes are used (e.g. methods which apply particle tracking). The groundwater system is firstly sub-divided into grid cells. Application of a finite element or finite difference computer code for transient groundwater flow results in a velocity field (direction and magnitude). In each grid cell a number of particles is positioned. An equal amount of salt is allocated to each of these particles in such a way that the total amount agrees with the actual concentration of salt in the groundwater system at that location. With the velocity field, the position of the particles can be computed after a certain time interval. The concentration of salt in a certain cell can then be computed from the particles present and the amount of salt they represent. This changes the density profile so that the hydraulic heads have to be adapted each time step (e.g. Lebbe (1986) and Oude Essink (1993)).

Salt water intrusion in coastal aquifers

In a first example, the distribution of fresh and saline water in the subsoil will be considered for a coastal phreatic aquifer with recharge (R). The saline water is stagnant, fresh water flow is horizontal and the transition zone between fresh and saline water is narrow. In other words, a rather sharp interface occurs and the BG-H approximation may be applied (see Figure 2.64). In this case Darcy's Law reads:

$$q = -k(h + H)\frac{dh}{dx} = -k\Delta(\Delta + 1)H\frac{dH}{dx}$$

The mass balance for fresh water gives:

$$\delta q = R\delta x$$

which, after integrating and combining with Darcy's law gives:

$$\frac{d(H^2)}{dx} = \frac{-2(Rx + C_1)}{k(1 + \Delta)\Delta}$$

which can be integrated to give:

$$H = \sqrt{\frac{-Rx^2 - 2C_1x}{k(1 + \Delta)\Delta} + C_2}$$

The integration constants C_1 and C_2 can be determined from boundary conditions.

For example, assuming a water divide at x = L m, where q = 0, leads to $C_1 = -RL$. The second boundary condition is at x= 0 m, where h = ΔH = 0 m. This results in $C_2 = 0$. These boundary conditions yield the following equation:

$$H = \sqrt{\frac{-Rx(x - 2L)}{k(1 + \Delta)\Delta}}$$

A second example deals with a confined aquifer discharging into the sea (Figure 2.67).

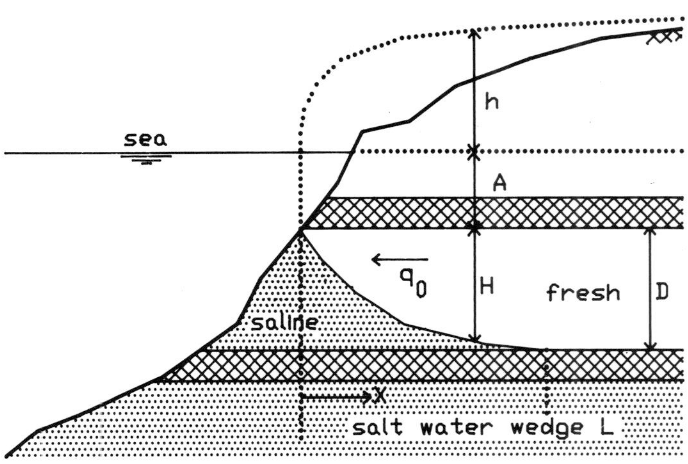

Fig. 2.67
Confined coastal aquifer discharging into the sea.

The magnitude of the (constant) discharge from the aquifer is q_0 m3/day per metre width. In this case, according to BG-H, h = Δ(A + H). Applying the same method as in the first example, leads to the following differential equation:

$$\frac{d(H^2)}{dx} = \frac{-2q_0}{k\Delta}$$

which can be integrated to give:

$$H = \sqrt{\frac{-2q_0 x}{k\Delta} + C}$$

For H = 0 metres at x = 0 metres, the integration constant is C = 0.

To find out how far salt water intrusion extends into the aquifer the condition will be H = D (where D is the thickness of the aquifer). The length of the salt water wedge (L) is then:

$$L = \frac{k\Delta D^2}{-2q_0}$$

The above equations indicate that at the outflow side (x= 0), the thickness of the fresh water body becomes zero. This is, of course, impossible in reality. Glover (1964) found a more exact solution to the problem by introducing a transformation of the x-axis:

$$x' = x - \frac{q_0}{2\Delta k}$$

For the second example this means there is a submarine outflow zone of 1.6 m and the thickness of the fresh water body is 3.2 m. For values of x > -10 $q_0/2\Delta k$, the differences in the calculated positions of the fresh-salt interface between the BG-H approximation and the Glover solution are smaller than 5%.

Human interventions

If fresh groundwater is abstracted from an aquifer, this introduces a drop in hydraulic head of several meters (see Figure 2.59). A simple calculation based on the BG-H relation shows that a drawdown in the well of s_w metres would lead to an upconing of the saline groundwater by s_w/Δ metres. This may easily lead to salinisation and abandoning of the well.

For a well in a confined aquifer the following equations can be derived (Schmorak and Mercado, 1969):

$$z = \frac{Q}{2\pi\,\Delta\,d\,k_{hor}} \cdot \frac{t}{t + \tau}$$

$$\tau = \frac{2nd}{\Delta k_{ver}}$$

where:

z	= upconing	(m)
Q	= extraction rate	(m^3/day)
d	= distance between screen and salt/fresh interface	(m)
k_{hor}	= horizontal hydraulic conductivity	(m/day)
k_{ver}	= vertical hydraulic conductivity	(m/day)
t	= time elapsed from start of pumping	(day)
τ	= time constant	(day)
n	= porosity	(-).

For t = ∞ these equations reduce to:

$$z_{max} = \frac{Q}{2\,\pi\,\Delta\,d\,k_{hor}}$$

Application of this equation may serve to determine a maximum permissible extraction rate such that the upconing does not exceed a certain critical height.

Reclamation of polders usually occurs in coastal regions. In most cases, the groundwater system may be schematised as a semi-confined aquifer in which fresh groundwater lies on top of saline groundwater. This is all over-topped with a semi-pervious layer. If the hydraulic head of the aquifer originally coincides with the water table, reclamation will demand a lower groundwater table. This disturbs the groundwater regime. The difference in hydraulic head of the aquifer and the reclaimed area will create seepage, and upconing of saline water will occur (see Figure 2.68). The magnitude of the upconing can be approximated from the lowering of the phreatic water table using the BG-H relation. This may eventually lead to saline seepage, which will restrict the land use.

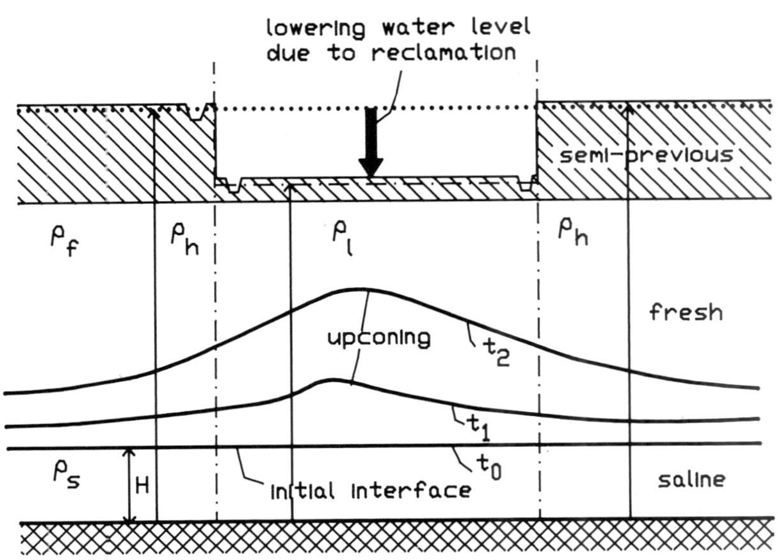

Fig. 2.68
Upconing caused by impoldering.

Lowering of the hydraulic heads and the effects of weathering may lead to land subsidence. This may necessitate a further change in the groundwater levels.

Another type of salinisation originates from irrigation, especially in arid and semi-arid zones. In these areas, rainfall is not sufficient to grow crops. Irrigation must therefore be applied. Irrigation also necessitates a good drainage system because of a build up in salt concentrations in the soil due to evapotranspiration. A surplus of irrigation water has to be applied to flush the salts. Drainage is necessary to export the excess water. This water may contain salt concentrations up to 10 times the concentration of the irrigation water. Obviously if there is no drainage system and the excess water recharges the groundwater system, the groundwater will eventually become unsuitable for further use. Decreasing the irrigation is no solution in this case, as the soil will become unsuitable for agriculture due to excessively high salt concentrations.

Climate change

During the last decade, much attention has been paid to climate change due to the greenhouse effect and global warming. A few effects related to salinisation will be discussed briefly here.

Changes in rainfall or evaporation, possibly up to 15% (Houghton et al.(1990)), will affect the fresh water balance. In general, an increase in rainfall will be advantageous, as the natural recharge of the groundwater will increase. This could lead to more extensive fresh water aquifers and less salt water intrusion. On the contrary, an increase in evaporation has the opposite effect.

A higher sea level will, in general, promote salt water intrusion into aquifer systems. Higher sea levels may also lead to a coastline retreat, thus decreasing the recharge areas (e.g. dune areas). The effect of this will be smaller amounts of fresh water and subsequently smaller permissible abstraction rates. Intrusion of saline water in unconfined and semi-confined aquifers will become more severe.

A few examples of salinisation in coastal areas will now be presented. A comprehensive overview of this subject is given by Custodio (1987).

Figure 2.69 shows an east-west cross-section through the Netherlands, near Amsterdam. The chloride distribution (in mg/l Cl^-) has been computed and calibrated by Oude Essink (1993).

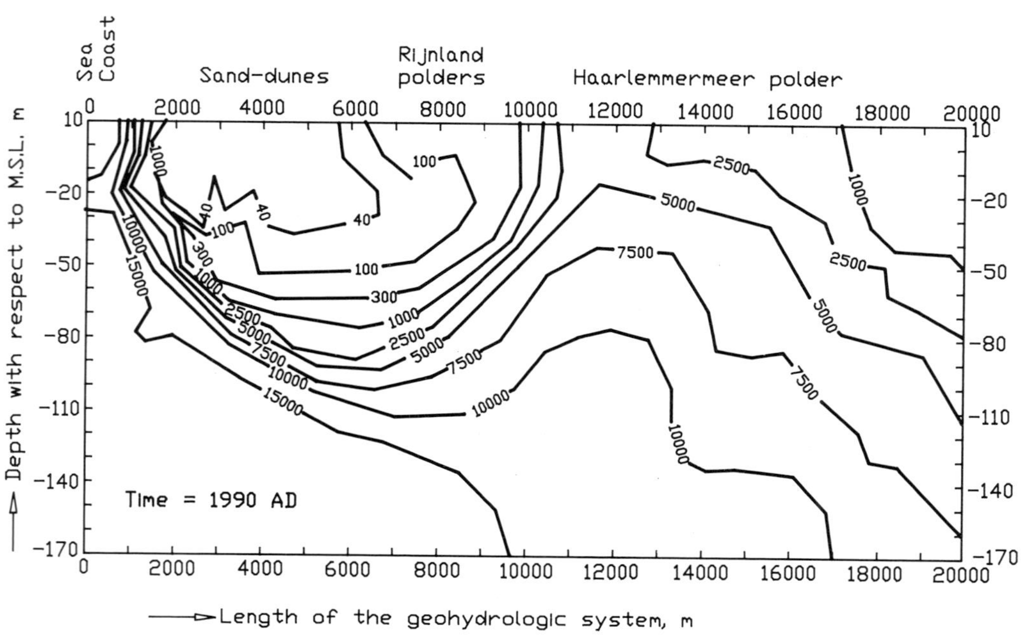

Fig. 2.69
Distribution of fresh and saline water in an east-west cross-section through the Netherlands, near Amsterdam.

Under the dune area in the west, a fresh water lens is present and salt water intrusion occurs in the deeper aquifers. Fresh water is extracted from the fresh water lens. To avoid salinisation due to the abstraction of groundwater, artificial recharge is applied. The Haarlemmermeer polder, a former lake, is situated east of the dune area. Due to the reclamation of this area a strong seepage occurs. The upconing of saline water as a consequence of the reclamation is clearly visible.

The next example shows the salinisation effects in the subsoil of the delta of the Llobregat in Spain (Figure 2.70). Under natural conditions the aquifers were in equilibrium with the river and salinisation was limited to the outflow zones. Extractions from the deep aquifers in combination with a decreased recharge from the river is now causing inflow of saline water into the deep aquifer. The recharge of the river deteriorated due to an increase of suspended solids in the river water and a decrease of flood discharges.

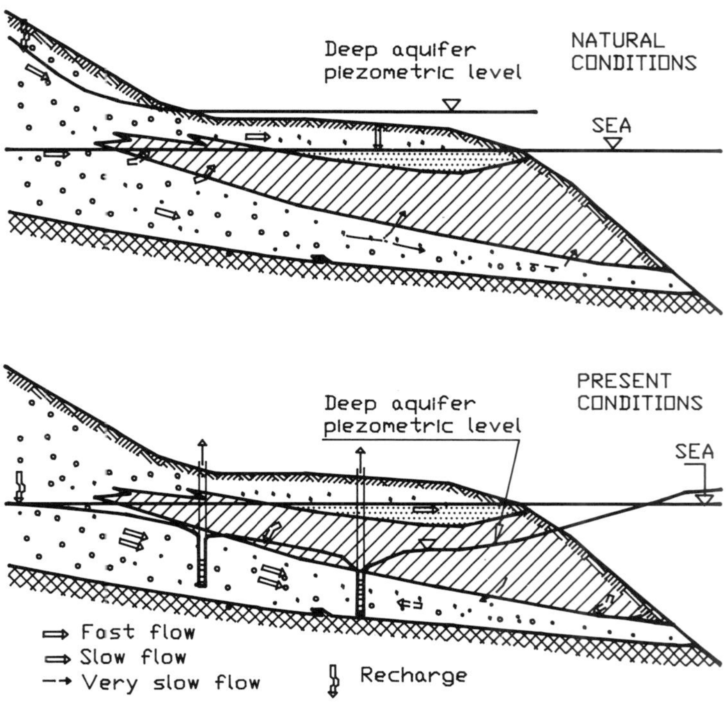

Fig. 2.70 Salinisation in the subsoil of the Llobregat delta, Spain (Custodio, 1987).

Possible solutions to this problem are a reduction of extraction rates, the use of surface water instead of groundwater and artificial surface recharge to the aquifers.

2.2.7 Groundwater modelling

Analytical solutions to problems of groundwater flow or seepage, as presented in the previous sections, are limited to situations where the region of flow, geological configuration and boundary conditions are simple and regular. For example, many analytical solutions require that the aquifer and aquitards are homogeneous and isotropic. There are, however, many practical situations for which analytical solutions cannot be obtained. This arises because seepage and groundwater flow problems have a number of non-linear features which are not amenable to inclusion in analytical solutions. These include the variable hydraulic characteristics (i.e. k, kD, S and c), moving or irregular boundaries, and long-term time-dependent effects. In that case, the differential flow equation can be solved numerically to find the distribution of head in the flow system.

Numerical methods are much more versatile than analytical methods of solution, but they are approximate and require the use of a computer. To apply these methods the flow region must be discretised. That is, the region must be divided into a finite number of blocks (cells), each with its own hydro-geological properties and each having a node at the centre, where the hydraulic head is defined for the entire block. Obviously, this type of numerical groundwater model is particularly useful in testing various groundwater management schemes and in predicting the effect of certain actions. As a predictive tool this type of model is also used to solve a wide variety of other problems such as travel times of groundwater, contaminant transport, heat transport, land subsidence, etc. The validity of these predictions will depend on how well the model approximates field conditions. Good field data are essential when using the model for predictive purposes. However, modelling a particular groundwater system with inadequate field data can also be instructive, as it may serve to identify those areas where detailed field data are critical to the success of the model. In this way a model can help to guide data collection.

Numerical groundwater models are based on a wide variety of techniques. In each case they are essentially alternative methods of solving simultaneous equations which represent the flow process in an aquifer. The most popular techniques nowadays are the finite difference technique and the finite element technique, both of which have their advantages and disadvantages. A third technique has recently been introduced, called the boundary integral equation method. The essential feature of this method is that solutions are obtained in terms of the functions at certain positions on the boundary of the flow domain (Liggett, 1977). It is beyond the scope of this text to deal with these techniques extensively. For a more detailed treatment, refer to the available textbooks on the subject such as Remson et al. (1977), Pricket & Lonnquist (1971), Zienkiewicz (1977), Pinder & Gray (1977), Rushton & Redshaw (1979), and Wang & Anderson (1982). A brief review of only some of these techniques is given here.

Finite-difference method

The basic differential equation which governs the unsteady two-dimensional flow in a non-homogeneous, isotropic aquifer is:

$$\frac{\partial}{\partial x}\left(kD_x \frac{\partial h}{\partial x}\right) + \frac{\partial}{\partial y}\left(kD_y \frac{\partial h}{\partial y}\right) = S\frac{\partial h}{\partial t} + Q_s$$

where:
Q_s is a source/sink term, which can include discharge or recharge by wells, induced recharge or seepage from rivers, drainage to rivers, evaporation or water uptake by vegetation, and leakage to or from other aquifers.

The finite difference method requires that both space and time be discretised. An example of space discretisation is shown if Figure 2.71 (see Pricket & Lonnquist, 1971). Darcy's equation is then applied to develop finite difference expressions for the flow at each node of the network.

This procedure yields a number of simultaneous algebraic equations (one for each node), with the head at each node as the unknown. The set of simultaneous equations can be solved by using, for example, an implicit finite difference scheme. With this method larger time steps can be used than with the explicit schemes. The coefficients of the resulting finite-difference equations form a tri-diagonal, N x N matrix, where N is the total number of nodes in the network. Such a matrix is easily solvable using a computer and one of the various algorithms available for this purpose.

The difference between analytical and numerical solutions to a particular groundwater flow problem should be noted. Analytical solutions are exact and have a physical basis. They allow the unknown quantity (e.g. hydraulic head) for any site in the flow domain to be calculated. Numerical solutions give only the values of head at a finite, pre-determined number of points in the flow region. Numerical solutions are non-physical and approximate. They do not give as much insight as physical models into the relative importance of various flow system parameters and input data. This insight is of great value in developing flow system models for analysis because, such models, no matter how complex they may be, are always drastic simplifications of the real situation (Bouwer, 1978). The technique of numerical groundwater modelling also has major advantages, as it allows aquifer heterogeneities and other non-linearities of the aquifer system, to be taken into account. Such non-linearity features are often poorly known. It requires great efforts of the hydro-

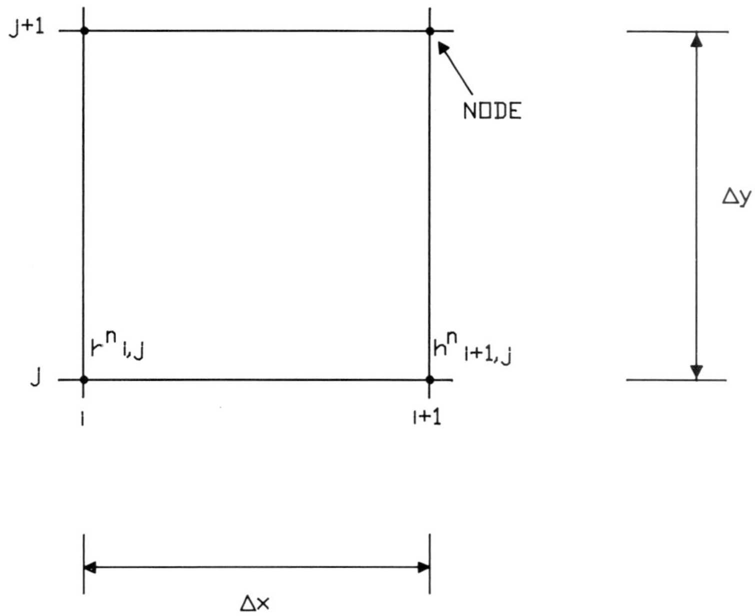

Fig. 2.71
Finite difference grid.

geologist to develop a conceptual model of the problem area and to provide reasonably accurate data on the magnitude and spatial variation of the aquifer's hydraulic characteristics and the nodal net recharge. These input values are tested in a process of calibration where generated heads are compared with measured ones. The process consists of adjusting the input data until computed heads match the field values (Boonstra & De Ridder, 1981). Calibration is necessary because the input data is not known exactly, and there will be a certain range of values that may be valid.

Finite-element method

With the finite-element method, the problem of the hydraulic head distribution within a flow region is solved through an equivalent variational functional (Verruijt (1970) and Wang & Anderson (1982)). The flow system is considered to represent a general system of energy dissipation, for which the solution of the unknown head is found as the head distribution which minimises the rate of energy dissipation. With the finite-element technique, the solution of the two-dimensional flow equation is obtained by finding a solution for h which minimises an equivalent variational functional of the form:

$$F = \iint \left[\frac{k_x}{2} \left(\frac{\partial h}{\partial h} \right)^2 + \frac{k_y}{2} \left(\frac{\partial h}{\partial h} \right)^2 + (S \frac{\partial h}{\partial t} - Q_s)\, h \right] dx\, dy$$

To find the solution, the flow region is divided into a number of sub-regions or finite elements which, unlike the square or rectangular grid for the finite-difference method, are triangular. The network of triangles should preferably be irregular, with small elements where the flow is concentrated (e.g. near a pumped well or near a strongly curved boundary) while larger elements can be used in areas of the flow region where hydraulic gradients are small and uniform or field data are scarce. For each time step, the parameters k_x, k_y, S, and Q_s are kept constant for a given element, but they may vary between elements. The use of the variational principle leads to a set of simultaneous algebraic equations that can easily be solved by a computer. However, the variational principle is restricted to specific groundwater flow problems only. Galerkin's method is much more widely applicable, as it is based on a particular weighted residual principle which turns out to be equivalent to a variational principle, if one exists for the problem under consideration. The basic difference between the two methods is that the weighted residual principle is expressed directly in terms of the governing partial differential equation without the need to resort to a physical quantity, such as the energy dissipation of a groundwater system. The residual at each point in the flow region is a measure of the degree to which the head does not satisfy the governing equation. If a particular weighted average of the residual is forced to vanish, the nodal heads are obtained as the solution of a system of algebraic equations.

2.3 Hydrological Data Collection

2.3.1 Hydrological information

Hydrometric information provides the basis for water resources development, surveillance of water quantity and quality, water apportionment and water licensing. Attention to human safety and the management of water quantity was important in the past, but during the last decade, rapid increases in industrialisation and population density have made the monitoring of water quality and the assessment of environmental impacts increasingly important.

The need for information for immediate operational use, water policy, planning and research, has brought about the necessity for extensive data collection. For immediate operational use, much emphasis is placed on rapidly available information (e.g. real-time control systems). For forecasting and warning (e.g. storm surges, high river discharges or the dispersion of heavy pollution), the amount of readily available information is more important than the accuracy and extension of such data. Actual information with respect to water-levels may be of essential economic importance. For navigation, the allowable draught on rivers is directly related to the quantity of freight and the accessibility of ports. Water policy and water planning require extensive data with a long time base. These data are then used for testing for compliance with water quality standards, water balances, determination of the natural variability of the phenomena, or the detection of long term trends.

There is a great need for accurate hydrological data for research purposes. In the calibration and verification of mathematical models, extensive data are often required with more emphasis placed upon accuracy and detail than on ease and speed of collection. Homogeneity and completeness of time series is important for setting up empirical relations and studies over long time intervals. For research in surface water, incidental and intensive measuring campaigns by vessels and measuring stations are required, where different parameters are measured simultaneously.

Operational use and water policy require routine data collection realised by networks of automatic measuring stations, or a preselected programme of point measurements (routine sounding) or point sampling (water quality) by vessels. To allow for a good comparison of data a fixed sampling programme and the use of standardised methods is a prerequisite.

2.3.2 Hydrological networks

Network measurements are employed to record the main elements of the hydrological cycle (i.e. precipitation, evaporation, groundwater and surface water). Both water quantity variables (e.g water-levels, flows and discharges) and water quality variables (e.g. oxygen, temperature, chloride, acidity, nitrate, turbidity, heavy metals, organic micro-pollutants etc.) are important in the case of surface water and groundwater. The character of a network may be quite simple or very complex, depending on the objectives, benefits and the available infrastructure. The network may be fully manual (e.g. daily readings of water-level gauges) or highly automated. Emphasis may be placed not only on the automation of data collection, but also data-analyses and data-transmission. In particular, when operational use is of great importance, automation can include on-line information supply and automatic generation of alarms at the user's address.

Where regulation of the water management system is possible, the combination of a measuring network with an automatic control mechanism, into one integrated system may be the next step. Complex mathematical computation modules and data-analysis techniques may be included in an integrated network. (see Abbott, 1991). In general, the available optimisation techniques are not the restricting factors for network optimisation. With the advanced technology available at present, it is rather the financial means which impose limitations. Networks should be realised at minimum cost. Often the significance of optimisation techniques is that they allow the realisation of the same information at lower costs.

Restrictions may also be imposed by regional conditions or the state of development of an area. The design of a hydrological network poses problems of maintenance and management of measuring stations, accessibility, power supply, data acquisition, storage and transmission, and the available organisation.

Design criteria

An important first step in network design is the specification of monitoring objectives and design criteria. This requires an analysis of the demand for information.

General objectives of the network can be easily stated (e.g. to provide actual information about water-levels in a coastal area or to detect long term trends in the water quality of river water). However, to deduce design criteria, a further specification and quantification of monitoring objectives is necessary. This can be done by considering how and by whom the information is to be used. Various users of the network may have different objectives. The *effectiveness* of a network is the measure of how well it fulfils the objectives.

The main aspects to be considered when analysing the need for information are the availability, accuracy, representativeness and homogeneity of time series. For operational use, an immediate on-line availability is also desirable. This imposes high demands on the organisation of the network.

Accuracy

In the case of water-level networks in deltaic areas used for forecasting, different levels of accuracy may be applied depending on the particular circumstances. With respect to high levels the requirements may be lower for uninhabited regions than near large cities and bordering lowlands of high economic value. On the other hand, navigational aspects may dictate the accuracy at low water-levels. The use of the network data for research purposes will place particular emphasis on accuracy. In this case, a maximum interpolation error between locations is often taken as a design criterion for network lay-out.

Representativeness

The temporal and spatial representativeness of the information requires an insight into the variability, in space and time, of the physical processes. In the case of a water quality network, sampling near the river-bank may suffice when trend analysis is the objective. If the objective is to trigger an alarm, phase shifts (i.e. non-homogeneity) over the cross-section have to be reckoned with. A water-level recording in an estuary should incorporate the influence of the tides, wind set-up and seiche, but swells and short waves due to wind and shipping should be excluded. Various monitoring objectives may have their own specific method of data-analysis and should therefore be carefully distinguished. In an automatic water quality network the exceedance of a limit value for a short time may be the reason for generating a network alarm. However, the same exceedance will be filtered out in a data analysis for trend detection.

Homogeneity and continuity

For research and water-policy objectives the availability of long time series with homogeneous accuracy may be required. A design criterion for network density may be that, in the event of a drop out of a measuring station, missing values can be estimated from neighbouring locations with equal accuracy (e.g. by regression).

In a cost-benefit analysis the value of the information obtained should be related to the network costs. The combined costs of the network and the costs incurred through lack of information should be totalled. A lack of information can lead to wrong measures being adopted or larger margins of safety, which result in economic losses. The information loss can be decreased by an extension of the network, which in turn will increase the costs. If the network is built up in such a way that the total of information loss and the network costs (e.g. construction, maintenance, equipment and personnel, data processing etc.) is minimal, the economically optimum solution is achieved (see Figure 2.72).

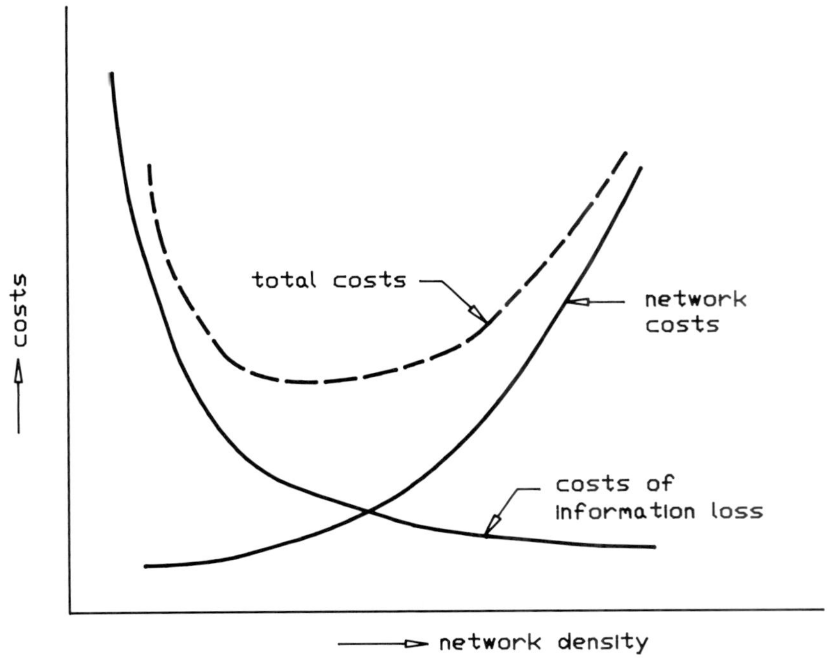

Fig. 2.72 Network costs as a function of network density.

An estimation can be made of the value of data for certain kind of uses (e.g. forecasting). For many aspects however, only a approximate value can be given. Assessing the real value of information for society is very difficult, if not impossible. As a result, a total cost-benefit analysis is not pursued in practice. Instead, design criteria are applied that start from an available budget, or a minimum level of effectiveness. In any case, the highest network efficiency should be pursued by optimisation of the network within the above mentioned restrictions (i.e. the available budget or minimum effectiveness level).

Design and optimisation

A measured variable forms part of a physical process that is dynamic and inhomogeneous in space and time. A monitoring network would be superfluous if the process was static and homogeneous. Generally, the physical process exhibits temporal and spatial variations about which users of the data wish to be informed. Irrelevant variations should be avoided, removed, filtered or averaged. On the other hand, relevant variations should be monitored as accurately as possible by the optimisation of data collection and data analysis. Insight into the heterogeneity and variability of physical processes is a prerequisite for network design. Often network data collected during some initial period are used for further optimisation.

Heterogeneity of the physical process may be caused by stratified flow or other gravity effects, restricted longitudinal and lateral mixing, and local effects (e.g. secondary flows, dead zones, etc.). Heterogeneity is important for the selection of sampling locations and for deciding on the positioning of sampling points at a certain location. Local knowledge and understanding of the system are also necessary because the nature and extent of the spatial heterogeneity may vary with time. The variability of a physical process originates from the variation in boundary conditions of the monitored system and the physical characteristics of the system itself. The extent of the variations in time and space (variance), and the time and distance scales of the considered variations (correlation structure), characterise the dynamics of the process (covariance structure). They greatly influence the optimal density of the network in space and time.

Monitoring objectives and design criteria for hydrological networks are usually defined in one-dimensional terms. Measuring locations are often situated near the river-banks. However, information is usually related to variables that are integrated over the entire cross-section of the river (e.g. water levels, concentrations). Locations should be selected based on suitability and representativeness. This requires insight into the heterogeneity of the physical process and the local conditions. Available sampling methods also have specific requirements, and thus play a role in the selection of appropriate locations. However, possibilities for the selection of locations may be limited. For example, measuring stations cannot be installed in the middle of the river. Sometimes political or historical reasons or different monitoring objectives result in sampling locations that are not optimal in all respects. It is then essential to investigate the degree of representativeness of the selected location and to describe how the collected data are correlated to the original data requirements.

Some examples will now be described to highlight some of the aspects of network design.

a) For a salinity network in a coastal area it is important to determine the degree to which the flow is stratified or mixed, and how far this characteristic varies in time, depending on tidal amplitude and river discharge. Stratification necessitates sampling at several points in the vertical, and at locations where sampling at great depths is possible. When the mean salinity over a cross-section is required, correlations should be established between the collected data at sampling points and the mean salinity over the profile. Through specific measuring campaigns the salinity distributions over the cross-section should be measured under different hydraulic conditions.

b) Due to restricted lateral mixing a pollutant drained near the river-bank may cause a non-homogeneous distribution in an otherwise homogeneous flow, even up to several kilometres downstream of the point of discharge. Special attention should be paid to the representativeness of water quality measuring stations at near-bank locations, especially when triggering an alarm is the monitoring objective.

c) For network sampling of suspended load (e.g. turbidity and contaminants), the vertical distribution of load content and particle size is important. Contaminants may be particle size selective. Near the surface or near the bottom, sampling points may not be truly representative under certain hydraulic conditions.

d) Locations for water level measurements should be free from local influences (e.g. groynes etc.). In river mouths and estuaries and near bifurcations and harbour entrances, local resonance and slow oscillations between the banks often influence the water-levels, implying a restricted representativeness of the collected data for the surrounding area.

Network density

The density of measuring stations in hydrological networks may be influenced by many factors. Data may have to be acquired at specific sites due to political considerations (e.g. water quality observation at international border crossings), direct operational use (e.g. navigation locks, intake points of fresh water), historical considerations (e.g. continuity of long time series), and the physical system (e.g. water-level recording at branching points in river deltas and in river mouths). In the intermediate area the density of observation points depends upon the design criteria of the network and the physical system. Nowadays, a criterion for design and optimisation of water-level gauging stations in a tidal river system is the maximum interpolation error between the measuring locations. If the accuracy of interpolated values, at intermediate points, is to be the same as that of the measured values then the standard deviation of computation, σ_c, must be equal to the standard deviation of measurement, σ_m. In fact, σ_m refers to both the accuracy of measurement and the representativeness of the location. Comparison of computed values at intermediate points with measured values at the same locations yields:

$$\sigma^2_{(c\text{-}m)} = \sigma_c^{\,2} + \sigma_m^{\,2}$$

The criterion for optimisation will then be:

$$\sigma_{(c\text{-}m)} \leq \sigma_m \sqrt{2}$$

Interpolation methods for intermediate points can be divided into three groups:
- mathematical interpolation (e.g. linear, non-linear, spline functions etc.),
- statistical interpolation (e.g. optimum interpolation, kriging),
- physical models.

Combinations of these methods in conjunction with tools such as Kalman filtering may lead to even better results. For example, multiple linear regression can be used for interpolation between water-level observation stations . Data for an intermediate point, y, are correlated to observations from a number of stations, n, such that:

$$y = A_O + \sum_{i=1}^{n} A_i x_i$$

where:
x_i = observation at station i,
A_i = correlation coefficient of station i,
A_0 = a constant.

This equation may be extended, if necessary, with observations at different times or with other physical parameters like discharge etc.

It is often possible to identify deterministic components in observations, which can be derived from physical processes. The deterministic part can be eliminated from the original data using a model. For example, in a tidal area, tidal predictions by harmonic analysis or by numerical model computations may be subtracted from measured water-levels. The residual data will have a smaller variance and smaller correlation scales. The monitoring network only has to provide information about the remaining stochastic part, and the network density can be reduced using the same design criteria.

When a mathematical model is available that gives a temporal and spatial description of the monitoring variable, the Kalman filtering method is a powerful tool for network optimisation in space and time. Essentially the state at a future level is predicted with the model and adjusted after comparison with measured values. The degree of adjustment depends on the specified uncertainties of the observations and of the model. Apart from the optimal estimate of the variables, the reliability of the estimates is also calculated. This reliability is determined by the measuring matrix, which indicates at which locations and at what moment measurements are made. An integrated optimisation of locations and frequencies can be realised by investigating the effect of different measuring matrices (i.e. different networks) on the resulting reliability. Optimisation of a network lay-out, and judgement of the information content of observations, is only possible with data from an existing network. Since these data are not available during network design, network density should be based on physical insight and preliminary research.

Sampling frequencies

To select the right sampling frequencies and measuring time of a network parameter, insight into the physical processes and the time scales of variability is of primary importance. Network parameters will vary with differing frequencies. For example, water-levels in an estuary will vary due to astronomical tides, seiches, basin resonance, and waves due to wind or shipping. On the other hand, storms may cause set-up of water levels and monsoon winds may cause significant seasonal variations in mean sea-level.

Different objectives require distinct methods of managing temporal variability. Certain variations should be eliminated while others should be highlighted in the frequency spectrum as accurately as possible. Measuring objectives may be:
- estimation of mean values,
- identification of a dominant variation,
- detection of the violation of critical limits (standards),
- trend detection (e.g. sea-level rise).

For the design of sampling intervals in time, three aspects are essential:
- the pursued network effectiveness in view of the monitoring objectives,
- instrumentation and observation errors,
- the variability (frequency spectrum) of the physical process.

Spectral analysis is a useful tool for the detection of dominant frequency components. Irrelevant frequencies should be removed by averaging over a sufficient number of independent observations. The ratio between measuring period (period of averaging) and irrelevant time scales should be sufficiently large.

In auto-correlated time series, increasing the sampling frequency will cause a higher degree of dependency of the observations. When the estimation of mean values is the objective, the effectiveness of each observation (i.e. the amount of independent information) will be reduced. The number of effective observations within a measurement period depends greatly on the correlation structure (time scales) of the physical process. Sometimes this implies that the desired level of effectiveness of the information cannot be met. The attainable effectiveness for the estimation of mean values has a maximum that cannot be increased by more frequent sampling. Where possible, the measurements should be extended over a longer period.

2.4 Features of Ecology

Plants, animals and other organisms generally occur in distinct biocommunities reflecting the abiotic properties of their environment. In ecology this phenomenon is investigated together with other aspects which determine community structure, such as competition among species, food relations, reproduction and longevity. Ecology therefore encompasses the total of structures and processes of plant and animal communities, their inter-relations -including those of man - and the relations with the abiotic components of the ecosystem. Therefore, especially in aquatic environments, the knowledge of hydrology and water management is essential in understanding ecological processes.

2.4.1 Ecology of a river basin

In this chapter, only aquatic environments will be considered, which means discussion will be restricted to the water-related components of the ecosystem. A short description of the essentials in river ecology will first be given. It is followed by a section on estuarine ecology. Finally, remarks are made on the effects of human activity in the deltaic environment.

A river drainage basin encompasses the surface water and groundwater together with the existing flora and fauna in the catchment area. For river ecology and management it is important to know the biological zonations. First however, a description of the sections of the river basin and aspects of abiotic zonation are dealt with.

River back-waters

Tributaries, or the lower order streams, are of great importance for the biota of the higher order streams. Fish, like salmonids, migrate into these back-waters for spawning and the juveniles re-enter the main stream. The chemical nature of the tributary may be different from the head stream. Hence it may influence the water quality downstream of the intersection and so determine the zonations.

If the water flow in a river basin is interrupted by a lake, the rheophilic flora and fauna is replaced by littoral, benthic and pelagic lake species. The role of lakes in relation to plankton will be mentioned below.

At the entrance of the lake, the affluent of the stream may form a so-called *internal delta* with gullies and moorland vegetation, by the deposition of eroded materials. Such an internal delta lacks tide-induced intrusion of salt water. An influence of a shallow river lake may be a rise in temperature of the downstream river water as the shallow water is warmed easily. In a deep river lake this may be less pronounced especially if the discharge is provided by the release of cold, deep water layers.

When the river lake empties in the river by means of a waterfall it may prevent the migration of fish to the upper reaches. Injury to fish or other organisms by the mechanical effect of the falling water on the rocky bottom has not been recorded. Possibly it is responsible for the occurrence of broken scales of diatoms in plankton samples.

Flood plain

Under natural conditions, the river foreland may be considered to be the land which is subject to flooding between the river-bed and higher riparian soils. In regulated river basins the river foreland is the land between the river-bed and dykes which protect the adjacent land from flooding. This land can be crossed by old branches of the river, which may or may not be connected to the river (oxbows) or are intermittent. During periods of high water such land is flooded, while in drier times bodies of stagnant water may remain. In its natural state, the foreland has a vegetation of trees and shrubs, but natural pastures may be present as well (e.g. in Africa such pastures are created by hippopotami). In man-regulated parts the foreland is cleared of woody vegetation components and used as pasture for cattle. During floods, suspended material from the river is deposited on the land and fish are spread throughout the flooded areas. They are trapped at low water in the remaining stagnant water bodies. Some species spawn specifically during these times. In the stagnant water bodies of the forelands, a diverse aquatic life develops. Oxbows and old branches are therefore very important for the biological richness of the main stream and they are a source of fish and plankton to the river. River forelands are also inhabited by many birds and some mammalian species.

In tropical rain-forests, forests adjacent to the rivers are flooded for long periods during the rainy season. Fish are spread amongst the trees, feeding on seeds and insects which fall into the water. In the dry period, they recolonise the main stream. As a result of temporary inundations, the character of the riparian forest, is different from the high inner forest.

Abiotic zonation

The importance of terrestrial input of coarse particulate organic matter (CPOM) decreases, and the transport of fine POM increases down the continuum.
For the classification of rivers, several criteria can be used. These may be based on the nature of their source, geomorphologic characteristics, or chemistry (e.g. calcium content, pH, and nutrient availability). Classification of river stretches in terms of bottom flora and fauna may be helpful in this respect

Biological zonations

Cummins (1979) presented zonations as illustrated in Figure 2.73. The flowing river is considered a continuum with flow-through being the main characteristic. The biological zonations in the different stretches can be treated as separate ecosystems, but they are not independent of each other due to the continuity of flow.

Headwaters are shown in Figure 2.73 as heterotrophic (i.e. the ratio of gross photosynthesis to community respiration, $P/R < 1$) because of restricted light, a consequence of shading by riparian zone vegetation. Mid-sized rivers are depicted as autotrophic, with a $P/R > 1$, through a combination of reduced riparian shading and relatively shallow and clear water. Large rivers are again heterotrophic because depth and turbidity-related attenuation creates a restricted light regime.

The passive transport of free floating zooplankton cannot be visualised in Figure 2.73 but conditions for development are present in parts of the river where the flow of water is minimal, as in the lower course. As a result of this, in many rivers plankton is not indigenous but originates from back-waters such as river lakes and oxbow lakes connected to the river. For example, the river Rhine carries diatoms and blue algae originating from Lake Zürich and other still waters. Eutrophication processes, initiated around 1900 in Lake Zürich, were reflected in the composition of plankton in the Rhine, starting with increases in diatoms followed by blue algae. In tropical rivers in humid regions, the plankton developed during the dry periods when the river flow is minimal is carried downstream in the rainy period.

Fish populations move freely in the water, which is not the case with sessile (=attached), benthic (=bottom-living) invertebrates and rooted water plants. Therefore, zones characterised by sessile organisms provide optimal information on local conditions of the river stretches.

The dominant macro-invertebrate functional groups shift from shredder-collector in headwaters to collector-grazer (scraper) in mid-sized rivers to collectors with plankton communities in large rivers. Fish populations shift from cold to warm-water invertebrate feeders. Mid-sized rivers have piscivorous forms as well, and large rivers have both bottom feeders and planktivorous species. In the upper, middle and lower course of the river, different fish zones can be discerned, characterised by trout, grayling, barbel and bream, respectively.

The presence and extent of the various stretches varies from river to river. For example, cold, alpine head-waters may be absent if the river originates from the plains. The middle course may be

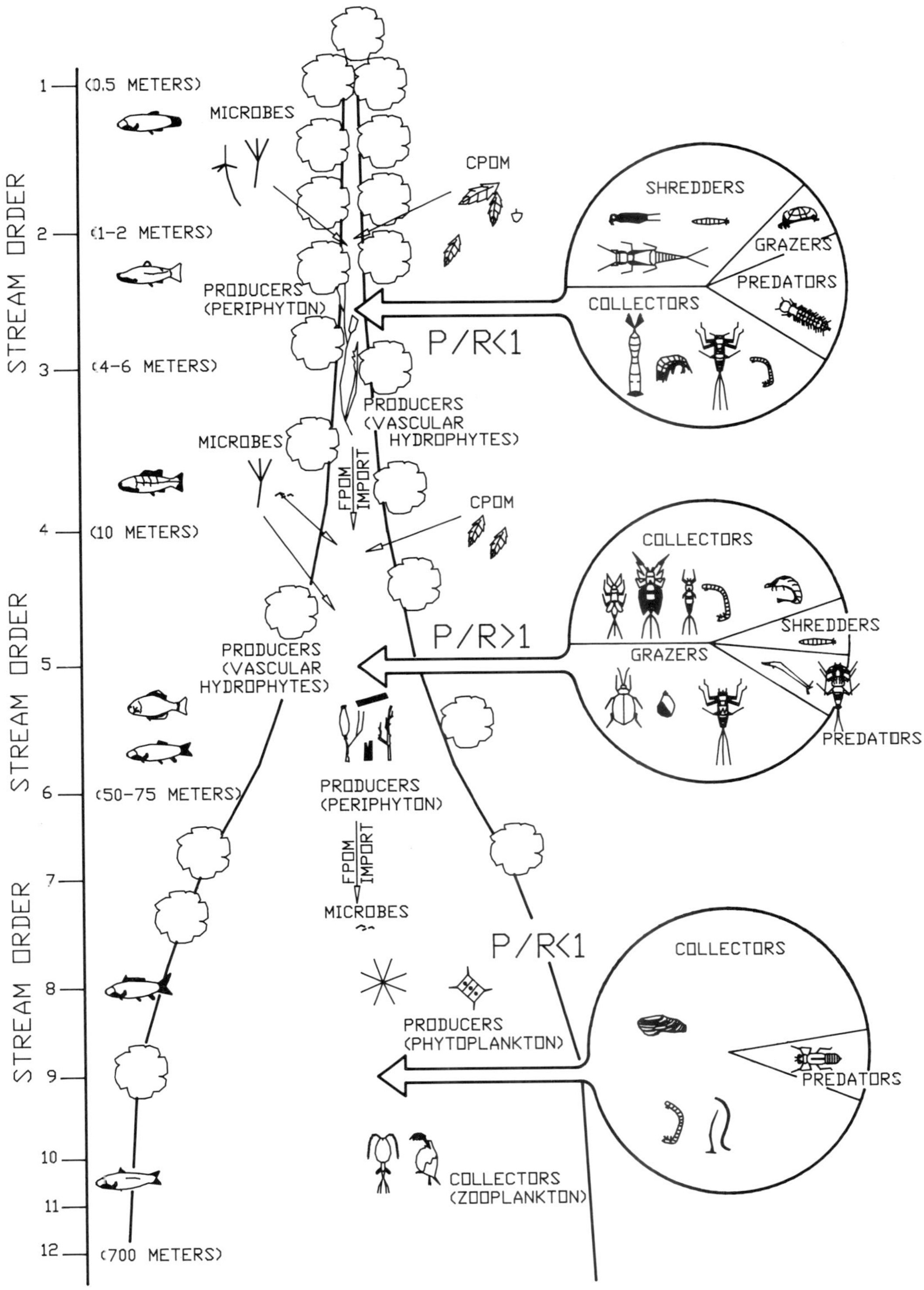

Fig. 2.73
Zonations of the river continuum.

the most extensive stretch or may be very short. In arid regions only a few species remain when long stretches dry up for a considerable period of time. Statzner & Higler (1986) presented a scheme as shown in Figure 2.74 based on stream hydraulics and species numbers.

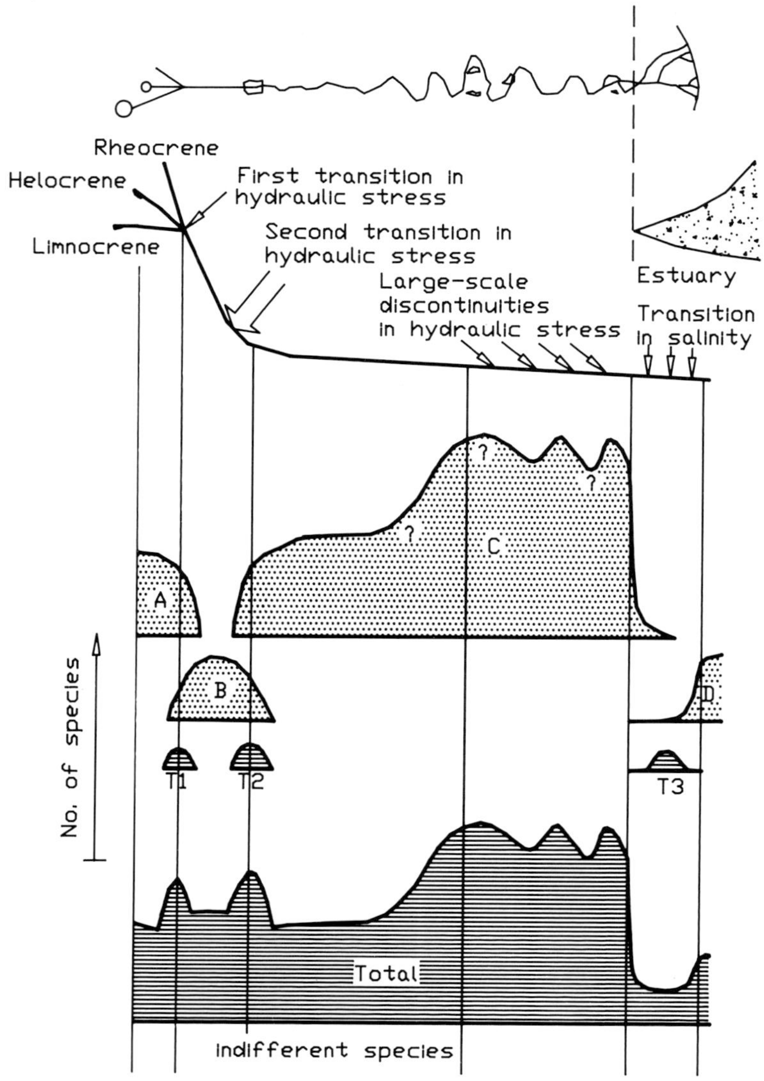

Fig. 2.74
General faunistic zonation pattern in pristine streams.

The various source types referred to in Figure 2.74 are:

rheocrene	- discharges directly into a channel,
helocrene	- discharges into a marshy pond,
limnocrene	- discharges into a pond.

Not all of the components mentioned can be, or must be, present in a stream. The species distribution in running water starting with a helocrene and ending with an estuary is indicated in Figure 2.74. Species occurring in the spring (A) and in the reach of high slope (B) overlap at the first transition in hydraulic stress. Species of group B and species occurring in the stream after it has entered the flood-plain (C) overlap at the second transition in hydraulic stress, where pristine streams are frequently braided. Large-river patterns are rather speculative due to sparse information. In the brackish zone a third overlap is found between species of group C and marine fauna (D). In all three transition zones a few species are confined to the transitions (T1, T2, T3).

Quality of a stream

As the water quality of a river depends on the properties of the soils in its catchment area, precipitation, and use of the basin etc., it is clear that it will vary not only in the different stretches but also among the river systems as well. Quality standards relate to the use of river water and have been developed for drinking water, agriculture, fisheries and industrial purposes.

Ecological quality is subjective. It may be considered as a relative quality (in a bad-good continuum) but it may also be assessed as an absolute quality. In order to come to grips with ecological quality, several independent ecological views may be employed. Key words in these visions are integrity (the degree in which natural dynamics and changes occur), diversity (species richness), scarcity (the uniqueness of species or ecosystems) and distinctiveness (e.g. does it represent a situation of approximately 1900?). In particular situations it is important to define which view will be taken as well as the spatial (local vs. mondial) or biological-integration (populations - species - ecosystems) level to be used in the different views.
In an integrated ecological quality assessment a combination may be made of the different visions. For an environmentally sound management of a river system, an integrated approach is necessary, in which consideration is given to the quality of the aquatic and terrestrial parts of the total catchment area.

The quality of a stream changes in the different stretches. Differences in nutrient contents, organic matter, minerals, nature of the bottom, hydraulics etc., all of which determine the ecological quality, are known to vary. This has to be considered in determining water management measures. Moreover, as most the rivers are affected by human activities it is difficult to find the natural ecological quality.

2.4.2 Ecology of the estuarine section

Where the flow of the river meets the flood of the tide, an aquatic environment exists which has characteristics of both fresh and sea-water. The boundary of this estuarine section has been described and defined by several authors. The landward boundary is found at the fresh-water tidal zone of the river. It is a dynamic boundary moving upstream and downstream as the river flow varies. In this fresh-water zone, the organisms are subjected to changing water movements. The seaward boundary of an estuary is found where salinity no longer changes due to river input. Hence, it may stretch out far into the sea but it is also a dynamic boundary with a changing position due to the existence of deltaic structures.

Fresh water, which floats on more dense coastal sea-water, can be traced as estuarine water at great distances from the geographical boundary of the estuary. This top layer carries fresh-water organisms originating from the estuary. Salt water from the sea enters the estuary in the lower layers, introducing marine organisms in the process. Tidal movements cause fluctuations in salinity and mix the fresh and sea water in the estuary. In general, conditions in the estuary vary more than those in the river or the marine environment. The salinity fluctuations in the estuarine zone limit the living conditions for marine as well as fresh-water organisms. Specific organisms have adapted to the brackish estuarine zone. Some species may occur temporarily as migrants or for reproduction.

Water movements

It is impossible to depict one environmental factor as being responsible for regulating ecosystem properties. In the case of a river, stream velocity and hydrodynamics have been indicated as major factors in this respect. When considering an estuary or deltaic waters, these factors from the river continuum are superimposed on the tidal movements.

While plankton organisms are transported passively in the water, bottom living (benthic) organisms such as worms, crustaceans and molluscs resist hydraulics as much as possible by their ability to burrow or fix themselves. As a result, the benthic fauna is dependent on water quality parameters such as salinity, oxygen content, nutrient supply and pollution that vary with the tidal movements. The result is that species richness is lower in estuaries than in fresh or marine waters, where a greater environmental stability exists. Figure 2.75 shows this effect in the Tay estuary, Scotland. The minimum number of species is found about 20 kilometres upstream of the point with minimal tidal variation. The degree of bottom variation declines together with the intensity of tidal changes. Therefore, regulating factors other than salinity fluctuations, such as hydraulics and related bottom structures, may also control species distribution.

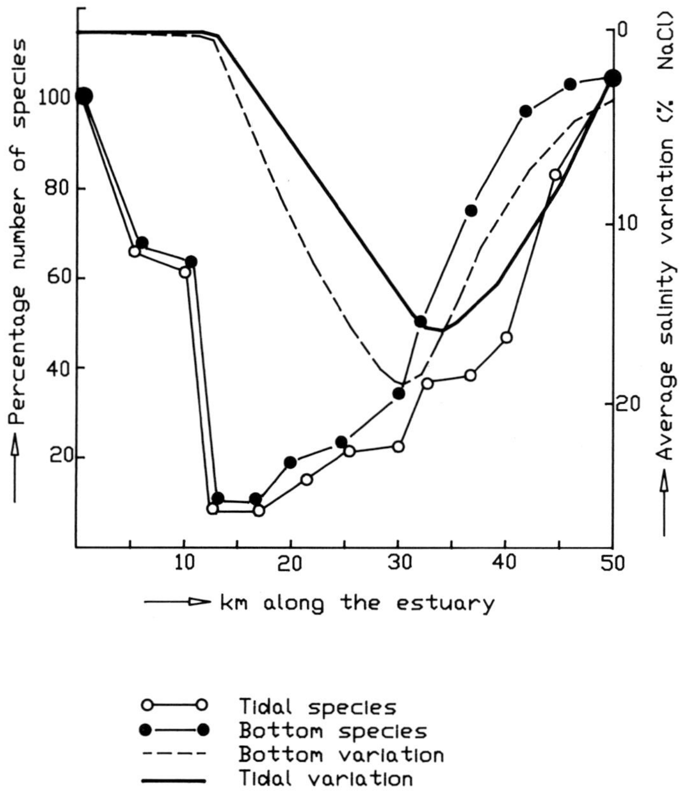

Fig. 2.75
Average salinity variation and percentage number of species.

Salinity gradients

In an estuary, a salinity gradient is present along the river to the sea, with the main characteristic being that the water is brackish. Several authors have tried to define brackish zones based on biological criteria. The system of Redeke (1922) and also the Venice system (Segerstråle, 1959) are widely used (Table 2.5). As with all biological system approaches, these classifications show gaps and overlaps, but they can be helpful and are often used in practice. The difference in the lower salinity limit used in defining fresh water in the systems (i.e. < 0.1 ‰ and < 0.3‰ Cl^-

Redeke [‰ Cl^-]	Water type	Venice system	
		‰ salinity	Equivalent ‰ Cl^-
< 0.1	fresh water	< 0.5	< 0.3
0.1 - 1.0	oligohaline	0.5 - 5.0	0.3 - 3.0
1.0 - 10.0	mesohaline	5.0 - 18.0	3.0 - 10.0
10.0 - 17.0	polyhaline	18.0 - 30.0	10.0 - 17.0
	euhaline	> 30.0	>17 (next to euhaline sea)
> 17.0	sea-water	30.0 - 40.0	17.0 - 22.0
	hyperhaline	> 40.0	> 22.0

Table 2.5
Brackish water classifications.

respectively) is probably due to the different areas of focus of the investigators. Perhaps this difference is based on a limit for inland water (limnological) against marine (oceanographic) waters. Wolff (1973), studying more specifically bottom fauna, questioned classification based on the hydrographical structures of the estuary, and proposed the neutral terms upper, middle and lower reaches of the estuary.

A characteristic of the brackish area of estuaries is the scarcity of benthic species. Remane and Schlieper (1971) describe the concept of species-salinity relationships for the Baltic brackish waters. They found a minimum number of species in the brackish zones compared to the fresh and marine environments. They also stated that the species minimum was consistent with the largest salinity fluctuations rather than with low salinity values. In the middle of the salinity gradient, where the largest salinity fluctuations usually occur, the species minimum contained fewer species than in more stable brackish water. The reason is that marine species, as well as fresh water species, do not penetrate as far into the unstable brackish estuarine waters as they do in stable brackish waters. As a result of its shallowness, the temperature of the estuarine water is also unstable, which restricts the species number as well. The pollution effect cannot be neglected as high turbidity may also be a factor (see Figure 2.76).

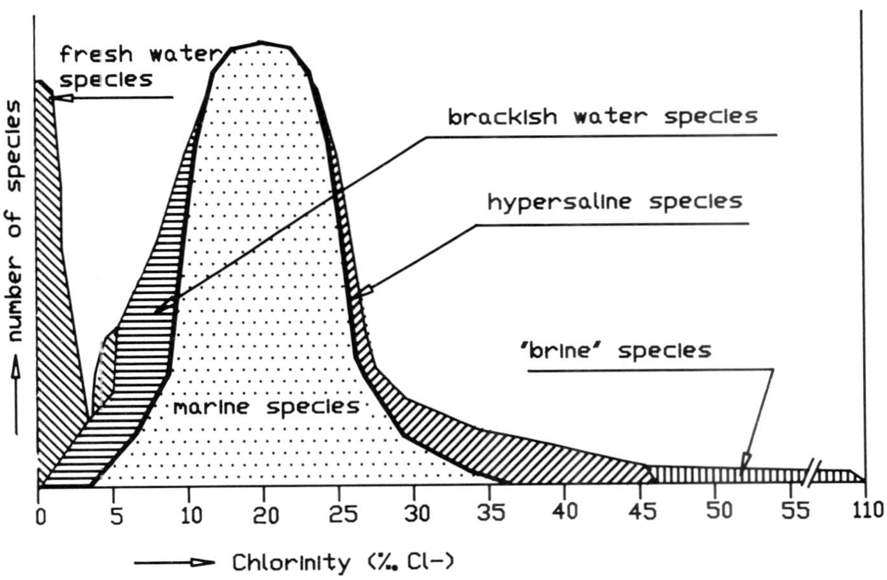

Fig. 2.76
Species richness as a function of chlorinity.

Although low diversity of benthic species may point to a low ecological quality, brackish environments and their characteristic inhabitants (e.g. mudskippers and mudshrimps) may be very valuable due to their restricted presence on a global scale. Moreover, the brackish environment cannot be considered to present a species minimum for species assemblages which are not directly dependent on salinity changes, such as birds.

Low species numbers are also found in hypersaline conditions as found in (sub)tropical shallow isolated parts of coastal regions with high temperatures and evaporation rates.

Turbidity

It is known that the species minimum in a salinity gradient also occurs in clear brackish waters, thus eliminating turbidity as a single cause.
Turbidity is often high in the brackish part of estuaries. This is caused by flocculation, a process

wherein dissolved matter encounters more saline water. It inhibits photosynthesis of plants by light extinction. On the other hand, all marine flowering plants act as silt traps, thereby reducing turbidity. The deposition of silt on the substratum restricts the occurrence of marine and fresh-water sessile species which prefer hard substratum. These organisms cannot find suitable substratum when their larvae settle, or too much energy has to be allocated in the clearance of silt. On the other hand, the deposition of silt favours species which have adapted to these circumstances and which use the supply of organic material as food.

Estuarine communities

The occurrence of estuarine communities is related to several factors such as currents, oxygen content, salinity gradients, temperature fluctuations, supply of food, etc.
Tidal movements, currents and wave action determine sediment composition and distribution. Sedimentological characteristics, such as median grain size and mud content, are important parameters for species composition.

The benthos in the fresh water tidal zone is mainly formed by Tubificid worms. They are present in the mud in spite of the low oxygen content in the bottom water layers.

In estuaries, several ecosystems may be found. Salt marshes and coastal marshes are found at the high tide water mark. On sand and mud flats, which are exposed at low tide, rooting vegetation may be absent or a cover of seagrass or algae (e.g. *Ulva)* may be present. The flats are inhabited by an infauna of polychaete worms, shellfish and crustaceans. In autumn, the intertidal biomass is greatest and, especially in this season, migrating birds feed on these grounds during exposure at low water. At high water, the soft substratum of the intertidal zone acts as a nursery for flat fish and shrimps.
Also, sand grains may be cemented into banks by polychaete worms such as *Sabellaria* . In tropical areas, mangrove vegetations (mangals) may develop. Seagrass vegetations are found in gullies and lagoons which retain water at low tide. Subtidally, oyster beds and mussel banks may develop. The pattern of gullies and flats creates different habitats for benthos and fish, which attract great numbers of waterfowl such as waders, ducks, gulls and mammals (e.g. seals). In deltas, the combination of the different ecosystems guarantees a great diversity of biological life.

Production and nutrient cycles

Although turbidity may be a limiting factor for bioproduction, estuaries are considered to be the most productive areas in the world. This is a result of the constant supply of nutrients by the river, run-off from the land, and clay deposition. Nutrients are used by primary producers such as macrophyte vegetations, phytoplankton and filamentous algae mats on mud bottoms. In the transition zone, the matter formed by dying fresh water and marine organisms also contributes to high production. Table 2.6 shows the gross primary production of estuaries compared with other ecosystems (Odum, 1971).

Estuarine vegetations such as salt marshes, sea grass beds and mangals are sites of higher biological productivity than the open water adjacent to them. This is probably also true for the secondary production (e.g. shell fish and crustaceans) to the benefit of bottom living organisms.

Tidal flats are exposed to the air during low-water periods, which favours oxygenation of organic material at the surface. This is mineralised and added as nutrients to the water, to the benefit of phytoplankton growth.

The example of Figure 2.77 (Saeijs, 1982) shows the subtlety of synchronised ecological mechanisms. In this case, the phenomenon of bird migration is triggered by light, through hormonal secretion and by the seasonal periodicity of regional food production.

ECOSYSTEM	AREA [10^6 km^2]	Gross Primary Productivity [kcal/m^2/yr]	Total Gross Production [10^{16} kcal/yr]
MARINE			
Open ocean	326.0	1,000	32.6
Coastal zones	34.0	2,000	6.8
Updwelling zones	0.4	6,000	0.2
Estuaries and reefs	2.0	20,000	4.0
Sub-total	362.4		43.6
TERRESTRIAL			
Desert and tundras	40.0	200	0.8
Grasslands and pastures	42.0	2,500	10.5
Dry forests	9.4	2,500	2.4
Boreal coniferous forests	10.0	3,000	3.0
Cultivated land (no energy subsidy)	10.0	3,000	3.0
Moist temperate forests	4.9	8,000	3.9
Mechanised agriculture (fuel subsidised)	4.0	12,000	4.8
Wet Tropical and sub-tropical forests	14.7	20,000	29.0
Sub-total	135.0		57.4
TOTAL for biosphere (excl. icecaps)	500.0	2,000	100.0

Table 2.6
Estimated gross primary production (annual basis) of the biosphere and its distribution among major ecosystems.

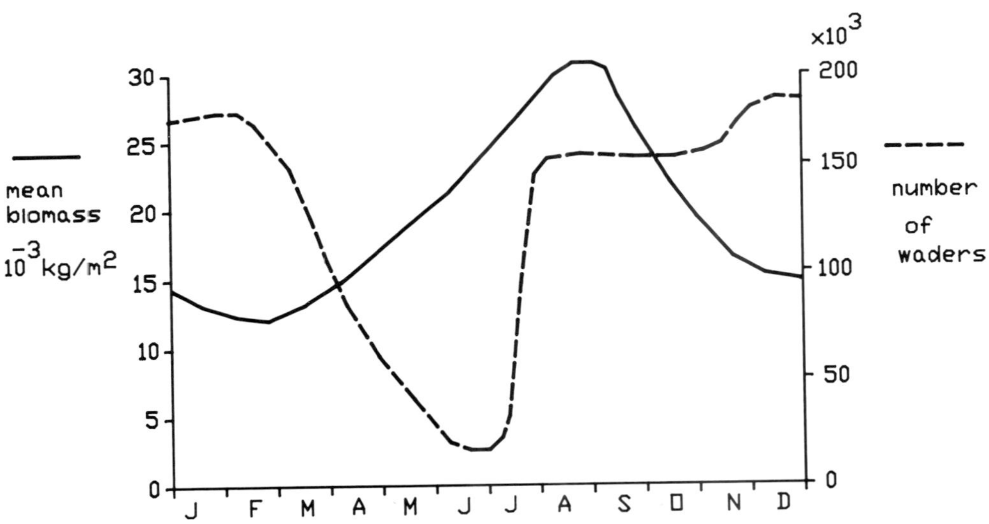

Fig. 2.77
Variations of the intertidal biomass and variations of wader population in the Eastern Scheldt.

2.4.3 Use of river basin resources

Water in a river basin is used by human societies for various purposes such as drinking, irrigation, fish culture, navigation, recreation, waste discharge, etc. Also, the exploitation of sand, gravel, timber, minerals, raising of cattle and poultry, fisheries and numerous other socio-economic demands must be mentioned. At present, one of the most important purposes is the generation of energy by hydropower or electricity plants. There is hardly any large river without dam constructions. Exploitations must be adapted to fit the ecology of the river basin whose presence is indispensable for our own existence.

The use of river basin resources depends on the development and structure of society (ethnologic history, religion, social aspects, economy, culture, technology, health and nutrition). It is a task to evaluate the interactions between nature and man and to develop guidelines which integrate negative and positive influences, in order to achieve optimum living conditions for human and plant and animal groups.

In the context of this book, consideration will be restricted to the use of water in a river basin in relation to technical constructions or manipulations. Impacts from other activities such as land use, industry and urbanisation processes, which affect water quality and the scenery of the basin, are the subject of a broader ecological approach. The impact of water uses, its hydrologic or hydraulic effect, and its general effect on biota, is briefly outlined in Table 2.7.

Types of hydraulic works	Possible effect on hydrology	Possible effect on biota
dams in river valleys	stagnation	loss of running water species
	changing groundwater table	loss of hygrophytes
lowland reservoirs	isolation	managed populations
decapitation of rivers	decreasing discharge	loss of spawning grounds
canalisation	changing of flow	loss of biotope diversity
locks, barrages	intermittent flow	migration disturbance
diversion of rivers	changing drainage	alteration of biocommunity
communication of waters	water quality affected	biotic forgery
flushing measures	water quality affected	removal of biota
level manipulations		
a. groundwater drawdown	loss of volume	loss of hygrophytes, wetlands
groundwater-rise	gain of volume	gain of hydrophytes, wetlands
b. surface water	shore erosion	loss of shore biota
c. fixed tidal water level	stagnation	changing brackish biota
bottom lowering (mining)	increase of volume stratification	loss of bottom fauna, fish and birds
transport from low to high lands	water quality	loss of specific biocommunities
recharge of aquifers	quality and quantity	alteration of vegetations
drainage of soils	loss of sources	loss of hygrophytes
polders	controlled system	loss of aquatic system
		gain of aquatic systems (ditches)
agricultural irrigation	flow regulation	harm to wild flora and fauna
agricultural sprinkling	surface and groundwater balance	affects wild flora and fauna
agricultural drainage	quality and quantity loss of ditches	loss of aquatic habitats
heating (cooling water)	stratification	fish damage
dredging	favours transport	resists moorland communities and favours aquatic biota
specific industries	acidification of rain	damage to specific species
salt water wedge repulsion	desalination	loss of salt marshes
control of floods	flow regulation	loss of spawning grounds and changing vegetation
dumping	turbidity and decrease of volume	depends on quality and quantity of dumped material
bank protection	wave energy dissipation	loss of biota depending of kind of artificial substratum used
shore and bank profile	circulations in the water	limits use of substrata by flora and fauna

Table 2.7 Hydraulic works and their possible impacts.

Table 2.7 lists only the types of technical works and not the manifold variations in magnitude, frequency or actual occurrence in a given river basin. Information on the aim of the technical work and whether or not this aim includes or specifically focuses on ecological interests, is not given either. If the latter items are incorporated, the basin resources are said to be wisely used. Although the demands and behaviour of human society remain difficult to predict the implementation and maintenance of a wise management practice must be pursued.

2.4.4 Effects of upstream interventions

The continuity of the river means that management measures cannot be taken independently. The ecological quality of a delta depends to a great extent on the upstream activities and hydraulic constructions.
The effect of human intervention in the upstream areas on the downstream stretches of a river is reflected in the quality and quantity of water and sediment. In river valleys with a dense population, the discharge of wastes results in pollution which affects the downstream biocommunities, human society included. The river Rhine is a well known example in this respect. In this case, the self purification capacity of the upstream and middle-stream areas, is overruled by the overloading of toxic, biodegradable particulate and soluble matter, which reach the downstream delta area. Several species and groups of organisms decrease in number or disappear completely. Sanitation measures are necessary for the improvement of the environment along the entire river. This process of increasing pollution is a trend related to growing societies in which, unfortunately, the cumulative and long term effect on the environment is easily neglected.

Several examples are known of the environmental effects of man-made lakes in river valleys. Deforestation is also considered to be one of the most important interventions. In moderate climates, the problem of barrage lakes filled up with eroded material and silt, forces managers to regulate the discharge of the sediment which is carried downstream, covering aquatic vegetations and fauna that die off in the process. Also, the release of anaerobic hypolimnion water from the lake may affect large downstream stretches of the river.

In the tropics, examples can also be found of the downstream effects of dams on deltas, and beyond that in the open sea. In the Lower Volta, this resulted in sea-water penetration upstream along the river bed which reduced the spawning areas of a river clam and so affected fishery economics. Aquatic weeds became more common as well as snails as a vector for schistosomiasis. The effect of the High Aswan Dam in the river Nile is a classic example, as it also affected the sardine population in the Mediterranean.

In Asia and Australia, estuaries are famous for their prawn concentrations, which is a food source for fish. As prawns are dependent on silty sediments, dam construction upstream will interfere with the food supply and the total food chain in the estuary. Recent research on the Solo delta, East Java, carried out on the sedimentation history, revealed that deforestation in the upper course of the Solo for sawah cultures, resulted in accelerated erosion and an increase of sediment in the delta. Larger areas of the delta no longer became flooded and mangroves thrived there. The loss of fishing grounds was compensated for by artificial aquacultures of fish and prawns.

The Solo delta example shows the ecological importance of forests in water management and control. Forestry should be considered as a water project. Water is stored in forests and other vegetation and released slowly, which is essential for the existence of the river system. Reference is also made to the problems of the river Ganges and the deforestation of the Himalayas. Forestry is not to be considered only for timber winning but must be managed in combination with water management interests in order to arrive at a sound environmental control. Deforestation without compensatory measures leads to disaster as evidenced by the irreversibly devastated areas of the Amazon rain forest.

Deltas which have developed where rivers enter into lakes in arid regions are particularly vulnerable. Damming the upper course of the river decreases river discharge and may cause salinisation of the lake water due to the high evaporation rates (e.g. Lake Chad, Africa).

In general, very little is known about the ecological effects of upstream interventions on tropical deltas. Mostly, problems become apparent after technical works are completed, and remedial measures must then be undertaken. This has been done, for example, after building the W.A.C. Bennett Dam on the Peace river in British Columbia, Canada in 1967. There was a significant change in the flow regime immediately downstream of the dam. Little attention was given to the study of the downstream effects during the planning stages. During the filling of the reservoir, potential problems became apparent some 1,200 kms downstream at the Peace Athabasca delta. This delta, with an area of 6,000 km2, is one of the largest fresh water deltas in the world. The purpose of the project was to produce electricity and the reservoir was designed to regulate the natural flow at a constant discharge. In 1970, the lack of flooding in the delta was deemed to be detrimental to the natural biological system, and a study was initiated for remedial measures. The results of this study are summarised in UNESCO (1984).

2.4.5 Effects of delta development

Estuaries with a high degree of man-made modifications, resulting from land reclamation and engineering structures (e.g. dykes, dams, sluices and reservoirs), may be found in the Netherlands. The reservoirs may be fresh, brackish, salt or even hypersaline, depending on their use or location. The Rhine-Meuse delta in the Netherlands has been modified to such an extent, that it might be considered to be a new, man-made land and water ecosystem. The natural components have to be managed. These include Nature reservations like the Wadden sea and Wadden islands. The deployment of the semi-permeable Eastern Scheldt Storm Surge Barrier in 1986, marked the completion of the Delta Works in the Netherlands. Planning and construction of this dam were carried out recognising the ecological values of the environment, while the primary purpose - safety against high water during storm surges - was guaranteed. The completion of the Delta Works as a water engineering product, does not signal the end of human interventions. The human society continues to develop and so do the natural forces inherent to the modified ecosystems in the delta.

The effect of the closure of the open water of an estuary on bird life is illustrated in Figure 2.78 (Saeijs and Baptist, 1976), where the response of two representative birds in Lake Grevelingen is given. As most of the delta area has been modified by impoldering for agricultural purposes, and the water balance is controlled by pumping devices or other regulatory works, the natural aquatic or semi-aquatic flora and fauna have been replaced by species adapted to terrestrial conditions.

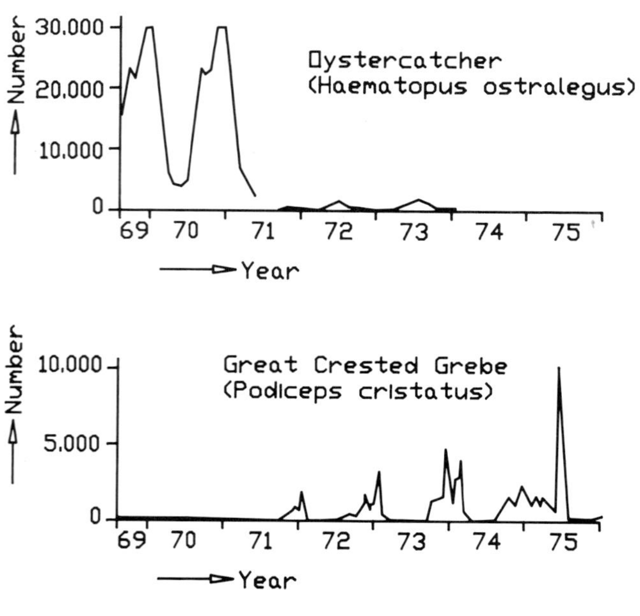

Fig. 2.78
The response of two representative species of birds in Lake Grevelingen on the closure of the Brouwers Dam in 1971.

It appears that in the new habitats, some known as wetlands, a rich and valuable diversity of plant and animal life comes into existence. The ditches and canals in polders are habitats for invertebrates, fish, birds and aquatic plants, while on the grasslands, other animals find a living which is quite different from the original deltaic life. Water management in these polders should be carried out considering the value of these wildlife elements. As deltas are very important for migrating birds, and for the migration of fish and some crustaceans that visit the river for reproduction, it is desirable from an ecological point of view, to create and preserve facilities for their needs. Delta development must therefore be undertaken by an integration of the ecological and socio-economic interests, in order to avoid natural hazards and to promote a wise use of the environment.

2.4.6 Management of the modified delta

If it is accepted that human activities modify the natural developments of ecosystems, then the fact that natural forces are suppressed to a certain extent, must also be realised. This necessitates management of the modified ecosystems, even if the functioning of an ecosystem will never be completely understood. The effects of human activities will often result in unexpected responses by Nature which make predictions difficult.

For the ecological management of a delta, which may be subdivided into several different ecosystems (e.g. open water bodies, polders and tidal waters), the key factors which govern the functioning and structure of the system, must be assessed. In the delta the hydro-environment is most important, reflected in the surface and ground water quantity and quality. For the use of this water, routine examinations are carried out to assess the values of the chemical and physical parameters which must be controlled in accordance with the standards developed for drinking water, industry, agriculture, fishery, recreation etc. In order to avoid surprises and to profit from the natural or the suppressed natural potentials as much as possible it is also necessary to set ecological standards. This means that not only must the chemical and physical parameters be evaluated in this respect, but the biological components as well. In ecological research, routine examinations have to be carried out for the evaluation of the ecosystem and its value. This includes monitoring the following basic parameters.

Diversity

If there are many species, the diversity is high. A low diversity is found in algal blooms which have a low number of species but a high number of individuals. A high diversity is considered valuable.

Density

The number of individuals per species expressed in numbers per volume or surface area, is used in relation to species, to assess competition, population dynamics, and use of environment. In routine examinations the assessment of the density may be restricted to the dominant species as only the species present in samples will be counted. Regular sampling may indicate seasonal fluctuations.

Biomass

The total amount of living substance may be used in estimates of productivity. In routine examinations biomass may be expressed in dry weight Carbon or chlorophyll.

Bioquality

The abundance of bio-indicators - key factors which reflect the specific quality of the environment (e.g. saprobic, oligotrophic, hypertrophic) - is important for the evaluation of the effectiveness of measures taken to manage an ecosystem.

Trophic status

Nutrients such as phosphate and nitrogen are the most important key factors in the regulation of the productivity of an environment. In water, phosphate and nitrogen compounds are examined in routine samplings. The trophic status, however, must be evaluated in relation to several other environmental conditions. The relation to specific species and biomass must also be considered.

All factors mentioned above are the minimum needed to evaluate an ecosystem and to set up guidelines for the ecological management of such a system. As the local environmental conditions are different and more specific, the general parameters must be complemented by specific local components. For example, calcium and iron may locally function as key factors which determine photosynthesis and the bioquality. In the deltas of Western Europe the bio-indicators are very similar, but they are different from deltas in other parts of the world. In conclusion, it can be said that for the aspects of ecological management, the above-mentioned general ecological parameters must be assessed everywhere, in addition to some local specific key factors.

2.5 References and Bibliography

Abbott, M.B., 1991, Hydroinformatics - Information Technology and the Aquatic Environment, Avebury Technical, Aldershot, U.K.

Abbott, M.B., 1992, Computational Hydraulics - Elements of the Theory of Free Surface Flows, Ashgate Publishing Ltd, Aldershot, U.K.

Abraham, G., de Jong, P. and van Kruiningen, F.E., 1986, Large-scale mixing processes in a partly mixed estuary, Delft Hydraulics Laboratory, Publication no 371.

Allersma, E., 1968, Mud on the oceanic shelf off Guyana, Symposium on Investigations and Research of the Caribbean and adjacent areas, Willemstad.

Anderson, H.R., 1978, Hydrogeologic reconnaissance of the Mekong delta in South Vietnam and Cambodia. Geol. Survey Water Supply, Paper 1608-R, Washington.

Badon Ghijben, W, 1888, Nota in verband met de voorgenomen putboring nabij Amsterdam, Koninklijk Instituut van Ingenieurs, The Hague, The Netherlands.

Bear, J., 1972, Dynamics of Fluids in Porous Media, Elsevier, New York.

Bear, J., 1979, Hydraulics of Groundwater. McGraw-Hill, New York.

Bird, E.C. and Schwartz, M.L., 1985, The World's Coastline, Van Nostrand Reinhold, New York.

Bishop, A.W. and Blight, G.E., 1963, Some aspects of effective stress in saturated and partly saturated soils, Geotechnique, 13, pp. 177-197.

Boonstra, J. and De Ridder, N.A., 1981, Numerical Modelling of Groundwater Basins, Publ. nr. 29, International Institute for Land Reclamation and Improvement, Wageningen.

Bosch, H., 1951, Geohydrological investigations at Bergambacht. Hydrologisch colloquium HC(S) no. 225A. The Hague (in Dutch).

Bouwer, H., 1978, Groundwater Hydrology. McGraw-Hill, New York.

Breeuwer, J.B. and Jelgersma, S., 1973, An East-West Geohydrological Section across The Netherlands. Kon. Ned. Geol. Mijnb. Gen., Vol. 29, 105-106.

Bruijnzeel, L.A., 1983, Hydrological and Biochemical Aspects of Man-made Forests in South-central Java, Indonesia, Ph.D. thesis, Free University of Amsterdam.

Bruijnzeel, L.A., 1987, Environmental Impacts of Deforestation in the Humid Tropics - A Watershed Perspective, Murs, Paris.

Bruun,P., 1966, Tidal Inlets and Littoral Drift, Universitets Forlaget, Trondheim.

Bruun, P. and Gerritsen, F., 1960, Stability of Coastal Inlets, North Holland Publ. Co., Amsterdam.

Colenbrander H.J., 1962a, Review of the seepage investigations in the Tielerwaard-West (in Dutch), Report 8, Province of Gelderland, Arnhem, The Netherlands, pp. 80-87.

Colenbrander, H.J., 1962b, A calculation of the aquifer characteristics, assuming steady-state groundwater flow (in Dutch), Report 4, Province of Gelderland, Arnhem, The Netherlands, pp. 55-59.

Coleman, J.M., 1981, Deltas - Processes of Deposition and Models for Exploration, Burgess Publ. Co., U.S.A.

Cummins, K.W., 1979, The natural stream ecosystem structure and function, Ecology of Regulated Streams, Ward, J.V. and Stanford, J.A. (eds), Proc. 1st Int. Symp. on Regulated Streams, Plenum Press, New York.

Custodio, E., 1987, Groundwater problems in coastal areas. Studies and Reports in Hydrology 45, UNESCO, Paris.

Dagan, G. and Bear, J., 1968, Solving the problem of local interface upconing in a coastal aquifer by the method of small perturbations, J. Hydr. Research, 6, pp. 15-44.

Dam, J.C. van, 1977, Determination of horizontal and vertical groundwater flow from piezometric levels observed in groundwater of varied densities. Delft Progress Report 3, pp. 19-34.

Dam, J.C. van, 1992, Problems associated with saltwater intrusion into coastal aquifers and some solutions, Proc. ILT Seminar on Problems of Lowland Development, Saga, Japan.

Delta Committee, 1960, Considerations on storm surges and tidal motions, Rapport Deltacommissie, Vol. 3, Staatsdrukkerij, The Hague.

De Marsily, G., 1986, Quantitative Hydrogeology - Groundwater Hydrology for Engineers. Academic Press, London, Orlando (Florida).

DHL, 1973, Hydraulic research for water management, Publication no 100, Delft Hydraulics.

DHL, 1991, Koelwaterstudie Warmte-krachtcentrale Moerdijk, Delft Hydraulics (in Dutch).

Domenico, P.A. and Mifflin, M.D., 1965, Water from low-permeability sediments and land subsidence. Water Resources Research, 1, pp. 563-576.

Dronkers, J.J., 1964, Tidal Computations in Rivers and Coastal Waters, North Holland Publishing Company, Amsterdam.

Dyer, K.R., 1973, Estuaries: A Physical Introduction, John Wiley, London.

Eisma, D., 1978, Areas of present mud deposition in the North Sea area, NIOZ, Texel.

Ernst, L.F., De Ridder, N.A. and De Vries, J.J., 1970, Gelderland, Geol. en Mijnb., 49, 6, pp. 457-488.

Ferris, J.G., Knowles, D.B., Brown, R.H., and Stallman, R.W., 1962, Theory of aquifer tests, U.S. Geol. Survey Water Supply, Paper 1536-E, pp. 69-174.

Fischer, H.B., 1979, Mixing in inland and coastal waters, Academic Press, New York.

Fookes, P.G. and Vaughan, P.R., 1986, A Handbook of Engineering Geomorphology, Chapman & Hall, New York.

Freeze, R.A. and Cherry, J.A., 1979, Groundwater. Prentice Hall, Englewood Cliffs, N.J.

Geirnaert, W., 1973, The Hydrology and Hydrochemistry of the Lower Rhine Fluvial Plain, Ph.D Thesis, Leiden.

Glover, R.E., 1964, The pattern of fresh-water flow in a coastal aquifer. Sea Water in Coastal Aquifers. U.S. Geological Survey Water Supply Paper 1613-C.

Goldenberg, L.C., Mandel, S and Magaritz, M., 1986, Fluctuating, non-homogeneous changes of hydraulic conductivity in porous media. Quarterly Journal of Engineering Geology, No. 19, pp. 183-190.

Groen, P. and Dorrestein, R., 1976, Zeegolven, Staatsdrukkerij, The Hague.

Gumbel, E.J., 1945, Simplified plotting of statistical observations, Trans. A.G.U., No. 261.

Hantush, M.S., 1964, Hydraulics of Wells, Advances in Hydroscience, vol. I, Chow, V.T. (ed), pp. 281-432. Academic Press, New York and London.

Harleman, D.R.F., 1961, Stratified flow, Handbook of Fluid Dynamics, Streeter (ed), McGraw-Hill, New York.

Herzberg, A., 1901, Die Wasserversorgung einiger Nordseebaden. Zeitung für Gasbeleuchtung und Wasserversorgung, 44 pp. 815-819 (in German).

Higler, L.W.G., 1988, World-wide Surface Water Classification System, UNEP-EMINWA, UNESCO, Paris.

Houghton, J.T., Jenkins, G.J. and Ephraums, J.J. (eds), 1990, Climatic Change - The IPCC Scientific Assessment, Intergovernmental Panel on Climatic Change, Science Working Group I, Cambridge University Press, U.K.

Huisman, L., 1972, Groundwater Recovery, Macmillan, London.

Ippen, A.T. (ed), 1966, Estuary and Coastline Hydrodynamics, McGraw-Hill, New York.

Jacob, C.E., 1940, On the flow of water in an elastic artesian aquifer, Am. Geophys. Union Trans., 72, part II, pp. 574-586.

Jacob, C.E., 1950, Flow of Groundwater, Proc. 4th Hyd. Conf., Iowa Inst. Hyd. Res., Rouse, H., (ed), John Wiley, New York, pp. 321-386.

Jones, P.H., 1970, Hydrology of Quaternary delta deposits of the Mississippi river, Symposium on the Hydrology of Deltas, Vol.I, Publ. no. 90, AIHS, pp. 49-63.

Kamp, G.S.J.P. van der, 1973, Periodic flow of groundwater: A systematic study of wave propagation under confined, semiconfined flow conditions, Ph.D. Thesis, Free University, Amsterdam.

Karelse, M., van Os, A.G. and van Urk, A., 1993, Effect of geometrical modification in the harbour area of the Rhine-Meuse estuary on the tidal motion and salt intrusion, Journal of Geophysical Research (Oceans).

Keulegan, G.H., 1957, Form Characteristics of Arrested Saline Wedge, National Bureau of Standards USA, Report 5482.

Kruseman,G.P. and De Ridder, N.A., 1986, Analysis and Evaluation of Pumping Test Data (4th ed.), 11, I.L.R.I., Wageningen, The Netherlands.

Lange, W.J. de and Maas, C., 1986, On the negative phase shift in the propagation of tides in the Dutch IJssel near Gouderak, H2O, 19, 2, pp. 24-29, (in Dutch).

Lebbe, L.C., 1983, Mathematical model of the evolution of the fresh water lens under the dunes and beach with semi-diurnal tides. Proc. 8th Salt Water Intrusion Meeting, Bari, Italy. Geologia Applicata e Idrogeologia 18, Parte II, pp. 211-238.

Leendertse, J.J., 1967, Aspects of a computational model for long period water wave propagation, Ph.D. thesis, Delft University of Technology.

Liggett, J.A., 1977, Location of free surface imporous media, J. Hyd. Div., A.S.C.E., 103, HY4, pp. 353-365.

Massie, W.W., (ed) 1986, Coastal Engineering, Lecture notes Delft University of Technology, f11a.

McDowell, D.M. and O'Conner B.A., 1977, Hydraulic Behaviour of Estuaries, McMillan Press, London.

Meinardi, C.R., 1973, The occurrence of salty groundwater in the lower part of The Netherlands, H2O, 6, 18, pp. 454-460, (in Dutch).

Meinzer, O.E., 1928, Compressibility and Elasticity of artesian aquifers, Economic Geology, 23, pp. 263-291.

NEDECO, 1965, A study on the Siltation of the Bangkok Port Channel, The Hague.

Odum, E.P., 1971, Fundamentals of Ecology, 3rd ed., Saunders, Philidelphia

Oude Essink, G.H.P., 1993, Effect of sea level rise on the groundwater flow system through Amsterdam Waterworks and Haarlemmermeer polder, the Netherlands. Sea Level Changes and their Consequences for Hydrology and Water Management, UNESCO, Paris, IHP-IV Project H-2-2, pp. 33-50.

Perrels, P.A.J. and Karelse, M., 1981, A two-dimensional laterally averaged model for salt intrusion in an estuary, Transport Models for Inland and Coastal Waters, Ac. Press.

PIANC, 1993, Problems created by saltwater infiltration, Report of working group no 6 PTC 1, Permanent International Association of Navigation Congresses.

Pinder, G.F. and Gray, W.G., 1977, Finite Element Simulation in Surface and Subsurface Hydrology, Academic Press, New York.

Pricket, T.A. and Lonnquist, C.G., 1971, Selected Digital Computer Techniques for Groundwater Resource Evaluation, Illinois State Water Survey Bulletin, 55, 62.

Redeke, H.C., 1922, Biologie der Niederlandischen brackwassertypen, Bijdragen tot de Dierkunde, No. 22, pp. 329-335.

Remane, A. and Schlieper, C., 1971, Biology of Brackish Water, 2nd ed., Scheizerbart, Stuttgart.

Remson, I., Hornberger, G.M. and Molz, F.J., 1977, Numerical Methods in Sub-surface Hydrology, Wiley-Interscience, New York.

Rijkswaterstaat, 1976, Salt Distribution in Estuaries, Government Publishing Office, The Hague.

Rijkswaterstaat, 1991, Een rechtlijnig WAQUA-model van het Haringvliet-Hollands Diep: Modelbouw, afregeling en verificatie, Report 91.191X, RIZA, Dordrecht (in Dutch).

Rijkswaterstaat, 1992, Effecten van een open Beerdam en open Rozenburgse sluis op de zoutindringing in het benedenrivierengebied, Rotterdam (in Dutch).

Rockwell Geyer, W. and Dungan Smith, J., 1987, Shear instability in a highly stratified estuary, Journal of Physical Oceanography.

Roelfzema, A., Karelse, M., Struijk A.J. and Adriaanse M., 1984, Water quantity and quality research for the Rhine-Meuse estuary, Delft Hydraulics, Publication no 325.

Roelfzema, A. and van Os, A.G., 1978, Effect of harbours on salt intrusion in estuaries, Delft Hydraulics Laboratory, Publication no 204.

Roelfzema, A., Perrels, P.A.J., Schoten W.N.G. and van de Wekken, A., 1987, Developing a mathematical model system for the Rhine-Meuse estuary, Delft Hydraulics, Publication no 384.

Rushton, K.R. and Redshaw, S.C., 1979, Seepage and Groundwater Flow, John Wiley & Sons, New York.

Saeijs, H.L.F., 1982, Changing Estuaries, Rijkswaterstaat Communications no 32, The Hague.

Saeijs, H.L.F. and Baptist, H.J.M., 1976, Wetland criteria and birds in a changing delta, Biological Conservation, 11:4, pp. 251-266.

Santing, G., 1963, Practical formulas for flow through aquifers - steady flow, The Development of Groundwater Resources with Special Reference to Deltaic Areas, Bangkok Symposium, Part II, Wat. Res. Series, 24, United Nations, New York, pp. 33-36.

Savenije, H.H.G., 1992, Rapid Assessment Technique for Salt Intrusion in Alluvial Estuaries, Ph.D. thesis, Delft University of Technology.

Sawkins, F.J., 1978, The Evolving Earth - A Text in Physical Geology, 2nd Ed., Collier Macmillan, West Drayton, U.S.A.

Schijf, J.B. and Schönfeld, J.C., 1953, Theoretical Considerations on the Motion of Salt and Fresh Water, Proceedings Minn. Intern. Hydr. Conv.

Schmorak, S. and Mercado, A., 1969, Upconing of fresh-water sea-water interface below pumping wells - field study. Water Resources Research 5, pp. 1290-1311.

Schumm, S.A., 1965, Quaternary Paleohydrology in the Quaternary of the United States, Wright, H.E. and Frey, D.G. (eds), Princeton Univ. Press, Princeton, U.S.A., pp. 783-794

Schureman, P., 1958, Manual of Harmonic Analysis and Prediction of Tides, U.S. Government Printing Office, Washington.

Segerstråle, S.G., 1959, Brackish water classification - historical survey, Symp. on the Classification of Brackish Waters, Venice, 1958, Archo-Oceanographical Limnology XI, pp. 7-33.

Shahin, M.M.A., 1991, Assessment of Groundwater Resources in Egypt, IHE Report Series 23, Delft, The Netherlands.

Statzner and Higler, L.W.G., 1986, Stream hydraulics as a major determinant of benthic invertibrate zonation patterns, Proc. 2nd Int. Symp. on Regulated Streams, Fresh Water Biology 16, pp. 127-139.

Steggewentz, J.H., 1933, De invloed van de getijbeweging van zeeën en getij-rivieren op de stijghoogte van het grondwater, Ph.D. Thesis, T.U. Delft, The Netherlands.

Streltsova-Adams, I.D., 1978, Well Hydraulics in Heterogeneous Aquifers, Academic Press, New York and London, pp. 357-423.

Stroet, C.W., 1987, The behaviour of aquitards in periodic groundwater flow, M.Sc. Thesis, T.U. Delft, The Netherlands (in Dutch).

Stuip, J., 1982, Proceedings FURORIS Congress, University of Suriname/Delft University of Technology, Paramaribo.

Terzaghi, K., 1925, Erdbaumechanic auf Bodenphysikalischer Grundlage, Franz, Deuticke, Vienna.

Terzaghi, K., and Peck, R.B., 1967, Soil mechanics in Engineering Practice, 2nd ed, John Wiley & Sons, New York.

Thatcher, M.L. and Harleman, D.R.F., 1972, A mathematical model for the prediction of unsteady salinity intrusion in estuaries, Dept. of Civ. Eng., MIT, report no 144.

UNESCO, 1969, Proc. Symposium on the Scientific Problems of the Humid Tropical Zone Deltas and their Implications, Dacca.

UNESCO, 1977, Atlas of World Water Balance, Korzoun, V.I. (ed), Unesco Press, Paris (translated from Russian).

UNESCO, 1984, Hydro-environmental indices - a review and evaluation of their use in the assessment of the environmental impacts of water projects. Techn. Doc. Hydrology, IHP-II, Project A.3.2., Paris, pp. 7-24.

Vellinga, P., 1982, Beach and Dune Erosion during Storm Surges, Delft Hydraulics, Publ. 372, Delft, The Netherlands

Verbeek, H., 1992, Verificatie DISTRO-NDB, Nota nr 92.016, Rijkswaterstaat, RIZA, Dordrecht (in Dutch).

Verbraeck, A. and De Ridder, N.A., 1962, The geologic evolution of the Tielerwaard-West (in Dutch), Report 3, Province of Gelderland, Arnhem, The Netherlands, pp. 43-54.

Verruijt, A., 1970, Theory of Groundwater Flow, Macmillan, London.

Volker, A., 1961, Source of brackish groundwater in the Pleistocene formations beneath the Dutch polderland, Economic Geology, Vol. 56, pp. 1045-1057.

Vreugdenhil, C.B., 1989, Computational Hydraulics, an Introduction. Springer, Berlin.

Walton, W.C., 1970, Groundwater Resource Evaluation, McGraw-Hill, New York.

Wang, H.F. and Anderson, M.P., 1982, Introduction to Groundwater Modelling, W.H. Freeman & Co., San Francisco.

Wesseling, J., 1979, Saline seepage in The Netherlands, occurrence and magnitude. Proceedings and Information, no.26, Committee of Hydrological Research, TNO, The Hague, pp. 17-33.

Wesseling, J. and Colenbrander, H.J., 1962, Determination of the aquifer characteristics from the propagation of the tides in the aquifer (in Dutch), Report 5, Province of Gelderland, Arnhem, The Netherlands, pp. 60-65.

Wiegel, R.L., 1964, Oceanographical Engineering, Prentice Hall, U.K.

Winterwerp, J.C., 1983, Decomposition of the mass transport in narrow estuaries, Estuarine,Coastal and Shelf Science, vol 16.

Wolff, W.J., 1973, The Estuary as a Habitat, Ph.D. Thesis, Leiden University, The Netherlands.

Wolff, W.J., 1983, Estuarine benthos, Ecosystems of the World, Ketchum, B.H. (ed.), Elsevier, Amsterdam, pp. 151-182.

Zenkovich, V.P., 1967, Processes of Coastal Development, Oliver & Boyd, Edinburgh.

Zienkiewicz, O.C., 1977, The Finite Element Method, McGraw-Hill, New York.

3. water management and control

3.1 Water Management Planning

This chapter marks a difference in approach to the subject area of this publication. Generally speaking, the preceding chapters dealt with the existing scientific knowledge related to the natural conditions and circumstances in deltaic areas. This chapter will deal with the possibilities and danger of human intervention in these areas. These interventions are motivated either by basic human needs, or by the desire to have a more comfortable life. The latter need includes measures to preserve or improve nature, the environment and ecology. There is a growing feeling of discomfort with the havoc wrought in this field by the massive impact of modern technology applied to satisfy mankind's desires. In order to sensibly utilise natural factors and scientific and technological knowledge aimed at realising these goals, it is obvious that planning is an indispensable aspect of water management and control.

This chapter is policy-orientated and deals with instruments for arriving at formulations of policies for the application of technological possibilities. These instruments, which may also be referred to as planning instruments, are mostly managerial, administrative, legislative, social and economic in nature. Of course, the possibilities offered to society by science and technology, as well as their dangerous aspects, must be the framework within which the planning instruments are applied.

The following aspects will be dealt with in this chapter:

- the need for planning.
- the conditions for planning; comprising degrees of organisation, character of legislation, economic need, willingness to cooperate and compromise, degree of development, political system and many others.
- fields of water planning; describing physical conditions and the nature of problems regarding the quality and quantity of surface water and groundwater taking the water system and its surroundings into consideration.
- interests in water policy; covering the various activities in society that depend on water, such as agriculture, water supply, shipping, environmental protection, drainage, urbanisation, etc.
- instruments of water planning; where planning methods, related to the nature of the need for planning, will be described.

3.1.1 The need for planning

Planning is the process of preparing intentions or proposals for measures aimed at realising an objective or set of objectives. By nature, this infers a projection of the future, be it the near or distant future. This definition does not make a distinction between small-scale planning and that over a larger time scale (e.g. a five-year development plan for a national economy). In fact, small-scale and short-term planning is as old as mankind. Planning is a capacity and a basic need, confined to the human mind, excluding all other living beings.

A way of looking at the development of human societies and cultures from primitive to highly sophisticated, is to view the development of planning from simple and short-term, to intricate and long-term. Although hunting to obtain food involves planning, the steps to cattle-breeding and from there to agriculture, implicate much higher degrees of it. In the same way, navigation developed from fishing to trading purposes, and from lake or river navigation to coastal shipping and world-wide voyages. Obviously, the planning need grew in that sequence. Canal building has also been practised, from very ancient times, for irrigation purposes, involving sophisticated planning and even system thinking. Planning also occurs in every art form, from community chanting to compositions for symphony orchestras, from ritual performances to motion pictures.

The explosive development of technology and industry in the last one or two centuries has given entirely new dimensions to the very concept of planning. The building of factories, roads, hydraulic constructions, space-rockets etc., involves such complicated planning that the human mind can only grapple with it through the cooperation of many individuals. The drafting of economic and social plans may be even more complicated, having to cope with many uncertainties. In this context, it is interesting to note, that although the periods for which planning is carried out, have lengthened considerably (from minutes to days to years to decades), there seems to be a limit inherent to the human condition. Plans (except when purely technical in nature) become very uncertain if they cover a period of a year, let alone several years or decades. The latter period may even be regarded as involving nothing but pure speculation.

The preceding remarks were made for the sole purpose of giving some perspective to the concept of planning. This section, however, should shed some light on the need for planning in the water field, especially in deltaic areas. In a general context, the historical aspects of planning have been mentioned. However, with respect to the field of water management, and of water policies as a whole, a striking aspect is the very lack of planning which existed until very recently throughout the world. Of course, water related constructions have been carried out, sometimes on a large scale, and in historical times. Obviously planning was indispensable for the proper execution of these works. However, safeguarding water supply for consumption, agriculture or industry, protecting the population from the dangers of floods from rivers or the sea, maintaining an acceptable water quality, creating conditions for economic uses like shipping or irrigation, have mostly been carried out in a haphazard way. These activities were implemented with a defensive or passive attitude rather than with a clear objective or vision for the future.

There were two predominant reasons for this state of affairs. The first is that water, like air, is such a basic necessity of life, and that the right of access to it was regarded as being absolute. The second is that manifestations of the destructive force of water were, like volcanic eruptions, regarded as *Acts of God*, that one could try to ward off, and against which lasting protection was impossible. These attitudes cannot be laughed away as ridiculously old-fashioned, because they reflect a realistic assessment of the human condition. Water was abundantly available, and regions where this was not so were regarded as uninhabitable. Moreover, all kinds of ownership conditions and water-rights developed, which excluded any planning for public benefit. In many parts of the world, this is still the case. Another factor was that water quality deterioration and threats to the environment did not exist or were so localised that there was no need for special measures. Finally, neither the scientific knowledge of the occurrence of floods, nor the technological means to effectively prevent them, were available. Only recently (the past 30 to 40 years) have some of these circumstances changed and the need for planning become very urgent.

Although water is regarded as a renewable resource, the process of renewing began to lag seriously behind the growth of demand. The most obvious cause of this is the enormous increase in the world population. Not only do all people need drinking water (a demand that has no serious quantitative consequences, but the qualitative aspect is prominent) but they also require food. Producing crops imposes a great demand on good quality water. Also, the so-called water culture became more important with the higher degree of development reached by a country. Water, being regarded as abundant was more widely used for purposes like showering, sprinkling, car-washing, laundering etc. There is no feasible means of providing water of less than drinking quality, for those purposes, although such quality is not always really necessary.

Another important demand factor is industry. Many industries use large amounts of water in their production processes. Generally, not much water is included in the product itself, but processing and cleaning water may amount to large quantities. This also has an impact on water quality.

Water quality

In many countries the dramatic increase of water pollution has rapidly become the major cause of water shortage. Apart from aesthetic, environmental and ecological hazards, polluted water is not a useful commodity. It has to be disposed of, if that is possible, without creating new problems. The growing amount of polluted water can be attributed to roughly the same reasons as the increase in water demand. Population growth in itself creates waste problems, and the water used in industry is always discharged in a polluted form. Agriculture uses increasing amounts of fertiliser and pesticides, and the intensification of cattle-breeding seriously contribute to the pollution problem. Even salt-intrusion is increasing due to human interference (e.g. shipping entrances into low coastal areas), and threatens surface water and groundwater alike. In balance, the amount of water of sufficient quality, available in places where it is needed, simply does not meet the demand in many parts of the world.

Not surprisingly, modern day planning has its roots in the traditional planning techniques. Planning for civil engineering works is old, but once people decided that the hazard of flood disasters was to be terminated, very large, long term and multi-disciplinary plans had to be formulated in order to cope with sets of construction works, some of them on quite a large scale. Once it became obvious in some regions that local plans for agricultural water supply were sure to overstrain the common source, planning was initiated to bring about a sensible distribution. Where such a source became polluted, purification plans had to be drafted. This problem often turned out to be nationwide (or even larger), and the national economy then entered into the process, involving long term investment strategies and the establishment of priorities. This is only a small sample of the problem areas, that will be dealt with more extensively later on.

As a conclusion to these considerations of the need for planning, comments on three relevant subjects will be made:

1. In 1977, the United Nations organised a Water Conference in Mar del Plata, Argentina. The "Mar del Plata Action Plan" resulted, stating among many other things, that water shortage had grown into a world-wide problem, requiring concerted measures on a global scale. The result was excellent, in that it drew the governments' attention to a serious problem, but it is remarkable that the statement in itself was factually wrong. While water shortages did and still do occur in most parts of the world, the problem is not global in nature. Water basins or water systems are always confined to certain areas, large or small, but in no way do they influence each other. It is puzzling nevertheless, that the problem does occur more or less simultaneously all over the world, in countries rich or poor, developed or developing.

2. From the foregoing, it should not be concluded that every region or country can be considered separately from a water management perspective. Where there is inter-dependence, it should be taken into account, regardless of political frontiers. Deltaic areas are always parts of river basins, and thus upstream and downstream phenomena have to be treated in unison. It is not possible, therefore, to confine a chapter on water planning, to these areas alone. An integrated, comprehensive approach is necessary.

3. A river basin with its coastal regions is a water system. The concept of water systems and integrated water management planning will be used later in this chapter, and it should be borne in mind that it is not identical to the concept of river basins. A water system is a body or system of water, together with all its boundaries and surrounding conditions that can be quantified in order to study their inter-relations with the water. As defined, it is a modern concept that lends itself to a systems analysis approach.

3.1.2 Conditions for planning

It must be emphasised that the concept of planning in this chapter does not refer to the traditional fields where specific works or processes are concerned, but to the type of planning needed to arrive at a beneficial water policy (in which, of course, hydraulic or other works may be included). The conditions for the former type of planning are simple and straightforward. The will, the means and the skill must be available, and the planning process, based on long experience, stands a good chance of being successful. As will be seen, however, the latter type of planning depends on an impressive list of conditions, which in practice are never completely fulfilled. Whether this means that the concept of planning is futile is a question that will, hopefully, be answered in the course of this chapter. A number of aspects pertaining to the conditions for planning will now be discussed.

Degree of organisation of water administration units

A situation where the territory for which planning is considered comprises one administration, is undoubtedly the ideal one. An example of this is the United Kingdom, where ten Water Authorities have integral administrative powers over separate river basins. On the other hand, this becomes a disadvantage if the need for national planning arises. The situation in which administrative borders have no relation whatsoever with the physical reality pertaining to water, is far more common. For a national plan, this drawback can be overcome by instituting cooperative associations or committees, but again, in the majority of cases, planning for water systems of smaller dimensions is the more obvious choice. A favourable condition is the existence of waterboards, like those in the Netherlands. They constitute a sub-structure on which plans for larger areas are based.

For planning on a national scale it is good to have a ministry with overall power in the water field. In modern societies, this power and that of the environment ministry must be clearly defined. Overlap cannot always be avoided, as the environment is an aspect of water, and water is part of the environment. In such cases, strict agreement on common policy is necessary. The existence of government power over national water bodies does not, however, solve the problem of trans-boundary river basins. Striking examples are the Rhine in Western Europe, and the Ganges in South Asia. The lack of international river authorities has made sensible international river planning impossible. This is especially so in the case of deltaic areas e.g. in Bangladesh and the Netherlands. The only solution in such cases is to plan for the downstream country, taking into account as boundary conditions, the consequences of upstream policies.
For the Rhine, however, considerable progress has been made in water quality management. These efforts have been undertaken within the framework of the International Commission for the Protection of the Rhine Against Pollution. At present, the Rhine Action Programme is under execution, which aims at a 50% reduction in the discharge of pollutants to be reached in 1995 (taking 1985 as a reference). The plan also includes an ecological rehabilitation of the river. The navigation on the river is also regulated through an international commission.

In conclusion, organised power pertaining to water management is a pre-condition to planning. Where this power does not exist for the territory concerned, it must be instituted, if necessary on a temporary basis, by a higher authority. Where no authority exists, there can be no planning.

Legislation

Traditionally, written law pertaining to the use of water and to the care of its sources, is not at all common. Obviously it has always been necessary to adhere to rules and agreements when abstracting, discharging or distributing water, but these were based on silent understandings, rooted in tradition. Often, they were far from equitable. The 'Water-Rights' in the United States of America automatically made a landowner the owner of the water occurring on his property, and he could (and still can in many parts) do with it as he pleased, without regard to the downstream interests. In some countries, farmers are generally poor and are tenants of their land. The way water is allocated to them depends on the whim of the landowner. In most countries, it is understood that the civil authority has some say regarding the management of water bodies, regardless of the actual ownership. This, in effect, is a necessary though not sufficient pre-condition for planning. The degree to which the authority is aware of its responsibility, may differ very much, and it is only in some countries, and of recent origin, that water management is regarded as a policy field.

Written water-law does exist, but is often haphazard, and made to cover certain areas of interest For example, there are rules for shipping in rivers and canals and for maintenance of the route, rules to manage river discharge or coastal defence works to prevent disasters, and rules set up by water boards or other farmers associations to protect their interests. Planning on a limited scale is possible as a consequence of these laws, but not policy formulation in the wider sense.

In many countries, the needs and mechanisms described previously are resulting in modern legislation that covers many or all aspects of water management. There are now water quality and environmental laws for surface water and groundwater. Some of these cover surface water management or groundwater management, in a quantitative sense. At times, they combine several of these aspects, but the ideal would be an integrated water law. In most countries, there are several levels of administration (e.g. central government, provinces, municipalities, waterboards etc.). Good legislation will determine the powers and the inter-relationships of these levels.

Legislation is a condition for planning. In principle, a law provides rules that enable the authorities to attain certain objectives. To do this in a complicated policy field, planning is necessary, and the law provides the power to execute the intentions laid down in planning. Some of the laws therefore make planning obligatory, which has the advantage of providing a framework for the process of planning and the shaping of such a plan.This is not a book on legislation, and the subject therefore cannot be treated in its entirety (e.g. by explaining the instruments the law may use to arrive at its goals). Also, the degree of centralisation of the legislation, which might be optimal for planning, is beyond the scope of this section. However, in the sub-section on public participation, some consideration of this subject will be found.

Economic aspects

Imagine a not too densely populated country, with abundant rainfall and well distributed river flows, a relatively high and rocky coast, fertile soil and a reasonable degree of industrialisation. In the past, there were such countries, and today possibly the Republic of Ireland more or less approaches it. In such a country there would probably be little or no incentive to do a lot of planning in the water policy field.

In a deltaic area, salt intrusion threatens agricultural production and maybe even water supply. Water pollution from inside or from abroad threatens public health, fisheries and the ecologic equilibrium, and population-growth increases water demand to such an extent that supply in certain (dry) periods is not assured. These and many other developments tend to become a threat to quite a few economic interests, to income and employment of the people, and possibly to the stability of the national economy. Generally, it takes a long time for governments to realise that a radical change of attitude is required, and a great deal of pressure is needed to sway the policy. Economic decline, however, is a factor that is apt to cause considerable political impact. The point being made here is that the difficult, long-term, tedious and possibly expensive process of planning will not be embarked upon without considerable economic pressure, and that economic need is one of the conditions for planning.

Public participation

Obviously, any kind of planning is subject to the participation of everyone concerned. However, in what has been referred to previously as the traditional type of planning, there is generally a hierarchical organisation, causing every participant to recognise his or her personal interest in the success of the project. The planning considered here depends on the willingness of the population, or certain parts of it, to participate. The interests of such persons are not always obvious, and may even be contrary to the objectives. In this context, the remark on the seemingly unrelated character of activities pertaining to the execution of a plan mentioned previously, is well worth remembering.

The question arises as to whether a centralised or a decentralised administrative system is more beneficial for planning purposes. In a centralised (autocratic) system the planning approach resembles the industrial or business-like approach. The process can go on undisturbed and be carried out by the central authority in a rational, purposeful manner. The difficulties arise in the execution phase of the plan, which in the final analysis always depends on the worker (the farmer, the local water manager etc.). Regardless of how autocratic the system or the basic willingness to cooperate may be, the plan will fail if the people who depend on it fail to understand its purpose. No measure of explanation will ever compensate for that drawback.
In a decentralised (democratic) system, the planning process is more subtle. Planning then arises from the inter-relation between the needs and wishes of the population and interest groups on the one hand, and the obligation of the national or local government to serve the interests of the inhabitants, on the other. The objectives and, later on, the intended measures, are invariably the result of lengthy discussions on various levels. In this process, the explanation and the

comprehension of the issues grow in time, although there is a tendency of some groups to see only a limited number of aspects. Consequently, a plan once it is finalised and publicly presented, tends to be outdated. Even worse, in the phase of execution, circumstances and ideas may change, frequently leading to considerable delay, loss of capital and the prolonging of undesirable situations.

A solution to this is sometimes to produce a progressive plan, implying process-planning, which results in a plan that can be adapted at certain (previously fixed) intervals. Such a plan provides for a high degree of flexibility. Measures, like legislation or economic incentives, may be changed. If capital investments are part of a plan, it must be understood that works carried out belong to the system conditions of the next version of the plan. In conclusion, it can be said that public participation is an absolute condition for planning. A centralised administration may produce more consistent plans, but a decentralised system stands a better chance of getting at least some results.

The influence of interest groups

Vested interests and lobbying may be considered undesirable and even unethical concepts. In a well-organised society, however, it is likely that all interests in a policy area are somehow interrelated and may participate in the process of planning. The chances of an unequitable plan being adapted are thus minimised. It stands to reason that the mentality for eventual compromise and cooperation must prevail. If the nation is well organised, the majority rule should overcome such difficulties.

There are countries in which one interest group is so dominant, and has such access to power, that it can frustrate planning for the public benefit. In developed countries, these are often large industries which, for instance, seek to postpone their submission to regulations concerning water pollution. In less developed countries, there are examples of groups of landowners who prevent measures for enhancing the security or fertility of the land, at the cost of the farmers, from motives that may be irrational but relate to their own income situation. Obviously, in these situations planning efforts may easily be frustrated.

Success of a planning strategy is, at best, conditional and excessively high expectations of the outcome may lead to disappointments. The smaller the territory concerned, the better the chances will be, as the degree of organisation is better and public interest is more keen. Deltaic areas often form a part of a river system of which the upstream portions are situated in one or more other countries, whose interests generally diverge. The only way to plan for such regions is to acquire knowledge of upstream policies and take these into account in planning.

3.1.3 Fields of water planning

Water planning can be applied to problems of water quantity and water quality. For each of these, surface water and groundwater pose separate problems. It is so difficult to adhere to divisions between these fields and hence, in most circumstances, integrated planning is unavoidable. The discussion will focus on deltaic areas, but obviously the relation to upstream areas is a very close one.

Surface water quantity

In deltaic areas, the most compelling parameter to be controlled is the water level. Water tends to be stagnant, and hydrological units are often man-made, in the form of polders or similar entities suitable for level control. Level control is important to prevent flooding but the precise objective is determined by agricultural demands. Such demands are different for various crops and for cattle-breeding, and are also very much dependent on soil type. As level control is carried out by the supply and discharge of water, it runs into difficulties if the available supply is insufficient or if the possibility to discharge is blocked. Both phenomena may have qualitative aspects in that the quality of available water may be unacceptable, or the water to be discharged may be prohibited from entering the receiving water body. In some cases a drop in water level may have disastrous consequences, causing the authority to permit salty or brackish water to be taken in. If certain quantities of water are transported for the sake of level control, much greater quantities are moved about owing to various water demands, not the least of which is evapotranspiration. Increasing demands have consequences for the capacity of the distribution system, and shortages create problems for which solutions may be sought on the supply or demand side.

Andra Pradesh Anicut at the apex of the delta of the Krishna river, India. During the dry period the flap gates are in an upright position and head is created to supply water to the irrigation canals. In the flood season the gates are lowered. The lifting and lowering of the gates is done by a special truck.

River discharge is generally the most abundant source of water. In deltaic areas, it becomes more and more necessary to dam off tidal estuaries and create fresh water reservoirs, so as not to lose too much of this resource to the sea. Again, if the water quality, arriving from upstream, is less than desirable, it will be necessary for the authorities to take measures. Another problem is often caused by upstream regulation, storage or diversion of water.

A prominent aspect of water management is the safe flow of river water to the sea. The river configuration and its confinement (dykes) must be such that the country is safe from floods. These may occur during extreme river flows, or in the case of many countries, if the flow is blocked by ice. Measures to ensure safety are not independent of those to ensure water supply. The same is true for measures to prevent floods from storm surges at sea. Dykes, dams and other coastal works must protect the country from disaster, but at the same time, they must be adequately constructed to enable discharge and storage if and when necessary.

A problem specifically related to deltaic areas is salt-water intrusion from the sea. This occurs in huge quantities through the tidal movement and is often also caused by the operation of locks to enable ship movement. A more subtle way is the salt intrusion through the subsoil into the groundwater. It is often possible to keep the salt at bay by forcing it back, using, either directly or indirectly, fresh water available from the discharge of rivers. Owing to the stochastic character of river flows, this method is liable to fail in dry times, but even in more favourable conditions it is a heavy burden on the often scarce resource of fresh water. In this respect an economical means of combatting salt-water intrusion is by damming off estuaries, or building sophisticated types of shipping locks. These measures, however, are expensive, and for this reason are not feasible in many countries. Also, the physical circumstances may prohibit this kind of solution.

Surface water quality

A problem that has invaded human complacency during the last three to four decades is the pollution of surface water. Of course the phenomenon is not new, for life as such produces waste, and industrial wastes are not all that recent. However, for a very long time it seemed difficult to imagine that human activities could cause more damage than Nature was able to mend. This is why it took so long before the awareness grew that Nature might need human assistance, and that this would mean a substantial sacrifice of wealth.

Huge concentrations of population, intensive cattle breeding, excessive use of fertilisers and pesticides, industries that grew not only in dimension, but also in the variety of substances utilised and discharged, and in the generation of heat waste, all contribute to a very grave situation with respect to the quality of surface water. Pollutants often adsorb to suspended materials that settle in sedimentation areas as heavily polluted sediments. Deltaic areas in this respect are in a particularly unfavourable position. The waste that they themselves create is often difficult to get rid of. At best, it flows sluggishly out to sea, and at worst it has to be pumped and dredged out. In addition, the upstream areas of the basin may discharge their wastes into rivers whose water arrives at the downstream areas in a fairly undesirable condition.

Of the factors which determine the state of the environment, water and sediment quality is a major one. The pollution of water has unacceptable consequences for the biological balance, human health, and even for the very quality of life. This problem is now widely understood, and in most countries measures have been and are being taken to improve the situation. In these measures, a marked development is evident. The first approach was the emission approach aimed at stimulating or enforcing the purification of waste water. Then came the water quality approach in which standards were set for the quality of the receiving waters. The latest ideas point to an approach comprising the integrated management of water systems.

Groundwater quantity

Fresh groundwater has always been a desirable commodity, because it is clean, of rather constant temperature and easy to obtain, especially in areas where there is no close source of acceptable surface water. Where fresh groundwater is available, it is preferably used for the supply of drinking water, but industries also use it for processing and cooling purposes, and agriculture may use it for irrigation.

Groundwater has been utilised in ever growing quantities, until extraction has surpassed supply. The source of supply is rainfall and cannot be influenced. Excessive extraction leads to a lowering of the groundwater table, resulting in a new equilibrium condition. If this continues, there will come a time when the resource is no longer available. In low-lying or holocene soil, or in dunes, excessive extraction will result in salt intrusion, rendering the resource useless.

In some cases, artificial recharge of the aquifers may be a solution. For this, surface water has to be utilised which must first be purified. Long term ecological drawbacks may sometimes prohibit this technique. Another possibility is the selective use of groundwater. Drinking water and some industrial processes depend very much on the purity of the resource. Other uses will have to be discouraged, either by economic measures (taxes), or by regulations. Often, the law does not allow this kind of discrimination but, if necessary, laws can be changed.

Groundwater quality

Until recently, the quality of groundwater was not regarded as a problem. The quality of the resource was not in danger, and could not be allowed to be in danger, because a cure was judged to be impossible, owing to the extremely slow processes in the subsoil. Opinions in this matter may differ, but it can be regarded as a tragic aspect in the development of modern societies that practices were allowed to continue for many decades, which slowly poisoned the subsoil.

Two main causes can be identified for the deterioration of groundwater quality in developed countries. These are waste disposal by industries and excessive fertilisation in agriculture. Both can be discontinued, though not abruptly. Industries may adapt their processes or find harmless solutions for waste disposal. Competition and employment policies necessitate careful approaches in agriculture. Optimal fertilisation, to avoid unnecessary losses, demands great care and sacrifices in profits. Curative measures (e.g. removing polluted layers of soil) are often excessively expensive, and are certainly not applicable on a large scale. Very grave cases may be remedied, but in most cases, time must do the work.

3.1.4 Interests in water policy

Policy, in general, is meant to be a way of enhancing the quality of life. This would remain a rather philosophical concept, were it not represented by so-called interest groups. These occur no less in the water field than in other policy areas. Planning therefore, is a useful tool to formulate a policy that strikes a fair balance between the various interests. Interest groups and government authorities must be continually involved in the process of thinking that is to result in plans. If not, a plan may eventually fail, however well conceived it may appear to be. Some water-related interests will be mentioned briefly in what follows.

Agriculture

Generally, this is quantitatively one of the largest users of water. The quality aspect of the supply is not one of the most important factors (though obviously not negligible). The supply of an adequate discharge of water on demand is of primary concern.

Horticulture

Depending on what is produced, this branch may have very high quality standards. Many plants are very susceptible to specific substances, like salt or certain organic micro-pollutants. The quantity aspect is often of less importance.

Public water supply

In most countries, and especially in deltaic areas, there is generally no problem of quantity. The quality aspects are, obviously, very important and water supply companies are apt to give priority to pure and fresh groundwater supplies. On the other hand, the importance of good quality water supply to society is so great, that eventually any amount of money may be spent to guarantee a high quality supply to the public, even if the source has unfavourable properties. This, of course, must be balanced in comparison with other interests.

Industry

Qualitative and quantitative demands are very much dependent on the type of industry concerned. It has been noticed that in cases where waste discharge is made very expensive or is forbidden, industries develop clean processes or prefer to invest in internal purification or water recycling.

Power

Hydro-power may have environmental drawbacks, but it hardly occurs in deltaic areas. However, thermal power generation creates a cooling water problem in these areas. Plants situated near rivers may experience quantitative shortages, but more often the critical problem is that of quality. The unnaturally high water temperature may create problems for the biological equilibrium.

Shipping

Shipping does not use up water, and does not cause quality problems on a large scale. It does, however, require depth for keel clearance, which may oblige water managers to direct water flows in a way that is harmful to other interests. It may also lead to canalisation of rivers, which is costly and not always beneficial to everyone.

Recreation

Swimming, beach recreation and boating demand water-level control, water depth and water quality. This may be contrary to interests such as water storage, waste water discharge or ecological demands for say, shallow water.

Environment, ecology, and nature preservation

These subjects are grouped together here, although their interests are not always identical. They are, however, generally opposed to various economic interests. Also, if the population demands that these interests to be carefully treated, the question may arise whether this in itself, does not constitute an economic interest. Whatever the case, the systematic neglect of these values has caused immense damage not only to plants and animals, but also to human life itself. It is difficult to define economic prices in these matters, which complicates weighing them against other interests.

Water management planning therefore has to take into consideration the surface water and groundwater systems at hand, both quantitatively and qualitatively. Also, the sediments, the surroundings, the constraints and the demands must be considered in order to arrive at a well-balanced and flexible plan.

3.1.5 Instruments of water management

To begin this section, reference is made to some of the conditions for planning described earlier. In fact, these conditions are equally valid for the execution of water management. Obviously it is no use planning if the conditions for carrying out such plans are not fulfilled. Special emphasis is placed on the importance for the population and interest groups to understand the problem and, based on this, to willingly participate in the process. No legislation, however well considered, will really work without this willingness. There should never be any doubt regarding the authority to enforce the legal regulations, but it is a bad sign if it has to be exercised too often.

The ways and means (instruments) of carrying out water management may be classed under four categories; technical, managerial, regulatory and economic. Generally speaking, all of these are necessary, but it is a matter of experience or taste, and also of the political and administrative system of a country, the relative importance that is attached to each. The meaning of these concepts will be illustrated by some examples.

Technical

A technical measure, from the point of view of water management, is generally a civil engineering project (e.g. drainage gates in a ditch, locks and weirs for canalisation of rivers or the distribution of water, barrier-dams in estuaries to guarantee security or create a fresh water reservoir). Large projects require high investments and can only be carried out in the context of flexible and adjustable, long-term planning (20 to 50 years). Apart from that, they do not differ in principle from other instruments at the disposal of water managers.

Managerial

The authority that is responsible for a water course, a water body or a water system will manipulate the infrastructure to achieve certain goals. The supply and discharge-gates (or pumps) of a polder may be operated in such a way, that brackish or polluted water in the canals and ditches is flushed away and replaced by clean water. The evacuation gates in a dam that closes off a former estuary, may be closed during such periods that the water-level in the man-made lake rises (or falls), enabling surrounding territories to be supplied with, or be drained of, water. In the branches of lower river reaches, weirs may be used to modify water distribution (e.g. force back intruding salt water from the sea, supply certain areas suffering from a shortage of rainfall).

Regulatory

This term comprises the entire system of water legislation of a country, from general national laws, to local rules and regulations. It may be forbidden to discharge polluted water without a licence. Limits may be imposed on the speed of ships, to protect the banks of canals. A polder may be prevented from pumping out surplus water if the water-level in the storage area rises above a certain mark. The discharge of waste heat or waste water may be subjected to standards related to the temperature and quality of the receiving water body. Extraction of groundwater may be maximised. A water authority may be obliged to receive and transport certain quantities of water from another authority.

Economic

Economic measures are generally understood to pertain to pricing, be it in the shape of incentives or of penalties. The instrument may be very powerful. If the price of discharging polluting substances is sufficiently high, an industry may decide to purify its own water or resort to clean production processes. However, the demand curve for drinking water is not very sensitive to price. If the economic correlations are sufficiently known, pricing may be an effective and subtle tool. However, gigantic mistakes are known to have been made in this field.

The process of planning begins with a study aimed at understanding how the system works, and which instruments should be used, and in what way, to make it work. Also, it is necessary to investigate in what way the various interests are served by the existing infrastructure and its management, and what aspects are unsatisfactory for each of the interests. From this, the need arises to attach weights to these interests, this being an important factor to take into account in the planning process. At a later stage, these weights may have to be re-evaluated by multi-criteria evaluation methods, which will be considered later.

The investigation into the functioning of the system may be time consuming and expensive, so it is essential to start with a search for available data. The investigation will include:

- physical and hydrological factors; e.g. geography, water system, morphology, water levels, flow quantities and flow patterns, frequencies, geology and hydrogeology, meteorological characteristics, infrastructure etc.;
- chemical and biological factors; e.g. the types and amounts of polluting substances arriving from upstream, data on the distribution and nature of the pollution inside the area under planning, present purification-capacity;
- economic factors; e.g. production outputs dependent on water, benefits or damages arising from limitations in supply (qualitatively and quantitatively), quantitative relations between various outputs, percentage of water-dependent production relative to national production, elasticity of demands etc.;
- political factors; e.g. the effectiveness of existing legislation in bringing about changes, the hierarchy of administrations and their powers etc.;
- social factors; e.g. the vulnerability of industrial and agricultural populations to changes in income or employment, traditions that determine the readiness to accept changes, levels of education and their influence on attitudes towards certain measures etc.

From the foregoing, it should not be concluded that in all cases the situation confronting the planner is so complicated that he has to resort to sophisticated methods for success in his planning effort. The territory for which planning is intended may be small and not of a very complicated structure. If a larger area, or possibly an entire country is considered, the issues involved may include only some of the aspects described above. Indeed, the first thing to seriously consider is whether the planning effort, without being superficial, can be realised by common sense, careful thinking and consultations and deliberations with experts and interested parties. If several courses of action emerge, decision-making may be aided by multi-criteria analysis.

It may not be possible to avoid modern computer-aided analysis methods in cases where many interests, authorities and policy makers are involved. This is referred to as systems analysis or policy analysis. Obviously, the application of these methods may also vary from relatively simple to very sophisticated and extensive. It is always wise to choose the simplest method that will do the job. A multi-disciplinary team has to be formed, of which the experts on science and technology may be available, but the methodology experts have to be hired, often from foreign consultancy firms. The descriptions of the methods that follow is intended to give an impression of the nature of systems analysis as a planning instrument. A more detailed description is given in the United Nations publication (1985).

Operational water management is enhanced by making use of Decision Support Systems (DSS). These provide the managers and operators with tools to facilitate the decision making in day to day operation. The DSS is often a mathematical model that simulates the relevant quantity and quality parameters in the water system. The model is run in order to try out possible operation strategies before actual execution. This helps the decison maker in selecting the best strategy in view of the specified objectives. Such models are accomodated with data assimilation techniques (e.g. Kalman filtering) to incorporate real-time field data. Further refinements in the DSS can be obtained through the use of expert systems or optimisation models. The position of a DSS in an operational management system is shown in Figure 3.1.

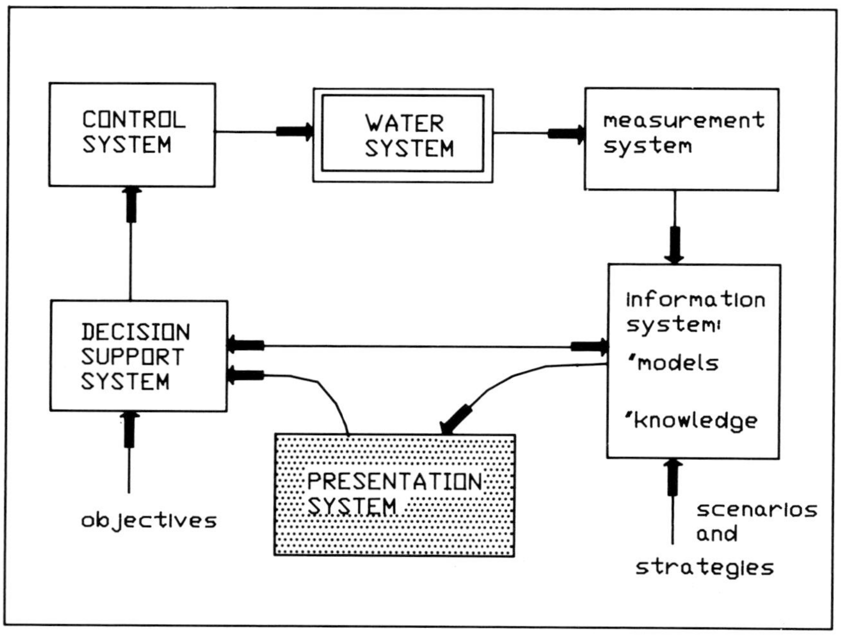

Fig. 3.1
Decision support system in an operational management system.

Systems Analysis

Analysis to support planning for water management has a simple aim. It is to provide systematic and quantitative information to decision makers to enable them to arrive at better decisions, or more specifically, to make a better selection out of proposed alternative measures (strategies). As a result of scientific and technological developments, there are now powerful computers, models and analytical techniques, that have very much improved analytical capabilities. They do not, however, in any way contribute towards solving the problems of obtaining basic data.

The changes described earlier in this chapter have their impacts on both the information required from an analysis, and the set-up and execution of the analysis. Systems analysis tries to respond to these changes and new requirements by providing a framework for analysis (a coordinated sequence of steps), and a computational framework (a coherent set of computational techniques) to facilitate quantitative analysis, or an analysis as quantitative as possible with the available data and tools.

The art of engineering and planning may be defined as drawing reliable conclusions from insufficient data. This may also be understood as introducing the right assumptions and simplifications into the analysis from which the conclusions are drawn. In the past, rather crude assumptions and simplifications had to be made owing to a lack of sufficient knowledge and of computational facilities. At present, planners are able to make better assumptions and introduce less severe simplifications. Reality today can be approximated more closely than a few decades ago. Also, without much additional effort, improved examinations can be carried out regarding the consequences of different conditions or assumptions, leading to an improvement of the assumptions. Computational capability is important because planning is always related to an uncertain future situation. Any analysis should treat these uncertainties explicitly, by examining different scenarios and by performing sensitivity analyses. A computational framework greatly facilitates such analyses. In a way it is fortunate that analytical capabilities have improved because problems have become more complex. For this reason, the analysis has to be structured in a sequence of steps or a set of procedures, in which the contributions of various disciplines and persons are coordinated and tuned to each other. Systems analysis can now be defined as a systematic process of generating, analysing and evaluating various strategies (courses of action). Through the application of coordinated procedures and computational techniques, systems analysis can reveal the multiple consequences of such strategies. Some of the tasks to be performed by the analysts are:

- observe and describe the behaviour of complex systems;
- develop models to explain the observations, test their validity, and predict behaviour under various sets of conditions;
- formulate measures and strategies and assess their consequences;
- compare various strategies;
- communicate the results to the decision-makers in a readily understandable way.

If applied properly, systems analysis will contribute to an efficient analysis of water management planning problems. It goes without saying that sound engineering and economic judgment and a good understanding of the decision problem remain essential ingredients for a successful analysis. It is important to emphasise that systems analysis is fundamentally a way of thinking about problems rather than some formal mathematical method. It is sometimes confused with the term operations research, which does connote a set of mathematical techniques. It is the capacity to integrate that is critical for producing useful results for decisions.

Framework for analysis

The framework for analysis structures the analysis procedure to provide the required information for the formulation and selection of water management strategies. This framework can be used as a reference for the design of problem-specific frameworks. It should be flexible to change and easy to apply in order to be useful for continuous analysis for planning. All steps in the framework, especially those steps that include the application of computational techniques, should be attuned to each other. They should produce no more and no less information than required in view of the total procedure, the accuracy of the different components of data, and the objectives. In many situations this internal consistency appears to be a dominant factor in the efficiency of the analysis. The framework presented in Figure 3.2 consists of two phases of analyses: SUTA (Setting Up The Analysis) and COTA (Carrying Out The Analysis).

SUTA

The SUTA phase is dedicated to the description of the problems, the identification of the objectives, and the specification of analysis conditions and the analytical approach. This phase may be described as the process which determines what types of analyses have to be carried out in the COTA phase. The major steps in the SUTA phase are discussed below.

1. Triggers
   Triggers that initiate a water management analysis have been described earlier.

2. Identification of planning objectives and criteria
   One of the fundamental premises underlying analysis for water management, is that the

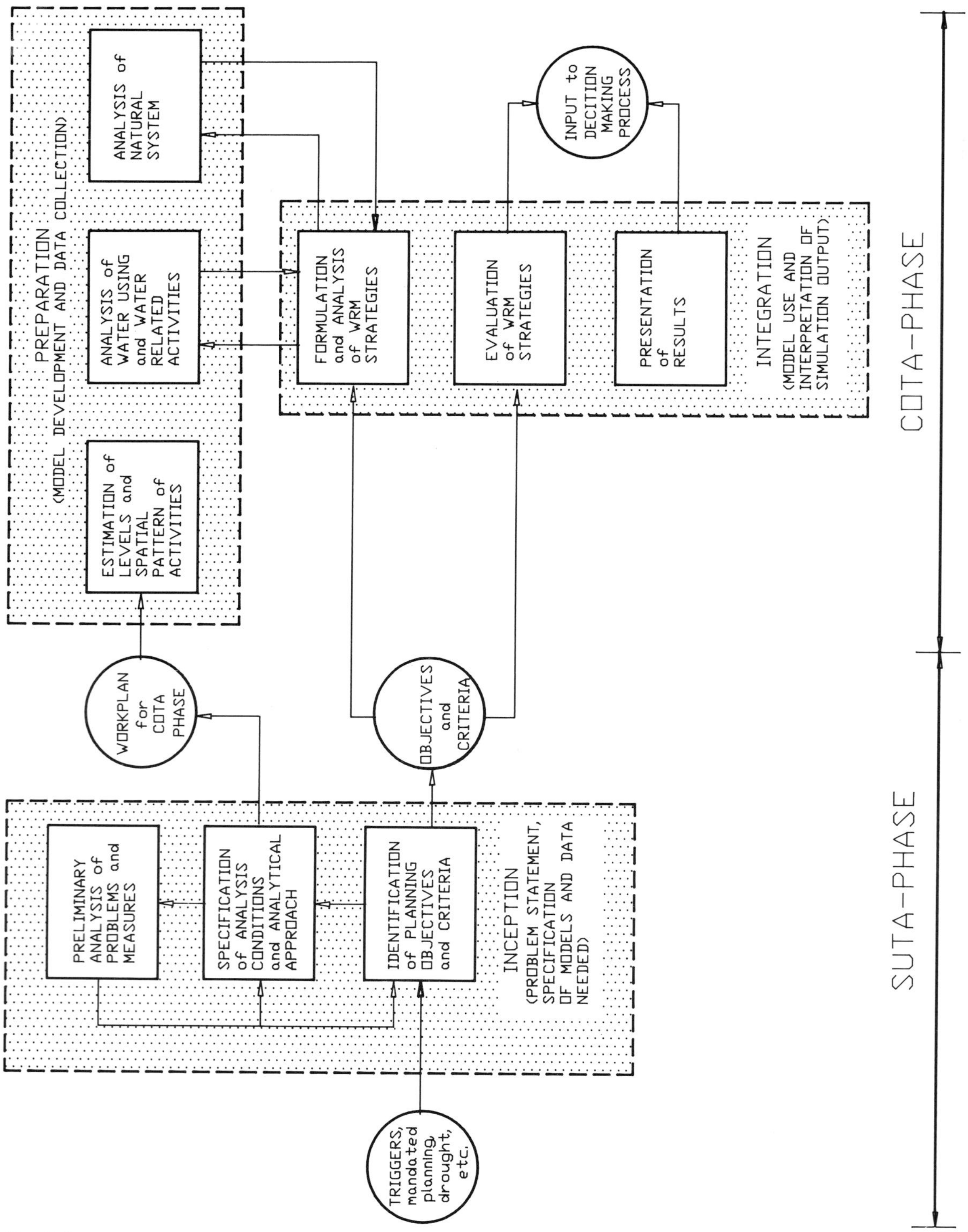

Fig. 3.2.
Framework for analysis for WRM studies.

analysis will be quantitative whenever possible. It is directed at the generation of alternative strategies. Criteria are needed to enable a comparison of strategies, with respect to the extent to which they could achieve certain objectives.

3. Specification of analysis conditions
   Analysis conditions refer to the complete set of conditions and assumptions under which an analysis is carried out. They include time horizons and area boundaries, discount rate and base conditions, hydrologic and meteorologic conditions, and various scenarios. There is always some uncertainty associated with the forecasting of economic, demographic, political and technological developments. Although such developments are exogenous to the system to be analysed, they do exert an influence on the behaviour of the system. To account for this fact, alternative developments or scenarios are developed and analysed.

4. Preliminary analysis of problems and measures
   The purpose of the preliminary analysis is to identify the relative importance of the various water-using and water-related activities, and to find out whether or not the water resources can meet the demands for water and water related outputs, in the present or future situation. The preliminary analysis aims at focusing the efforts during the COTA phase on the most important problems and the most promising measures. The specifications of the analytical approach will, to a large extent, be based on the results of this preliminary analysis.

5. Specification of analytical approach
   Based on the previous steps the output of the analysis can be specified (e.g. agricultural production, or the costs of measures). The level of desegregation (the amount of detail) is an important element in this specification. Given the required output an analytical approach can be specified (i.e. what data are to be collected and what models are to be developed). Also, other required resources should be specified, such as computer facilities, number of personnel and required disciplines and skills, time and budget. Finally the outcomes of the various steps should be combined and summarised in a work-plan for the COTA phase.

6. Feed-back in SUTA phase
   The previous parts of the SUTA phase must be checked with the decision-makers. Although such checking should be part of the process of SUTA, the final output of SUTA must be one to which the decision-makers agree. Particularly important in that respect is the work-plan for COTA. In case of inconsistencies, revision may be necessary or additional time and resources may have to be obtained.

COTA

The COTA-phase in its general form comprises six main activities.

1. Estimation of the levels and spatial patterns of activities.
   Any analysis must begin with an estimate of levels of population and economic activities, such as households, agriculture and industry, for the specified time horizons. Information on economic activities pertains to the final demand of society for the production of goods and services. These projections of national activity levels need to be desegregated into regional levels, accompanied by a further desegregation into sectoral activities.

2. Analysis of water-using and water-related activities.
   This main activity comprises the translation from levels of population and economic activities into quantities of water demanded and of residuals discharged. The output of an activity analysis is input into the analysis of natural systems and into the formulation and analysis of water management strategies. Input into the first analysis concerns the estimations of the demand for water and the residuals generated, which affect water quantity and quality in various bodies of water.

3. Analysis of natural systems.
   This activity includes the analysis of how the demands for water and water-related goods and services, and the measures to meet those demands, affect natural processes. The amount and the accuracy of available data are predominant factors in the overall accuracy of the analysis of strategies. Collecting, evaluating and processing natural systems data require a good knowledge of the represented phenomena, as well as of the purpose of the analysis. Inputs into natural system models involve the outputs of the activity analysis (i.e. the demand for water and discharged residuals, both at specific locations and in specific time patterns). The link between analysing activities and analysing natural systems is emphasised. The former must result in data inputs in the proper form for the natural system models used. Conversely, these models must be chosen and formulated in relation to the outputs which the analysis of activities can provide, the available data, and the desired outputs in terms of physical, chemical and biological indicators.

4. Formulation and analysis of water management strategies.
   This main activity includes various rounds of analysis. The first steps include a screening of individual measures, often based on criteria related to technical feasibility and benefit-cost relations. Subsequently, strategies are formulated and, in accordance with the criteria applied, impacts other than costs and benefits are estimated. These steps include an environmental impact statement. Finally, this activity results in a set of promising strategies and their impacts or consequences in terms of the criteria applied.

5. Evaluation of water management strategies.
   Decision-makers use multiple criteria in making decisions. Costs represent a major criterion but not the only one. The decision-makers should indicate the relevant criteria and attach relative weights to each. The analyst plays an important role in providing the framework for making explicit criteria and related weights. Analytical methods exist to contribute to the evaluation of strategies, such as cost/benefit analysis and multi-criteria analysis.

   A cost/benefit analysis represents a systematic enumeration and evaluation of all relevant, direct and indirect social benefits and costs of a given strategy. Benefit/cost analysis can be done in ways which provide information on objectives in addition to national economic efficiency. For example, the analysis can track the costs to those who will spend the money, both in terms of the geography of political jurisdiction and in terms of economic classes, and the benefits to those who will have the advantage, in the same terms. Benefit/cost analysis can be done under various types of constraints. For example, the objective function can be to maximise net benefits subject to preserving a particular free-flowing stream. If the analysis is repeated without the constraint, what is produced is the net economic benefit lost to society by preserving the stream. This does not represent the value, as such, of the free flowing stream; what it does represent is the minimum amount that society would have to be willing to pay in order to preserve the stream. In analogous fashion, benefit/cost analysis can be done under other constraints, such as specification of maintenance of some level of employment in a given region, or the provision of a certain amount of water to a reservation. Doing so indicates the trade-off between economic efficiency and the other objectives. All of these benefit/cost analyses can be done with multiple sequences of hydrologic events, in order to produce *expected values* of the outputs.

   Multi-criteria methods, such as expected value methods, permutation methods and concordance analysis aim at providing an instrument for a systematic evaluation characterised by the presence of multiple criteria or multiple attributes. The basic idea of these analyses is the fact that some of the project impacts are not necessarily translated into monetary units as is the case with cost/benefit analysis. A critical part of evaluating strategies is the financial analysis. The results of the financial analysis represent important inputs into the decision process for selecting a water management strategy.

6. Interaction with decision-makers and presentation of results.
During the SUTA and the COTA phases of the analysis a close interaction is required between the analysts and the decision-makers. Each individual main activity requires this interaction. The final stage of the analysis is the presentation of results. In this phase, explicit consideration is given to the kind of participants in the decision process and the nature and types of information needed or desired by those participants. Apart from the results of the analysis, information on the selection and definition of the characteristics of the system considered, on objectives and criteria, and on analytical approaches are also presented in this phase.

Computational framework

The SUTA phase includes an outline of the analytical techniques to be used in the different activities of the COTA phase. A computational framework aims at tying together those different analytical techniques. At present, a variety of techniques and models exists and more are being developed. The selection from this variety and the subsequent implementation should be:
- internally consistent (matching inputs/outputs);
- tuned to the objectives of the analysis;
- adequate in relation to the required level of detail;
- consistent with the fact that planning for water management should be considered to be a continuous procedure.

In water management analytical techniques from many different disciplines are needed (e.g. from hydraulics, hydrology, economics, statistics, agronomy etc.). Some examples of possible analytical techniques that may be included in a computational framework are:
- models to predict the demand for water (irrigation, drinking water);
- models to simulate water flows and levels in surface waters (flood routing);
- models to simulate water and contaminant movements in groundwater aquifers;
- response or impact models to predict the effects of water management strategies on specific water users (e.g. crop yield, industrial production, shipping movements);
- national or regional input/output models to provide the economic data base on national and regional economies for the analysis.

In the present context, a more elaborate overview of available techniques cannot be given. However, an example of a computational framework is given in Figure 3.3, showing the system diagram of the Policy Analysis of Water Management Study for the Netherlands (PAWN) (see Goeller et. al., 1983). This study was carried out from 1976 to 1979 and will be described in Appendix 7.

The analyst will use various analytical techniques to solve the individual models. As is the case with a framework for analysis, the computational framework also has to be problem specific. When carefully designed and rigorously applied such a computational framework may add considerably to an effective and efficient execution of an analysis. In particular, during the execution of the main activities of the framework for analysis, the development of a computational framework can be of great assistance in the integration of the individual contributions to the analysis. Moreover, the computational framework provides a valuable tool in the formulation and analysis of water management strategies, in particular with respect to impact assessment. It has to be applied to the water system at hand.

A water system is defined as the geographically demarcated, inter-related system of surface water, groundwater, underwater beds, banks and technical infrastructure, including the existing ecosystems and all the accompanying physical, chemical and biological features and processes. The demarcation lines of a water system of this kind are determined in the first instance on the grounds of morphological, ecological and functional relationships. Obviously, the less extensive such a water system is, the better will be the chances of an integrated plan. Problems should be approached on a regional or even local basis, without losing sight of the national objectives.

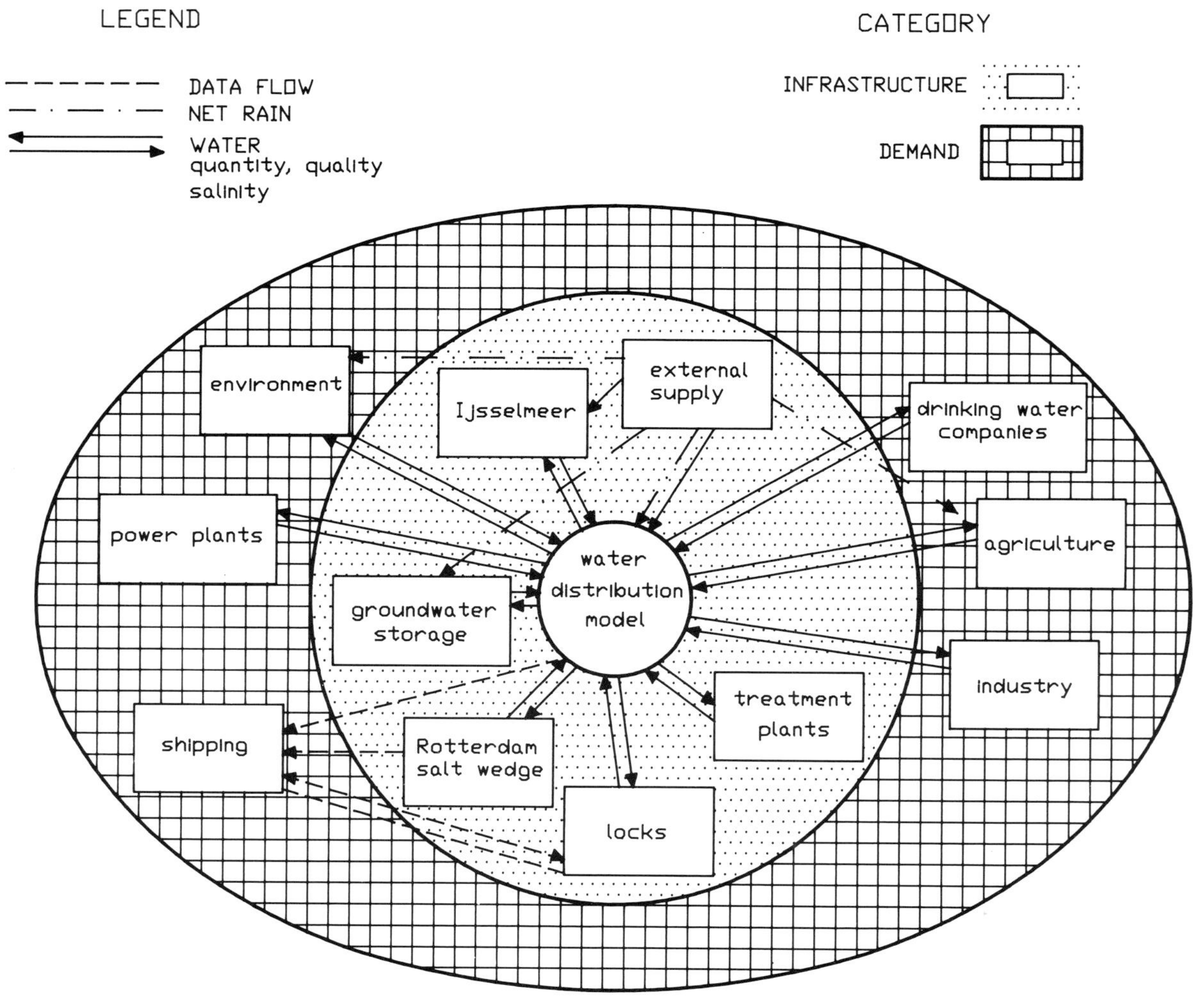

Fig. 3.3
System diagram of the PAWN study.

3.1.6 Practical consequences of an accelerated rise in mean sea-level

In Section 1.2.2, mention was made of a possible accelerated rise of the mean sea level as a result of global warming, caused by the so-called greenhouse effect. Such a rise would have many far-reaching repercussions for the water control and water management of deltaic and other low-lying coastal areas. When examining the practical consequences of a sea-level rise, subsidence should also be taken into account (see Section 1.3.4). The practical effects are determined by the *relative* rise of the sea-level. Water and land management will also be affected by changes in the local hydro-meteorological parameters, but these changes are the most difficult to predict.

The relative rise of the sea-level will have consequences in many fields, the most important being:
- the protection of the coast and lower river reaches against tidal levels and storm surges;
- the drainage, especially by gravity, of the low-lying coastal areas;
- the seepage of brackish and saline groundwater in these areas, both from the sea and local seepage;
- the sea water intrusion into open estuaries;
- the morphology and erosion conditions of the coastal strip, the estuaries and lower river reaches.

These consequences depend, of course, on the specific physiographic and socio-economic conditions of a particular low-lying area. In this respect, two areas can be identified, namely the areas where little human interference in the natural processes has taken place, and the areas where the hydrologic conditions have been substantially modified by man. In the former, which are often quite productive and densely populated, man has learned how to live with the flooding and the crops have been adapted to the prevailing hydrological conditions. Examples can be found in the large deltaic areas of South East Asia. The latter comprises areas where man has created an environment with full flood protection and artificially controlled surface water and groundwater levels so that the optimum conditions exist for a complete utilisation of the high potential of the delta. Examples can be found in Europe, Japan, China and North America.

A solution to the problems posed by a possible accelerated rise of the sea-level cannot be given. The following considerations lead to a reserved attitude with respect to measures taken on the short term.

a. Since about 1990, the predictions of the various General Circulation Models have been seriously contested by parts of the scientific world. Doubts have been raised about whether the observed increase in global temperature during the past two to three decades has in fact been caused by other effects than the greenhouse effect. It will take several more decades before this question can be answered with any certainty.
b. In principle, more reliable and more detailed models are conceivable but the amount of data required is so enormous that it cannot be expected that such models will become available within the next 20 or 30 years.
c. Continuous monitoring of the sea level in a large number of well-founded recording tidal gauges, well-distributed over all the oceans of the world, will clarify a number of questions, but it should be kept in mind that even the annual averages of tidal levels show considerable random fluctuations. These are due to variations in the wind patterns, temperature and salinity distributions in the ocean waters. As a result, neighbouring stations may show very different short term trends. Even rapid rises of the sea-level, caused by the greenhouse effect, can only be detected after several decades.

While a reliable prediction of a rise in sea-level will take 20 to 30 years, there will be another delay of 10 to 15 years before hydraulic counter measures can be executed. Nowadays, many years are required for the decision process, public hearings, environmental impact assessment, comparison of alternatives, authorisation, financing, etc. Abandonment of the threatened low-lying areas is not an acceptable policy from a socio-economic and political point of view. In these areas, with a clustering of population, the largest cities and ports of the world have been built and resettlement elsewhere would be unacceptable.

An accelerated rise of the mean sea-level may have far-reaching consequences for those areas where little interference in the natural system has taken place. These cases are mostly found in developing countries. A substantial relative rise of the sea-level will disrupt the natural processes of delta build-up and may make it necessary to install a flood protection system of embankments and an artificial drainage system with pumps. This will entail a change in the cropping pattern and agricultural practices and also a change in the way of life of the inhabitants. New possibilities for further economic development will be created but the realisation will require considerable investments and a change in human attitude.

The situation is quite different for the highly developed deltaic areas. A striking example can be found in The Netherlands where a relative rise of sea-level, or relative land subsidence, has been observed for many centuries. A land subsidence of three metres, and even more in large peat areas, has been recorded for more than 800 years. The consequences of this subsidence have been countered early on using primitive technology (e.g. manual earthworks, windmills, wooden box sluices, etc.). Later, these lakes and tidal embayments were reclaimed and are now at elevations of more than six metres below mean sea level. Storm surges on the Dutch coast may produce water levels of more than five metres above mean sea-level.

In areas of this type, a further rapid rise in the sea-level would not cause insurmountable technical problems and not upset the economy of the country. Figure 3.4 shows a possible adaption of the water management system in The Netherlands for a sea-level rise of several metres (Rijkswaterstaat, 1986). Essential elements of this plan are the construction of vanguard dams in the North Sea, parallel to the existing coast, and huge pumping stations to lift the waters of the Rhine and Meuse to the sea. The vanguard dams should take over the function of coast protection from the vulnerable sand dunes. The dams are located about 20 kilometres to the west of the existing coast. Between the dams and the (former) coast, new lakes would be created to serve as buffer basins for the temporary storage of flood waters from the rivers and also store saline seepage water from the sea. These lakes would be drained to the sea by pumping.

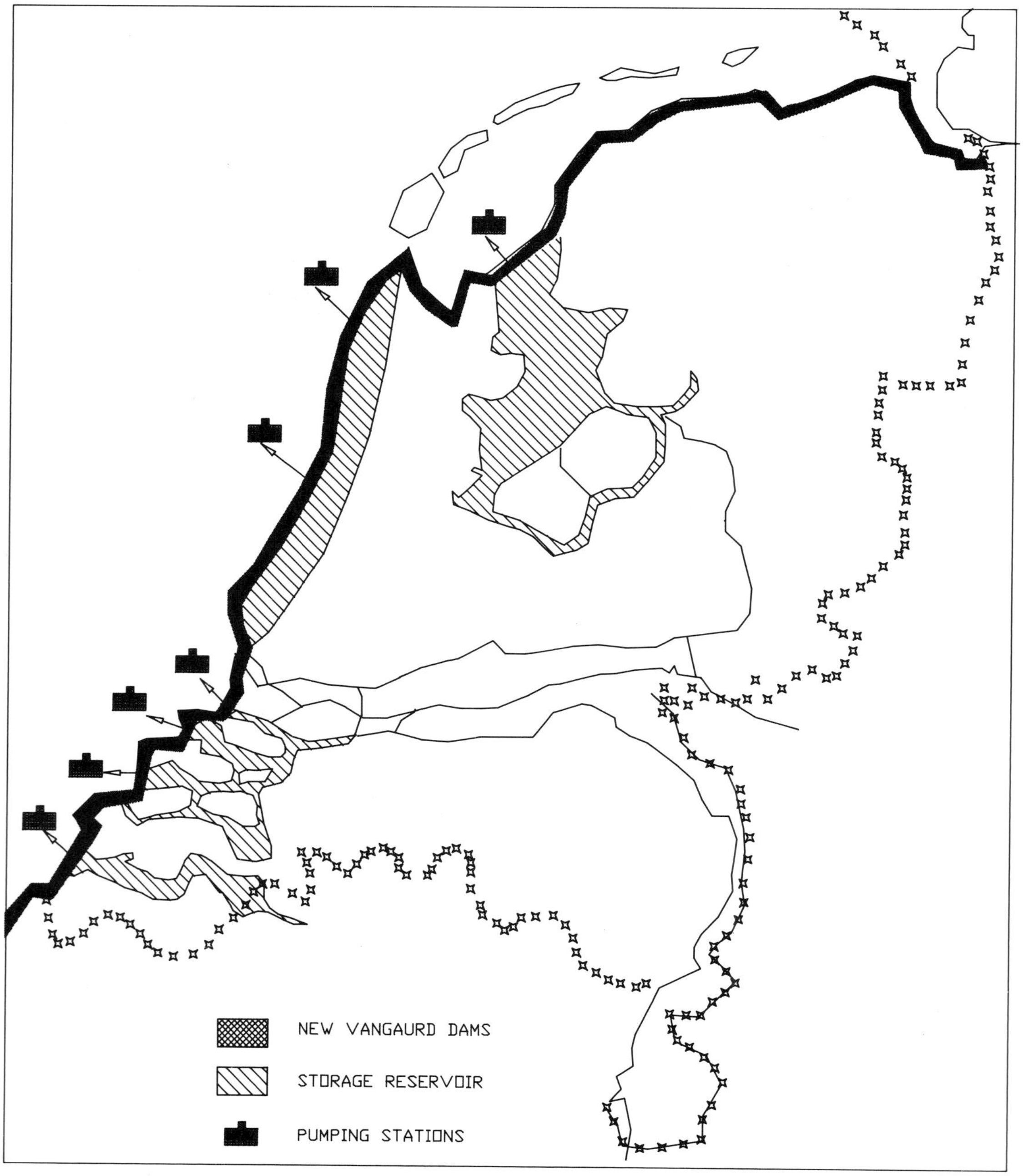

Fig. 3.4
A new water control system for the Netherlands.

3.2 Floods and Flooding

3.2.1 General

Throughout the world floods and flooding occur as natural phenomena which, in most cases, are not much appreciated by people living in the affected areas. Consequently, flood control and flood protection measures are introduced in many places to prevent the negative consequences of this flooding. These acts of nature as well as the reactions by mankind can be expressed in by a great number of words and expressions. Though most of these are well known to the hydraulic engineer acquainted with engineering works in deltaic areas, it is considered worthwhile to define some of the most common words and expressions before beginning to discuss the subject of flooding.

Flood: Flood is defined as the unexpected or periodic inflow into a defined region of more than average quantities of water due to storm surges, upland river discharges or local rain. They are characterised by increased current velocities and higher water levels. Floods in rivers generally show the characteristics of a passing long water wave.

Flooding: Inundation of areas which are normally dry, for a limited period of time, due to the occurrence of floods is referred to as flooding.

Flood control: Management of the inflow and outflow of flood-water into or out of a region, in such a manner that flooding, if any, is either kept to a minimum (return period extent) or occurs at a planned moment and during a planned period, is referred to as flood control.

Flood protection: This is defined as works especially designed and constructed to protect an area from flooding caused by floods.

Flood plain: This refers to the area which will be flooded periodically due to the passing of a flood wave or the temporary storage of flood water. In many cases the flood plain can be considered as a temporary extension of the flood discharge river channel.

Diversion channel: This refers to a natural or excavated channel which is used for the discharge of flood water during periods of high river stages (due to floods).

Bankfull stage: The stage (water level) at which the flood water in a certain reach of the river begins to flow over the natural levees or man-made embankments into a flood plain or polder is termed the bankfull stage.

Bank overspill: This is the volume of water which has spilled from a river into adjacent flood plains or polders.

Flood prone area: An area which will be flooded periodically and for which no flood protection measures have been taken, is referred to as a flood prone area.

Flood parameter: This is a parameter which characterises a flood, e.g. duration, or size (m^3) of a flood, highest water level, bank overspill (m^3) or river discharge (m^3/s) due to a flood.

Zoning: Legal and institutional measures aimed at restricting the use and occupancy (building, persons) of a flood plain in view of its flood discharge or flood storage function is called zoning. It is also practised to prevent the construction of permanent structures along the alignment of future or existing flood embankments.

Submersible dyke: This refers to a low embankment along a river or coastal water aimed at providing flood protection from low and frequent (flood) water levels, or during the early stages of a flood (resulting in delayed flooding).

Reasons for flooding

Whatever the type of flood or flooding, the phenomenon can always be described as follows: a volume of water enters a certain area and, as it cannot be discharged quickly enough, the water level rises and flooding occurs. Apparently, the elements volume and time play an important role. When dealing with floods people tend to forget that water is not compressible and that deltaic areas tend to be very flat. The fact that water is incompressible is known by everyone, but the consequences are not always fully realised. The huge masses of water causing flooding cannot be decreased. Aquifers cannot accept such huge masses within a short time span, and the water will not evaporate. Consequently, water must be stored locally, which is not very easy in deltaic areas. At most only a few metres of height will be available and large areas will be required to store significant quantities.

Apart from non-availability of storage space, flooding can also be due to the insufficient discharge capacity of the river system (river channels and flood plain). This can be due to several reasons:
- wet profile of the river system being too small;
- water depth on the flood plain being too small (hydraulic radius);
- bottom slope of the river channel and flood plain being too gentle;
- roughness of the flood plains being too large due to obstacles and bushes;
- high downstream water levels (sea-level).

In fact the relative importance of these factors follows from well known discharge formulae. Generally speaking, it may be stated that the flatness of coastal lowlands as well as their low elevation hampers the quick discharge of flood water towards the sea. The non-availability of storage space and impossibility of discharging flood water quickly towards the sea leaves the inhabitants of deltaic areas with high water levels during floods. When the embankments around polders are high enough and structurally stable there are no problems, as with time the water recedes and no flooding occurs. However, in many cases in the past, the safe embankment around a polder turned out not to be as safe as people thought. They were either too low and overtopping destroyed them or dyke breaches occurred for various reasons. In both cases the result was flooding and the formation of deep erosion gullies (scour pools) in the land behind the breach.

Flooding may occur because of overtopping of an embankment. When this is limited in height and time (say up to 20 cm during 3 hours), the embankment may be able to continue its function as a water retaining structure, while at the same time limiting the flooding of land inside the embankment. In many cases in the past, embankments have been destroyed along rivers during high water levels or in coastal defence works by high water levels in combination with wave attack.

Investigations in the Netherlands have shown that the collapse of an embankment (dyke breach) can be due to the following reasons:
- the crest height of embankment is too low; crest overspill destroys landward slope by erosion and the dyke body collapses (slips);
- the crest height of the sea defence is too low; the overtopping of water due to wave run-up has the same effect as above;
- the earth mass of the embankment is saturated (due to long lasting high water levels or rain); in combination with steep landward slope this condition is not stable and leads to failure;
- insufficient compaction of the embankment fill or wide and deep cracks after a dry spell provide easy access to rain and flood water; this leads to seepage and piping and ultimately to

the collapse of the embankment;
- erosion of the foreshore may result in a slip of the toe (the seaside base of the embankment), followed by slips in the dyke body or liquefaction.

The collapse of an embankment can be prevented by designing a good profile which is well constructed, and well maintained afterwards.

Types of floods and flooding

People speaking about a flood do not always know why it occurred and from where it originated. In fact the flood wave passing through a river and its adjacent floodplain is relatively easy to understand. Somewhere upstream it rained a lot or the snow melted and the water travels down the river towards the sea. Locally, the phenomenon is manifested by a gradual (or abrupt) change in water level. The water level first rises and, after reaching a maximum, it falls again.

The shape of the hydrograph depends on the characteristics of the particular catchment (hydrology, climate, topography, etc.). As already mentioned, interest is not only centred on the water level, but also on maximum discharges, shape, volume, duration and time of occurrence.

Flood waves travelling down a river may cause flooding. In principle this can be due to either high water levels or impeded drainage. Flooding due to high water levels is more common. As soon as the water level is high enough bank overspill will commence and flooding of riverine lands occurs. The high water levels in the main river may, however, have other consequences as well. They also give rise to backwater effects in tributaries which may result in flooding along these tributaries. In addition, high water levels will prevent natural drainage of riverine lands. When rainfall occurs in the period that a flood wave is passing, the rainfall cannot be discharged and flooding resulting from impeded drainage takes place.

Obviously, flooding due to rainfall will also occur when the drainage network is not sufficiently developed to remove rainwater from the field, through the network of drainage channels out of the area.

In coastal areas flooding can also be caused by storm surges. In this case it is the combination of high tides and water level set-up due to storm surges, as well as the specific shape of the estuary, which leads to extreme high water levels resulting in flooding.

In deltaic areas flooding is mostly caused by a combination of storm surges at sea and an increased river discharge due to rainfall on the catchment area. Each of these phenomena may not be very dangerous when taken separately but their combination leads to flooding.

Flood parameters and frequency curves

Selection of a flood parameter implies that sufficient data are available about floods and flooding which occurred in the past. In fact the whole phenomenon of floods and flooding must be thoroughly understood before a flood parameter can be selected and frequency curves drawn.
A flood parameter should represent a flood in such a manner that the different values of the parameter valid for the different floods, reflect the magnitude of these floods in relation to each other. Flood parameters can be:
- maximum discharge during flood at a certain location;
- maximum water level during flood at a certain location;
- maximum extent of flooding in a certain area;
- duration of flooding in a certain area;
- bank overspill in a certain area;
- duration of the actual flood wave travelling down the river or of a storm surge in a coastal area.

The results of hydrological and hydraulic studies should indicate which flood parameter best represents the flooding phenomenon. It is known from experience that the hydrological and

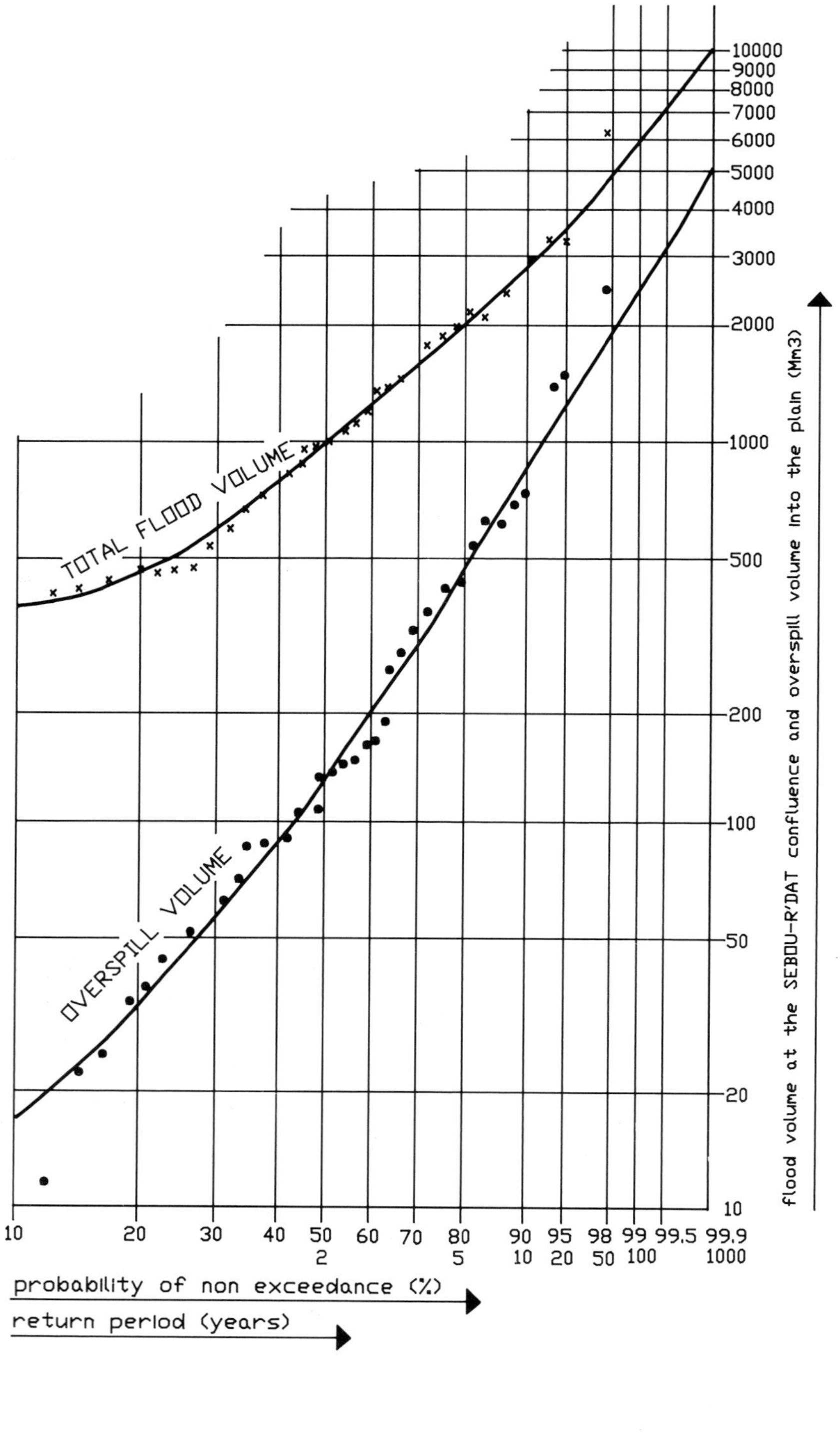

Fig 3.5
Frequency curves for flood and overspill volumes for the River Sebou in Morocco.

topographical situation as well as the size of the area, determine to a major extent the flood parameter to be used. During studies in Morocco, it turned out that flooding of the Rharb Plain by the Sebou and Ouerrha rivers was best represented by the overall bank overspill into the plain. In the case of the Paraná-Paraguay studies, which covered a much larger area, no less than seven locations were earmarked at which the maximum annual water level could have been considered as the representative flood parameter for a certain area (such an area was called a sub-area).

Frequency curves can be drawn for the values of various flood parameters attained during recorded historical floods. In addition, the relationship between various flood parameters can be established such as:

- total discharge over and above a certain flood level vs. the bank overspill;
- maximum discharge during a flood vs. the duration of flooding;
- maximum discharge during a flood vs. the bank overspill;
- maximum water level during a flood vs. the extent of flooding.

As mentioned earlier, the frequency curves for the selected flood parameter play an important role in the economic calculations. In Figure 3.5 the frequency curves for the total volume and the overspill volume into the Rharb Plain are presented. The data for the overspill curve were found by comparing the overspill volumes (calculated for a number of floods by means of a mathematical model), with the volume of the same floods above 2,100 m^3/s as obtained from the hydrographs at the Sebou-Ouerrha confluence.

The selected parameter for the Paraná-Paraguay studies for each sub-area was the water level. In Figure 3.6 the frequency curve for the highest annual water levels at Paraná (sub-area A7) is shown.

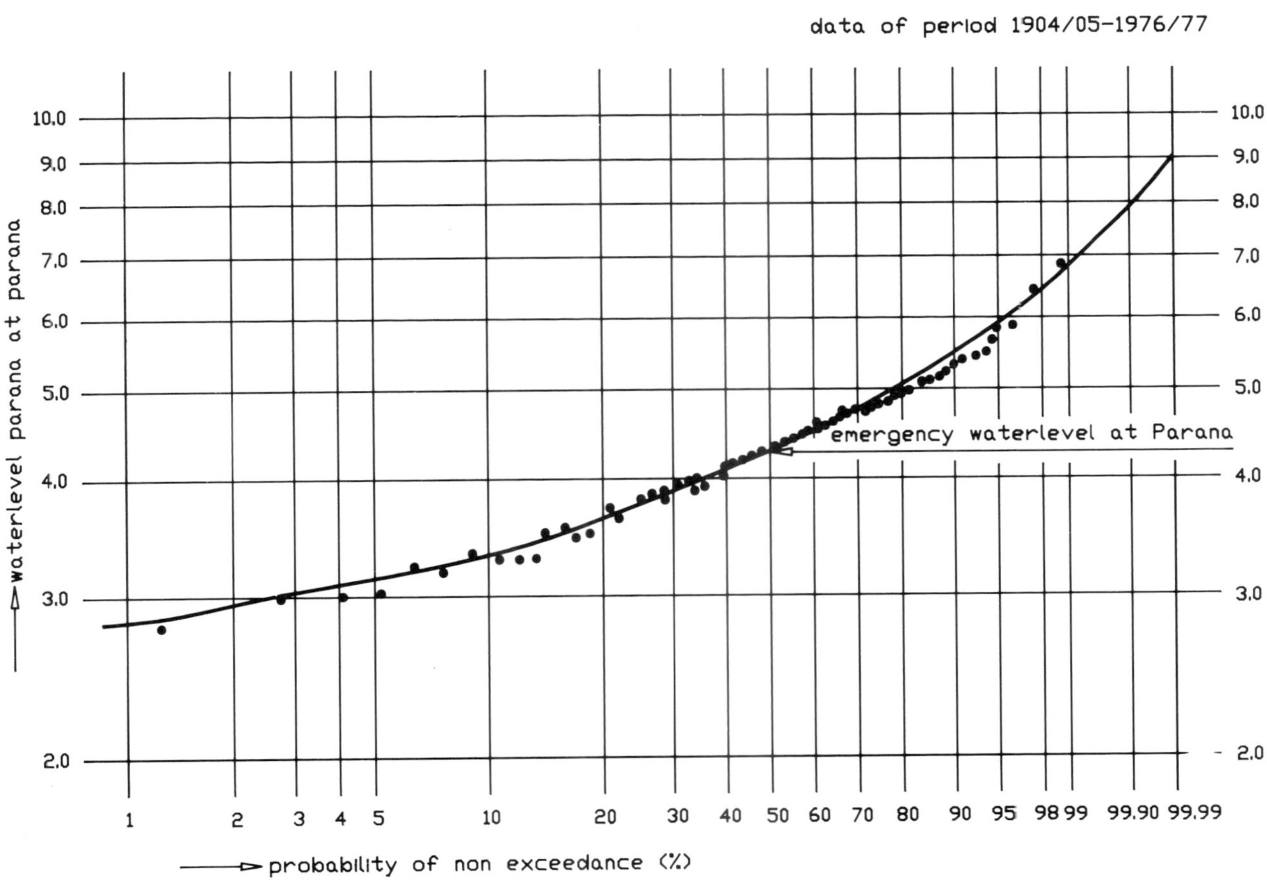

Fig. 3.6
Frequency curve of the highest annual water levels at Paraná, Argentina.

3.2.2 Flood control

In the foregoing sections the analysis of floods and flooding was discussed. Such an analysis is made by collecting first of all a large amount of data on topography, bathymetry, geology, hydrology, agronomy, agro-economy, regional and urban development plans, and last but not least, development plans for water resources of the river basin and its delta as a whole or, when there is no river, of the coastal strip. Floods, flooding and damage caused by flooding are then studied through modelling as described in the last section.

As soon as the flooding phenomenon is thoroughly understood the designer is in a position to formulate a number of physical measures for arriving at a certain degree of flood control. Obviously, some of these measures would have been already provisionally formulated, at an early stage, in order to provide a reference point for the study of the various disciplines. However, it is now possible not only to formulate more definite flood control works, but also to evaluate these on their technical and economic merits.

Basically, there are five different types of physical measures which can be adopted for flood control purposes:
- watershed management;
- temporary storage of flood water in reservoirs;
- creation of floodways (bypasses);
- increase of conveyance capacity;
- flood plains.

These five types of flood control measures will be looked at individually but a combination of two or three will also be discussed. In Figure 3.7 the various possibilities for flood control as well as for flood protection are shown.

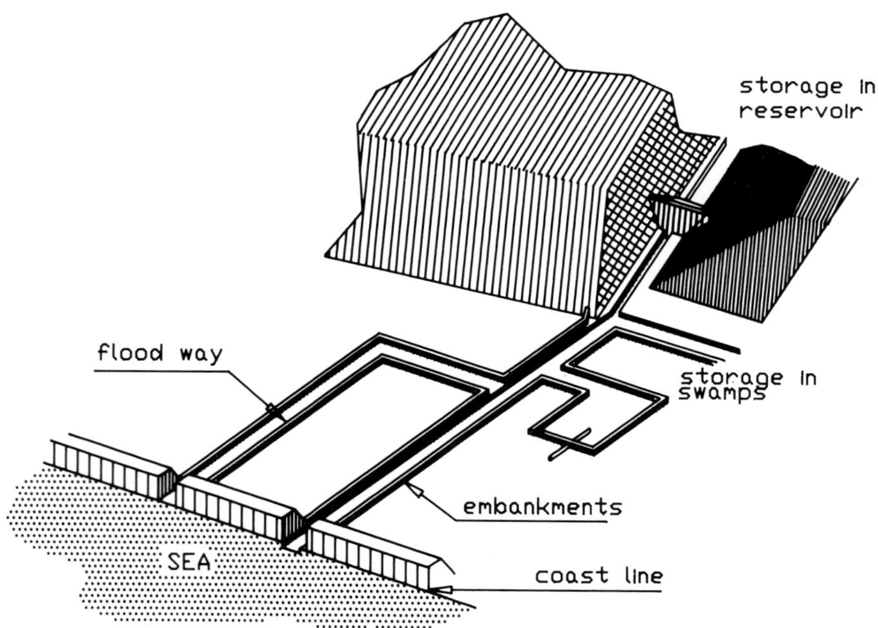

Fig. 3.7 Various possibilities for flood control.

Generally speaking, it may be stated that reservoirs and floodways have a regional impact on flooding while embankments are only effective locally. Also, the latter sometimes results in more serious and frequent flooding elsewhere.

Watershed management

Watershed management is the management of the watershed in a catchment such that erosion and flash floods are reduced to an acceptable minimum or disappear altogether. Watershed management in a catchment is executed by means of soil conservation and erosion control measures. Many countries suffer severe damage to their land area because of soil erosion of one form or another. Water erosion in which soil particles are loosened and detached by the impact of rain and flowing water causes heavy soil losses through sheet erosion, rill and gully formation and landslides. Although the eroded sediment can have a beneficial effect when deposited in low-lying areas (like deltaic areas), the more common effects of water erosion are the degradation of the upland areas, flooding of downstream river valleys because of flash floods, and rapid siltation of reservoirs. These are costly negative effects which have severe implications. The serious nature of soil erosion can be illustrated with some figures from India. About 175 million hectares (53% of India's total land area), are estimated to be subject to wind and water erosion and some 6,000 million tons of material are lost annually from the sub-continent.

From the point of view of floods and flooding, soil erosion produces the following negative effects:

- rapid run-off of rainwater as it can no longer be stored in the disappeared soil, and because of the pattern of erosion gullies which have been established towards the rivers;
- rapid siltation of reservoirs in which floods could otherwise be temporarily stored;
- increase in elevation of the river bed necessitating higher embankments along the river.

Soil erosion is encouraged by deforestation of large areas because of the need for fuel for an ever increasing population. As large areas are involved, social problems are manifold. Also, the positive effects of soil conservation and erosion control are only apparent after a number of years. Thus, in many countries, the results of watershed management on the whole, are not very encouraging so far. To complicate matters, preventive and curative measures to be taken, may differ between micro-catchments because of the many different aspects involved, like topography, soil structure, climate, vegetation, land use, tenurial rights, cropping patterns, etc. This in turn implies that many different surveys and studies are needed before effective measures can be implemented.

Finally, the problem is complicated by international borders. To prevent the arrival of high floods in certain deltaic areas, watershed management has to be carried out in many cases by other countries.

Temporary storage of floods in reservoirs

As such, it does not matter whether the reservoirs are located in the upper part of the river (in the mountains), or in the middle or lower reaches. In practice, however, reservoirs located in the middle or lower reaches are not very effective, as the land is generally too flat and the reservoir must be very large to provide substantial storage. Obviously, a large reservoir reserved solely for the storage of flood waves and, in many cases, occupying fertile lands is not always looked upon favourably. In addition, as the lifetime of such a shallow reservoir is generally limited, because of rapid silting-up, this solution is mostly disregarded. This leaves the reservoir located in the mountains as a possible solution for flood control. This would be feasible provided the community, aiming at flood protection in the middle and lower reaches of a river, has the power to carry out works in the upper reaches. However, this is not always the case, as large rivers are, in many cases, also international rivers. Examples of countries facing this problem are: Bangladesh, which can only hope to achieve a certain degree of flood control if large reservoirs were built in Nepal to check the floods of the rivers Ganges and Brahmaputra, and Argentina and Paraguay, who do not know what the effects of Brazil's works on the river Paraná and its tributaries will be.

For the Rharb study in Morocco, all possible reservoir sites, identified in the upper reaches of the rivers Ouerrha and Sebou and their tributaries (see Figure 3.8) were scrutinised for flood control purposes.

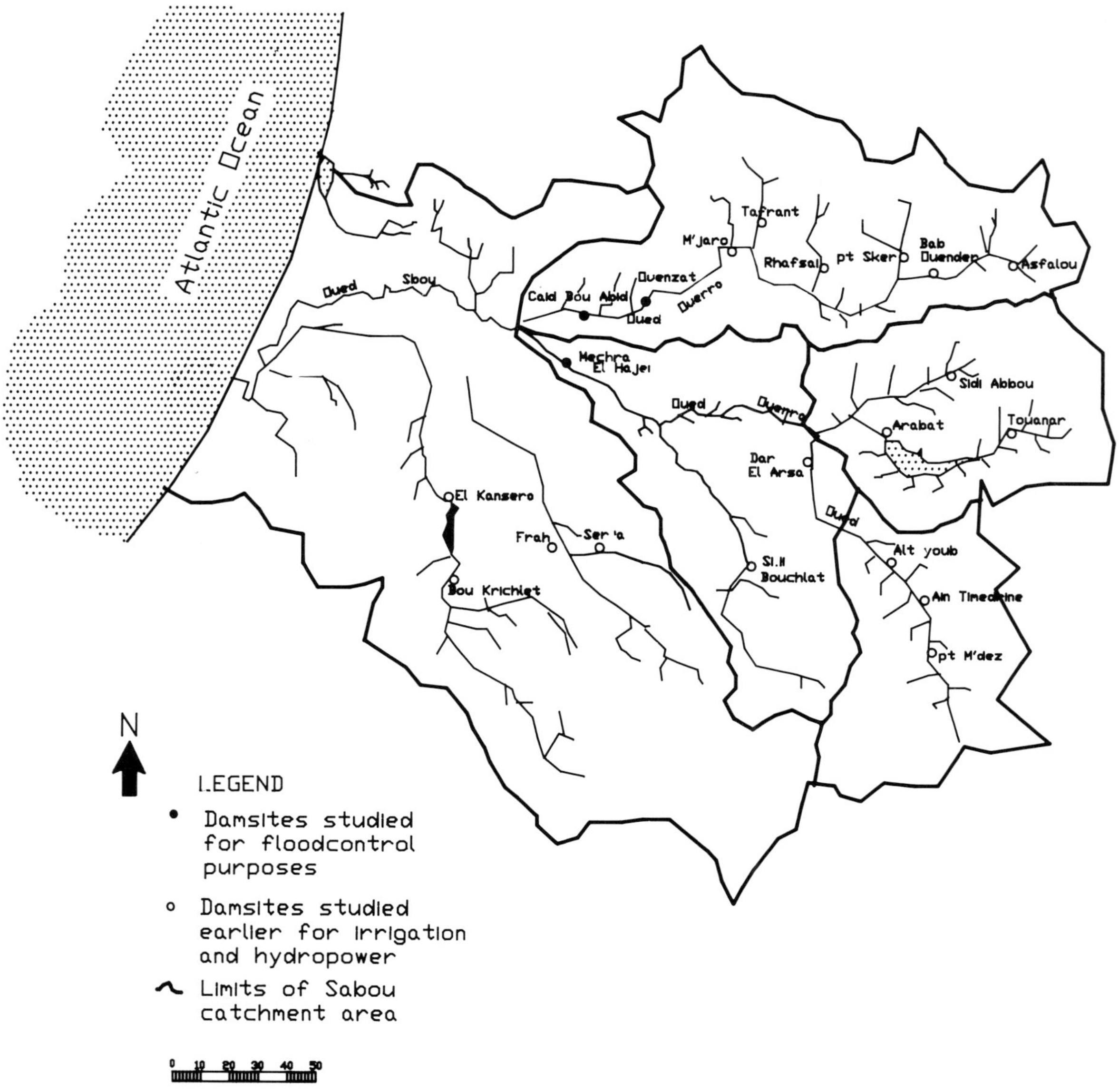

Fig. 3.8
Reservoirs planned in the Sebou basin, Morocco.

For each reservoir, the size (i.e. the height of the dam) was varied and its hydraulic effectiveness in preventing flooding in the Rharb Plain calculated. In Table 3.1 various particulars of this study are given. An interesting feature is the interdependence of the reservoirs as far as flood control is concerned. The hydraulic effectiveness of Rhafsaï (260) is 36%, and of Bab Quender (350) 43%, but if operating together, they have an effectiveness of 55-60% and not 79% as would be expected.

Reservoir	Code (Volume in $m^3 \times 10^6$)	Hydraulic Effectiveness in %
M'Jara	Rmj (800)	80
	Rmj (1600)	73-90
	Rmj (2100)	87-94
Mechra El Hajer	Rmh (200)	33
	Rmh (400)	41
Rhafsai	Rr (260)	36
	Rr (210)	34
	Rr (160)	31
Bab Quender	Rbo (350)	43
	Rbo (175)	35
Tafrannt	Rt (260)	37
	Rt (160)	32

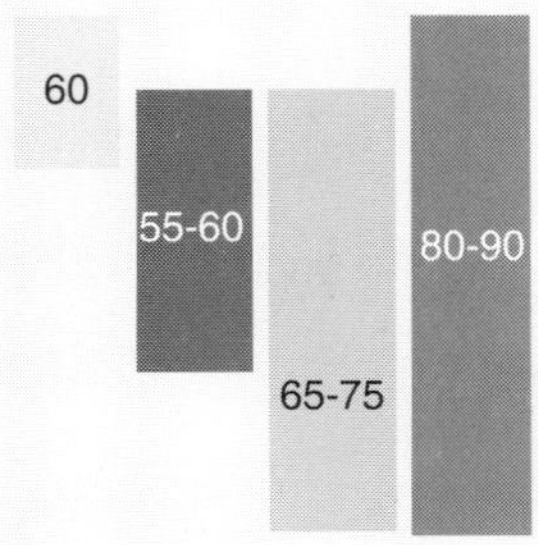

Table 3.1
Hydraulic effectiveness of combined use of reservoirs in Sebou catchment, Morocco.

As already stated, the idea of utilising reservoirs for flood control in the upper reaches, was disregarded in the case of the Paraná/Paraguay studies. However, it is possible to create a very large reservoir at the site of a natural depression, the Iberá marshes (see Figure 3.9).

Floods, coming from the Upper Paraná and entering the Yacyretá reservoir (soon to be constructed), would be diverted to the Iberá reservoir. The Iberá reservoir in turn, would discharge through the river Mirinay into the river Uruguay and through the river Corrientes into the Middle Paraná. Though, from an ecological point of view, the siting of a large reservoir in the Iberá marshes is not favoured, it is, technically speaking, a sound solution as will be shown below.

Floodways and flood plains

By creating floodways, flood plains, or alternatively, by fully embanking a river, an attempt is made to enlarge the discharge capacity of the conveyance system. These flood plains are mostly situated adjacent to the river, as is the case in Argentina and the Netherlands. This however, is not always possible and there are also situations where the flood water can be diverted through an old river course (bypass) to a point downstream where it re-enters the river, or to a second outlet into the sea. Such floodways are also called flood diversion schemes.
In Argentina, it was proposed to divert the flood water to another river basin while in Morocco, the combination of a movable weir (already required for facilitating the intake of water for irrigation), and a floodway to a newly constructed outlet into the sea seemed to be attractive, as well as a diversion channel (bypass) on the left bank (see Figure 3.10).

Again, generally speaking, it can be said that floodways are only economically justified when formed by natural processes. In this case extensive excavation works are not required and agriculture is not seriously affected by periodic flooding of the land. The rural society of the past had learnt to live with floods (for instance in the Nile delta in Egypt), and agricultural practices were adapted accordingly (e.g. the *floating rice* in Bangladesh).

At present it may be difficult to introduce a floodway where none already exists. Advanced agricultural production will not allow such flooding, or alternatively, the landowners will exaggerate the annual damage to crops. All the same, from a technical point of view floodways are an efficient means of flood control.

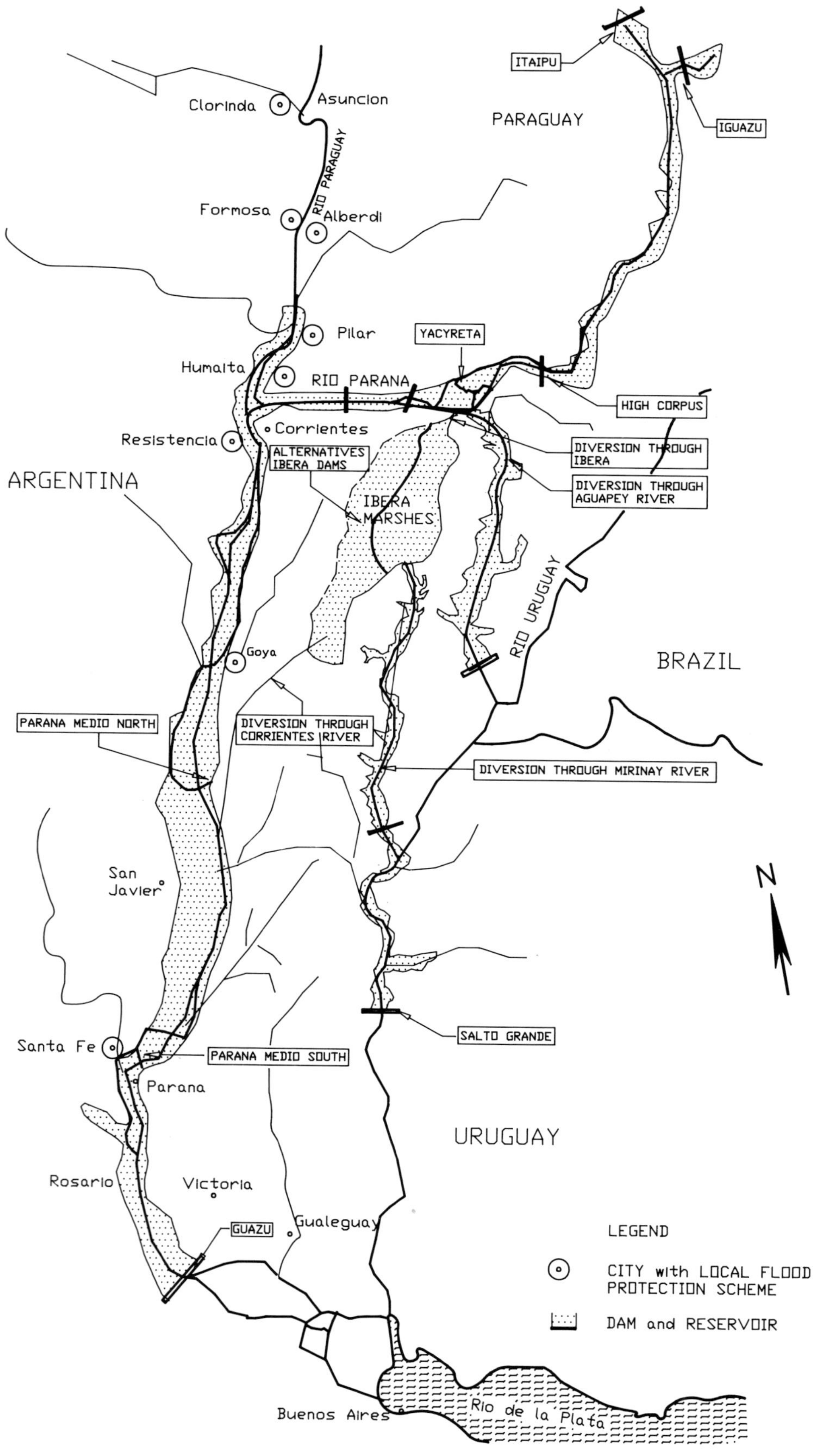

Fig. 3.9
River Paraná and Paraguay and diversion possibilities to River Uruguay.

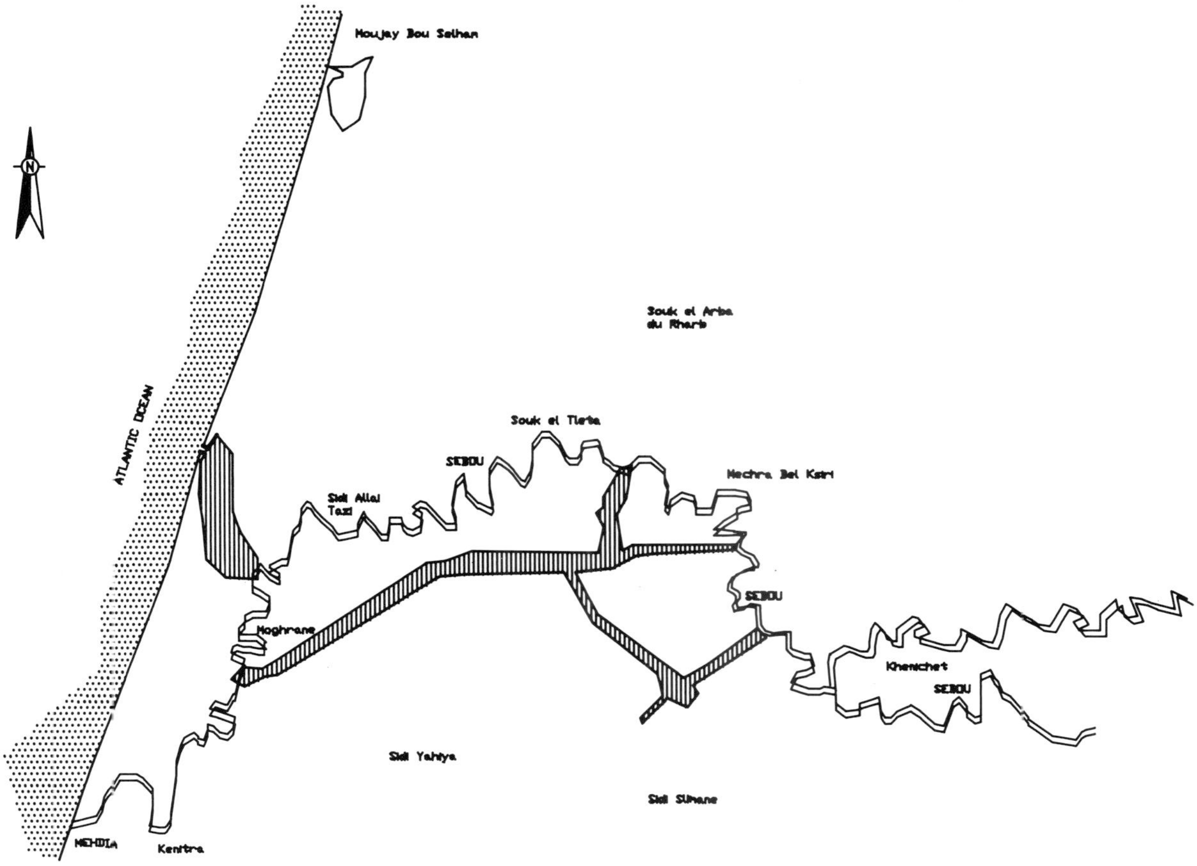

Fig. 3.10
Diversion of flood water from the River Sebou, Morocco.

Increase of Conveyance Capacity

The increase of conveyance capacity in a river or estuary to achieve a certain degree of flood control, is mentioned here for the sake of completeness. As previously mentioned, the enlargement of the wet profile of the river, widening of flood plains, increase of the slope of the river bottom and decrease of the roughness on the flood plains, will individually and collectively contribute to an increase in conveyance capacity.

Relatively cheap improvements like removing local bends, and clearing the flood plain of obstacles and bushes, will no doubt be worth considering when contemplating flood control measures. Other measures like lowering of the flood plain or deepening of the river channel are either costly or have to be repeated periodically (dredging), and the result (degree of flood control) is often marginal. Moreover, in the tidal zone the increased conveyance capacity will increase the inflow of tidal waves and wind set-up caused by storm surges. Thus, its effect on flood control in the coastal zone would be negative!

Combination of measures aimed at flood control

When combinations of measures aimed at flood control are discussed, it must be realised that these govern mainly the measures which can be taken to control flood waves arriving at the project area from upstream. The floods arriving from downstream (tidal waves, tsunamis and storm surges), cannot be controlled in the real sense. However, flood protection works are possible in this case.

By combining different flood control measures, it is possible to achieve a high degree of flood control which is not always easily obtainable by uncombined means. Moreover, such a combination may comprise the most economical solution for flood control in the circumstance. There are three reasons for this:

- to obtain the desired high degree of flood control by a single protection measure is not always physically possible or is too difficult and costly;
- the shape and magnitude (duration, maximum discharge) of the flood wave sometimes require the combined input of storage and floodways in order to control it;
- the complexity of the river system:
  - flood waves originating from different sub-basins but ultimately travelling on the same downstream reach of the river;
  - impossibility to divert (part of) a flood wave to another basin during a certain period, because of floods occurring in the latter basin at the same time.

It is impossible to give clear-cut solutions for such situations. It goes without saying that the larger the rivers and floods, and the more complex their hydrological situation, the more difficult it will be to obtain a high degree of flood control. This statement can be illustrated by presenting in global terms, the hydraulic effects of different flood control works for the river Paraná and Paraguay. Before this illustration, the limits of diversion to another catchment as well as the combined effect of storage and diversion are explained.

All discharges exceeding the discharge capacity of the river downstream of a diversion point will be diverted. Diversion capacity is usually not the limiting factor but during periods of flooding in the other catchment, diversion may not be possible and flooding will occur (see Figure 3.11). It is evident that, if storage capacity is available,it should be reserved for storing the peak of the flood wave that cannot be diverted because of floods in the other catchment.

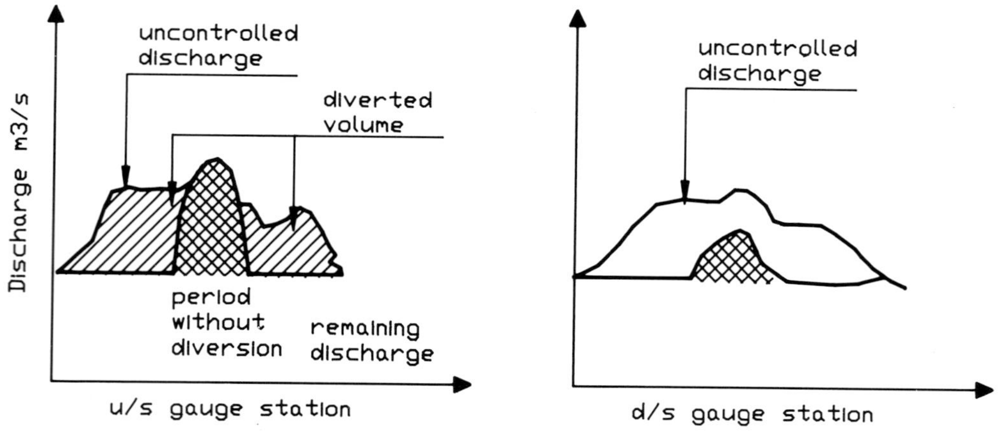

Fig. 3.11
Hydrographs upstream and downstream of diversion point.

The floods travelling down the river Paraná, downstream of the confluence of the Paraná and Paraguay rivers, are very large: the one in 50 years flood, for instance, has a maximum discharge of 42,000 m^3/s, a volume of 80 km^3 above the emergency level of 5.50 metres, and a duration of

170 days. The annual hydrographs at Corrientes are the result of superposition of floods originating from the Alto Paraná (Brazil), the river Iguazú and the river Paraguay. The floods cannot be controlled easily due to the large volumes, high discharges and large areas involved. In fact only a combination of storage, diversion and embankments will provide the desired degree of protection. After looking at a large number of possibilities with respect to siting as well as sizing of storage schemes, diversion schemes and embankments, the following basic schemes were examined in more detail for varying sizes (see Figure 3.9):

- diversion of flood water from the Middle Paraná towards the river Uruguay through the river Aguapey;
- storage reservoirs on the river Iguazú;
- reservation of storage for floods in the planned Corpus dam on the Middle Paraná;
- storage/diversion schemes in the Iberá marshes (diversion to river Uruguay through the river Mirinay);
- embankments around the towns.

Combinations of these five basic schemes finally resulted in sixteen typical flood schemes, which have been evaluated on the basis of various criteria. The hydraulic effects, for three of these sixteen schemes, are presented below:

Scheme no. 040 would consist of a diversion weir in the Yacyretá reservoir, an excavated canal from the weir to the Aguapey river, and finally, 4 dams in the river Aguapey. Diversion would start as soon as a discharge of 14,610 m^3/s is exceeded in Posadas. This scheme would have an effect on frequent floods, but this effect would sometimes be reduced by simultaneous floods in the Uruguay river. Economically speaking, 34% of the flood damage in the non-project situation would be avoided. The investment cost would be US$ 354 million (1979).

Scheme no. 410 would include all three elements of flood control: diversion, storage and embankments. The latter were introduced to provide flood control along the Paraguay river. Though it is now possible to divert 10,000 m^3/s and store 14.6 km^3 of flood water in a reservoir on the river Iguazú, the damage avoided only amounts to 61% while the investment cost is increased to US$ 1,801 million (price level 1979). The hydraulic effectiveness of this scheme, measured at Corrientes, is 60%.

Scheme no. 351 By constructing a large reservoir (area 6,000-11,000 km^2, gross volume 65-117 km^3) in the Iberá marshes, it is possible to divert large volumes of water from the Yacyretá reservoir. After temporary storage, these can be conveyed to the Uruguay river through the Mirinay river.

After having studied a large number of alternatives, the scheme finally selected for more detailed evaluation was:

- diversion capacity Yacyretá - Iberá 30,000 m^3/s
- storage in reservoir at a maximum storage level 73.5 metres 117 km^3
- available volume for the storage of floods 75 km^3
- diversion capacity to the Mirinay river 10,000 m^3/s

If constructed, together with the embankments around the towns along the Paraguay river, the investment cost would amount to US$ 2,595 million (1979), while 68% of the damage would be avoided in the project area. In Table 3.2 some of the effects of these three schemes on floods have been summarised.

Code no. of scheme	Volume diverted annually		Decreases obtained at Corrientes		
	average (km^3)	maximum (km^3)	average annual discharge (m^3/s)	av. discharge of 10 largest floods (m^3/s)	max. water level of 1 in 10 year flood (m)
040	15	60	2,544	3,510	0.5
410	17	77	3,488	6,224	0.8
351	21	85	4,874	10,000	

Table 3.2
Results of flood control on Paraná-Paraguay system.

3.2.3 Flood protection

The difference between flood control and flood protection follows from the definitions given earlier. Flood control aims at controlling (completely or partly) the flood, but in designing flood protection works the flood is considered as an input, a basic data which cannot be changed. Floods may originate in the catchment of a river (upstream floods) or at sea. At times, floods from both sources may arrive in deltaic areas simultaneously. Unable to carry out flood control measures the inhabitants of these areas have limited options at their disposal. They may either construct mounds, structures on piles, or build dykes or dams. Each of these flood protection measures is elaborated upon in the following sections.

Mounds

Mounds constructed by hand and consisting of clay or sandy material found in the neighbourhood, are undoubtedly the most primitive means of flood protection. At present they are still in use in parts of rural Bangladesh where annual flooding can be expected because of a combination of heavy rainfall and bank overspills from large rivers. Provided they are high enough, these mounds will be relatively safe as they cannot be breached like dykes. Also, unexpected flooding will not occur unless they are built in a cyclone prone area.

Bangladesh, south-western region. Rehabilitation of 3500 kilometres of coastal embankments by manual labour.

Cyclone shelters

Cyclones, originating in the Bay of Bengal, can induce a water level rise of several metres within a few hours. The coastal embankments which in the rural areas of Bangladesh will be overtopped once every 20 to 30 years, are usually too low to retain the high water levels of a cyclone. The stormy weather together with the high water wave travelling over the low-lying flat lands of the coastal area, does not provide much time and opportunity for travelling inland to higher places or to temporary accommodation in boats. As such, the idea has been conceived to build so-called cyclone shelters. In principle, these cyclone shelters could be earth mounds, built specially for this purpose, on which some shelters are erected. In practice, however, they are concrete and brick structures on stilts, consisting of a building or buildings of several floors and a large open area (for cattle). Obviously, stairways or ramps must be constructed to allow cattle to climb to the upper levels. A cyclone shelter should be large enough to offer shelter to all people and cattle living in a radius of a few kilometres around it. It should have a food store, a safe drinking water supply, and sanitary facilities for a large group of people. It should be possible to cook meals and it should be operated and maintained by the local authorities as part of a flood response planning programme.

A cyclone shelter can serve as a community building (village centre, dispensary, mosque, school etc.) in normal times as long as the purpose it was built for is not jeopardised. Cyclone shelters are only useful as a non-structural measure against flooding if (a) they are large enough for people and their belongings, including cattle, (b) the network of shelters is dense enough in a flood prone area to give people the opportunity to reach a shelter with a minimum of warning time, (c) everybody has access to it and (d) the mentioned facilities are indeed operational in times of distress.

Submersible dykes

The submersible dyke is a low embankment (1 to 2 metres high), between a river and its flood plain, or around a polder. The dyke will be overtopped and subsequently submerged almost every flood season. Its function therefore, is definitely not to retain high water levels during a long period of time. Material used for the inside and outside slopes, and slope "protection" (in most cases grassing on a clay top layer will suffice) should prevent serious damage during periods of overtopping. The idea behind the use of submersible dykes is to postpone the flooding until the crops on the land have been harvested.

Reliable hydrological records on daily water levels at the start of flood seasons can indicate how high submersible dykes should be in relation to harvesting time and rising (flood) water levels. In many cases (e.g. in the Netherlands) these submersible dykes will never become full fledged flood retaining embankments, as the lands which they temporarily protect against flooding are needed during the flood season as flood plains, i.e. for the storage and discharge of flood water.

Embankments around flood prone areas

Embankments are also called levees, seawalls, sea defences or dykes. In this section the word embankment will be used to illustrate the flood protection function along rivers. Only when the protection against wave attack and high tides is part of the function of embankments would the term sea defence be used. These sea defences will be discussed in a later section.

Embankments around flood prone areas are part of what is called impoldering, and have been the standard solution for local protection against flooding through the centuries, in many river deltas throughout the world. There is nothing wrong with this solution provided that the river retains sufficient space (floodway, floodplain) for the discharge and storage of flood waves, the embankments are well maintained and flood levels are monitored. However, these points also indicate the weakness of the system. In flat low-lying areas, the river may require its storage and large flood plains at the time of floods (as is the case in Bangladesh), in which case only limited impoldering is possible.

In addition, the continuous periodic inspection and maintenance of embankments, together with a fool-proof flood warning system, require a mentality of both peasants and local authorities, which can only develop in time. Assuming there is nothing wrong with the embankment (slopes not too steep, no seepage underneath, no danger of slips, no settlement, no lapse in maintenance), it is mainly the height of the crest which determines its risk of overtopping. For economic reasons, such overtopping is generally not allowed in urban areas, but is considered to be quite acceptable in rural areas. Embankments surrounding urban areas are therefore designed for overtopping once every 1,000-10,000 years (the Netherlands and Argentina), while the peasants, living in the rural areas, may expect to see the embankments overtopped once every 20 (Bangladesh) to 1200 (the Netherlands) years. There are designers who maintain that a high frequency of flooding is to be preferred, as the alternative (once in every 50-100 years) would create a false sense of safety, and consequently, when flooding occurs, it would incur a great loss of life, both human and animals.

Coastal embankments ("village bunds") in the delta of the Ganges in India. Embankment to keep out high spring tides offer little protection against cyclone levels. In the foreground the borrow-pits for the clay can be seen. These borrow pits may form a danger in the future by instigating erosion at the foot of the embankment.

A good example of a flood protection scheme consisting of embankments, is the project developed for the protection of irrigated areas in the Rharb Plain. The aim was to protect 43,000 hectares against flooding. This land was to be brought under irrigation and it had been demonstrated that the damage to irrigated crops and to the irrigation infrastructure justified the investment in flood protection. A mathematical model of the Rharb Plain provided information on the discharge capacity of the Lower Sebou at bank-full stage, and on the location and magnitude of bank overspills along the course of this river during the higher stages of a flood. For the solution finally adopted, the locations of major bank overspills were fixed. These were channelled into floodways, which diverted the peaks of high river discharges to a point 200 kilometres downstream, where the water re-entered the Lower Sebou. The various primary units comprising the irrigated area were completely surrounded by embankments. In fact, some primary irrigation units had to be reshaped to accommodate the floodways. As the impoldered areas were located mostly on higher ground, the storage capacity of the Rharb Plain for floods was not seriously affected. Detailed flood routing calculations were carried out to determine the crest levels of the different sections of embankment around the irrigated areas. The overall length of the embankments required for flood protection was 250 kilometres.

One of the more difficult decisions the designer has to make when fixing the horizontal alignment of an embankment is the rate of meandering of the river. Obviously, the peasant would like to have the levees as near as possible to the river, while the authorities responsible for maintenance do not like the idea of frequent rebuilding of embankment sections. In Morocco, the designer used aerial photographs, taken in different years, to determine his approach. In Bangladesh, the 200 kilometres long embankment, along the right bank of the Brahmaputra river was constructed at a distance of 0.5 to 1.5 kilometres from the bank. However, this mighty (braided) river is continuously shifting to the left or right resulting in no less than 10 kilometres of embankment being rebuilt in the last 14 years. Clearly, the cost of such periodic reconstructions has to be included in the economic calculations. It must be added that river training works can never be economically justified for rural areas, so the meandering of the river and the realignment of the levees have to be accepted.

It was stipulated above that the crest height of embankments is a matter of economics and psychology. The horizontal alignment, apart from considerations related to river engineering and flood control, is also a matter of economics and human interest. Apart from the crest height and horizontal alignment, the design of the cross-section of the embankment should be a major point of study. Design in this case is closely related to the fill material to be used, the subsoil, the standard of construction to be expected and the maintenance after construction. It would be beyond the scope of this book to describe and discuss all design features in greater detail. Only a few significant remarks will be made:

- fill material should always be borrowed in the neighbourhood of the alignment to limit costly transport;
- borrow-pits must not be located alongside the embankment otherwise the danger of scour to the toe or heel of the embankment by currents during flood cannot be excluded;
- width of the dyke body depends on the soil mechanical stability, duration of high water levels (seepage and saturation), and minimum width of crest in view of transport;
- good grassing is a vital tool in the battle against erosion by wind and rainwater, and the formation of cracks;
- animals should not be allowed to graze on the embankments;
- a freeboard of around 0.5 metres is required to allow for unreliable hydrological records, mistakes in (schematised) hydraulic calculations, unexpected local settlements (dyke body and subsoil), etc.;
- when calculating seepage patterns the subsoil (the foundation) of the embankment must be included in the calculation;
- maintenance entails frequent inspections and continuous minor repairs, it should not comprise large re-construction contracts say once every ten years;
- certain local animals may cause piping by digging tunnels through the embankments in which case local control measures must be taken.

Sea defences

The main differences between embankments along rivers and sea defence on the coast are:

- embankments have to retain high water levels for long periods (1 to 4 months), while sea defences are only confronted by high water levels during limited periods of time (1 to 3 days);
- sea defences have to be designed for wave attack, wave run-up and overspill of water from breakers, while river-side embankments are usually subjected to very little wave attack;
- sea defences are built and maintained in a salt water environment which is not conducive to grassed slopes and various construction materials; river-side embankments do not have this problem.

These differences are sufficient to arrive at a completely different design for sea defences in comparison with those for riverside embankments. When designing the cross-section of sea defences the following major aspects have to be considered:

- design water level (frequency of water level including wind set-up);
- design wave (or wave pattern and consequent wave run-up) in relation to crest height, seaward slope, roughness of the revetment and permitted overspill of spray from breaking waves;
- revetment design in relation to fill materials used, materials available for revetment design, seaward slope(s), wave attack and maintenance requirements;

- design of toe protection in relation to the level of the foreshore, the average low and high water levels, wave attack, subsoil characteristics and fill material used in the dyke body.

For the design of a sea defence the choice of the type of revetment is very important as it is a major factor in determining the cost. In this respect, it is recalled that in deltaic areas stony materials are not always available and must either be hauled over considerable distances or imported. Alternative solutions like pre-fabricated concrete blocks, or revetments incorporating bitumen are also expensive.

Coastal embankment near Chittagong. Testing a revetment consisting of bricks and brick-blocks.

As the dimensions of the revetments (thickness, length along slope) and specific weight of revetment materials are mainly governed by the expected wave attack, it can be advantageous to select the horizontal alignment in such way that a (high) foreshore exists in front of the sea defence. On this foreshore the waves will be reduced in height. Moreover, it is better to keep the tidal channel at some distance in order to avoid scouring of the toe protection, which eventually can lead to complete collapse of the sea defence. As in the case of embankments, a comprehensive discussion on all design features of sea defence works would be beyond the scope of this book. In this respect interested readers are recommended to consult publications on coastal engineering.

A last remark is made regarding the levees along the banks of estuaries. These are located in what can be called a transition zone between the river and the coast. This is reflected in the design of the levees (also called coastal embankments) where the revetment can be of much lighter design than for sea defences. However, crest heights sometimes have to be higher because of higher tidal water levels and wind set-up, while tidal differences are also larger than on the coast.

Downstream measures

The word downstream is used to indicate coastal protection works located near the sea, aimed at protecting areas situated more inland, where water levels are influenced to a great extent by tidal motion. These downstream measures relate to either estuary closures or storm surge barriers. The latter are in fact mobile weirs which are operated during storm surges when they function temporarily as an estuary closure. Both types of downstream measures will, to some extent, render obsolete the function embankments and sea defence have as water retaining structures and wave walls. The magnitude of the tidal prism determines whether or not the closure of an estuary or tidal inlet, or the construction of a storm surge barrier, should be undertaken. Generally speaking, one may say that apart from the tidal volume, the distance to the open sea as well as the water depth, determine the difficulties encountered by designers in collecting necessary data, designing the tidal closure, constructing the works in open sea, and operating and maintaining these afterwards.

Estuary closures

Though the title of this sub-section indicates very clearly what type of work is meant, it is not completely correct. The term estuary closure refers to closures of lagoons, tidal creeks, tidal inlets, estuaries and tidal waters like the Eastern Scheldt in the Netherlands. Obviously it applies to the closure of a coastal water at its mouth to the sea creating a lake behind the dam (also called a barrage). In this lake the water can either be salt, brackish or fresh, and the tidal regime and water level set-up due to storm surges no longer exist. Such coastal barrages sometimes separate the coastal lake completely from the sea. In many cases however, facilities are needed in the dam for:

- flushing purposes: intake of salt water and the discharge of waste water;
- drainage purposes: sluices for discharging river floods;
- navigation: locks for pleasure craft or commercial vessels;
- tidal power: powerhouse with turbines and generators to allow the inflow and outflow of seawater.

The idea of shortening the coastline by the construction of major coastal barrages has now been accepted throughout the world. During the last 40 years major coastal barrages have been constructed in France, Japan, Korea, Germany, Bangladesh and the Netherlands. Others are planned in Canada (Fundy), Great Britain (Severn) and India (Sundarbans). It must be emphasised, however, that in the past, flood protection was not always the major purpose of the barrages. Other purposes were:

- tidal power: La Rance (France);
- fresh water storage: Feni (Bangladesh), Nakdong (Korea);
- preventing salt water intrusion: Brielse Maas (the Netherlands).

Obviously, a multi-purpose function of such an expensive construction makes its economic justification a lot easier. The technology and techniques applied in the design and construction of coastal barrages will not be discussed here as it is well covered in literature.

Storm surges barriers

When a coastal barrage is contemplated for flood protection purposes, but other interests or ecological aspects would be greatly affected by its construction, a storm surge barrier can be built. Such a barrier is normally open allowing the tide to flow in and out of the estuary. When exceedingly high water levels are expected, due to storm surges, the barrier is closed until the water level at sea has fallen to an acceptable limit.

Storm surge barriers have been built and are operating:

- in the Eastern Scheldt (the Netherlands) for ecological purposes;
- in the Thames (Great Britain) for river discharge, sanitation and navigational purposes.

One is under construction near Leningrad (USSR), while another is earmarked for protecting Venice (Italy) from floods.

Non-structural flood protection measures

So far only structural measures regarding flood protection have been discussed. However, there are also measures which do not require infrastructural works but place emphasis on the operation of existing works, or are of an institutional nature. Such measures are:

- operation of multi-purpose storage schemes aimed at decreasing peak discharges;
- zoning of flood plains in order to prevent damage to persons, infrastructure and industries;
- flood forecasting and warning.

Multi-purpose river basin management

In the foregoing sections, only the development in a river basin aimed at flood control was discussed. Embankments and most diversion schemes will, in fact, have as their sole purpose protection against floods. However, the situation differs when storage schemes are considered. The creation of a reservoir in a river basin makes it possible to regulate river flows. This can be of interest for irrigation, while the potential energy of the stored water can be used for the generation of electricity.

The combined use of the reservoir for a number of purposes is always attractive from the point of view of optimum use of water resources. Also, the economic feasibility of a storage scheme is more easily demonstrated when multi-purpose utilisation of the reservoir is intended.
As the focus here is on flood control, the subject of multi-purpose use of reservoirs is only briefly discussed. The technical and economic features of these types of projects are very interesting but a detailed review would divert too much attention away from the main subject.
Nevertheless, it must be emphasised that nearly all major storage schemes include multi-purpose elements which, in turn, determine their siting, dimensions, technical features and economic feasibility.

A simple example of water management is the choice which has to be made in the operation of certain reservoirs, from either:

- storage of an arriving flood by utilising all available storage space in a reservoir and consequently, accepting that the next flood arriving cannot be stored and will give rise to flooding downstream; or
- allowing the non-interrupted passage of high flows which will cause damage, while reserving the available storage space for flood peaks or extraordinary and extreme floods.

By introducing rule curves (derived from statistical values obtained from the simulation of flows), it is possible to make optimum use of storage space. This was demonstrated in the case of the M'Jara reservoir in Morocco when the economic optimum size had to be determined. It was realised that the following factors, as well as a few others, played an important role in determining this optimum size:

- the integrated operation of three reservoirs (including M'Jara) in the same catchment, as opposed to independent operation where each dam served a separate command area for a single purpose;
- the joint use of available storage space for irrigation, hydropower and flood control purposes, rather than reserving entirely separate reservoir pools for each purpose.

Hydraulic effectiveness of flood protection

Measures aimed at flood control and flood protection of a certain area can differ in nature, size and location. Designers as well as decision makers now require knowledge about the extent of flood control offered by certain measures or by infrastructural works. It is therefore considered best to express this as a coefficient. As this coefficient ignores possible economic (damage) or financial considerations only considering the effectiveness of flood control works in keeping water out of a certain area, it is called the hydraulic effectiveness. Although the hydraulic effectiveness does not include any cost element, it provides a relatively simple method for comparing the quality of various flood control measures. As will be seen below, it also shows whether two or more flood control measures can be taken independently of each other.

Finally, if the construction cost is introduced, the relationship between construction cost and hydraulic effectiveness, of various sizes of a certain flood protection measure, will give an indication as to the optimal solution for the flooding problem.

Evaluation of flood protection schemes

The evaluation of a flood control scheme solely on the basis of the hydraulic effectiveness and cost is, however, not enough. Such an evaluation only shows "what you get for money" but it gives no indication as to whether the investment is worth it. The so-called economic evaluation is also not sufficient. Though the economics of flood protection are vital when external financing is required for the implementation of the scheme, other factors turn out to be equally important when decision makers have to choose from a range of schemes which differ from each other in cost, benefits, degree of protection obtained and other aspects.

Designers who, in the context of a pre-feasibility study, have defined a number of alternative flood protection schemes, must therefore present the key figures and relevant effects due to these schemes, in such a manner that decision makers have all the available tools at their disposal to draw conclusions and to make their decisions. It must be admitted that the significance of relevant effects differs from country to country. There is also a rapid development in the number and type of aspects considered to be important when evaluating a flood protection scheme. Twenty years ago, an effect like environmental impact was already known but practically no one was inclined to take it into consideration when making decisions.

Evaluation is often hampered by the fact that many effects cannot actually be expressed in figures. These effects are therefore called intangible effects. For flood control schemes, the following aspects would appear to be relevant for evaluation purposes:
- financial and economic figures;
- safety aspects;
- various intangible aspects such as social, regional, ecological effects and possible phasing of a scheme.

Financial and economic analysis

For a financial and economic analysis of a flood control scheme, costs and benefits must be calculated. Costs comprise the following:
- investment costs related to the works to be constructed;
- operational and maintenance costs;
- periodic costs for the replacement of worn-out equipment;
- lost agricultural benefits due to the use of agricultural lands for floodways or floodplains.

Benefits consist of:
- damage avoided due to the execution of a flood protection scheme;
- induced benefits in rural areas because of the possibility of more profitable and intensified agricultural use of the reclaimed lands;
- benefits in urban areas, derived from the possibility of having the former flood prone area urbanised;
- other benefits such as tourism, fisheries, less frequently blocked transportation systems, etc.

Macro-economic growth rates must be introduced for each region, both for the damage avoided and the induced benefits. A distinction must be made between rural (agricultural and advanced agricultural), and urban (industrial and mixed industrial-residential) areas when applying individual growth rates. The period of evaluation, commencing at the start of construction, is generally taken as 35 to 50 years depending on the interest rates taken in the calculation of present day values (NPV).

The calculation of benefits due to damage must be done on the basis of damage curves. These curves must reflect the relationship between flood damage and a selected flood parameter, for a certain reference year. This flood parameter must be representative of the floods in a certain sub-area. The occurrence of flood must, in turn, be laid down in a frequency curve for the selected flood parameter. By combining this frequency curve with the flood damage curves (for one sub-area but for various reference years) it is possible to calculate the mathematical expectation value of damage during each year of the evaluation period in the non-project situation. This non-project situation is, in fact, represented by the frequency curve which applies to the non-project situation.

Safety aspects

In this paragraph, when safety is mentioned, it refers to the external safety, which is the safety which follows logically from the combination of all mathematically calculated risks of flooding. There is another type of safety, the internal safety, which reflects the subjective feeling one may have concerning the risks and dangers of flooding.

In reality, there are many factors which endanger safety. These can be grouped as follows:
- factors due to natural conditions;
- factors due to the actual flood protection works;
- factors due to human behaviour.

Within the context of this paragraph, it is not possible to discuss all possible risks in detail. The main issues are:
- natural conditions in the river basin or coastal area deteriorating in time such as:
  - higher river discharges due to deforestation,
  - silting up of storage reservoirs,
  - scour leading to the collapse of protection works,
  - higher water levels due to the silting up of a river course,
  - higher water levels due to sea-level rise or due to the deepening of tidal channels;
- deviation of values of hydrological parameters, derived from statistical analysis of data on limited observation periods and applied to short life spans of projects;
- human actions in the upper regions of the catchment area (e.g. the construction of reservoirs), or in the coastal area (e.g. the construction of harbours, dredging of fairways, reclamation and impoldering works);
- risks arising from the actual flood protection works, caused by:
  - insufficient field investigations,
  - inadequate design,
  - bad construction,
  - bad construction materials,
  - lack of maintenance,
  - malfunctioning of equipment;
- risks created by human beings such as:
  - mistakes in operation of spillways, weirs, drainage sluices or storm surge barriers,
  - acts of sabotage,
  - acts of war.

It is possible to analyse each risk in detail and to take measures to lower a particular risk and increase the safety of the system under consideration. At times this is not possible, and in such a case, the total abandonment of the envisaged flood control works could be considered. A few years ago the possibility of embanking a large rural area (24,000 ha) in a flood prone area in South Asia was investigated. Flood control would have entailed 80 kilometres of embankments 6 to 7.5 metres high, and annual water levels (outside the embanked area) of 3 to 4 metres above the land during a flood season lasting 5 months. In view of the expected low quality of construction and lack of maintenance, the risk of a breach of the embankment, with the consequent loss of many lives, was considered to be too high and the project was not recommended for implementation. The client was not happy with this recommendation but the international financing agency sponsoring the study agreed, and subsequently refused to finance implementation of the works.

In general, it can be said that a false feeling of safety must always be avoided when planning and executing flood protection works. Commonly accepted departure points for flood control works are, for instance:

- provision of a high and continuous (in respect of different types of floods) protection against flooding;
- provision of a higher degree of protection in urban and industrial areas than in rural areas where people and movable properties can easily be evacuated in case of floods.

More than by financial and economic figures, safety will be conceived by the people in terms of the consequences a certain flood has: how much land is flooded, how many lives are lost and how many houses are destroyed. With this in mind, the safety of flood protection schemes in the Paraná-Paraguay studies (see Keller et al., 1981) was measured by calculating for a range of floods (from a flood having a return period of 10 years, up to the probable maximum flood (PMF) and dambreak flood) the percentage of avoidance of flooding (area), persons affected (number) and monetary damages (US$). This was referred to as the degree of protection obtained.

Intangible aspects

Intangible aspects are effects which are observed after the implementation of flood protection works and which cannot be expressed in economic figures. In this respect, it is noted that safety can be considered to be an intangible effect. Other intangible effects are:

- social effects, i.e. the influence of flood protection on the social structure and social behaviour in the region concerned;
- regional effects. Here the balance is held between the areas benefiting from a certain flood protection scheme and the areas which have to pay for the investments;
- ecological effects. Broadly speaking, these are all changes in the ecosystem, which are due to the implementation of a flood control scheme;
- phasing effects. These effects concern the possibility of constructing a flood protection scheme in parts, having the option to disregard, in the future, the construction of the last phase(s) of the scheme without rendering useless, the parts already completed.

Obviously, the list of intangible effects can easily be extended while the social, regional and ecological effects can be subdivided into numerous specific items. It may be possible to relate some items to a parameter. Values given to this parameter for different flood protection schemes, can be a useful indication to decision makers when selecting a scheme or schemes for further study (full feasibility stage). Examples of intangible effects of flood protection schemes will not be given, as the effects themselves, as well as the importance allotted to them, depend heavily on the particular circumstances (nature, population, culture, politics) of the region studied. Contrary to the technical and economic aspects, the intangible effects cannot be evaluated by using internationally recognised criteria or yardsticks. Therefore, the significance of certain effects would be difficult for readers living elsewhere in a completely different environment, to understand.

3.2.4 Flood forecasting and warning systems

The safety of an area is commonly expressed in terms of a probability or a return period of a certain flood, e.g. the "100 years' flood". This does however not mean that the particular flood will occur only once during the next 100 years and that it then will reach the water level as calculated. One conclusion from this stochastic phenomenon is that the feeling of safety is not an absolute certainty. This knowledge has led mankind to undertake action instead of sitting on the river banks and waiting until the flood will come. Flood control and flood protection measures can be undertaken to reduce the frequency of flooding, a fool-proof protection against all types of flooding, however, is impossible. Therefore further measures are required to be able to look ahead and foresee whether a flood is likely to develop in the near or medium future and to undertake appropriate measures to mitigate the negative impacts of flooding. The monitoring of the river basin in order to forecast the development of a flood at a certain moment in time and at a certain location in the river basin is called flood forecasting. The series of measures that can be undertaken to prevent or mitigate negative impacts of flooding are commonly known as flood warning.

Flood forecasting

Flood forecasting enables responsible authorities to anticipate risks of floods and flooding that are expected in the near or medium future. The measures that will be taken will depend on the relative importance of the area of the river basin under consideration as well as the hydrological conditions. Such measures can be simple routine actions (enforcement of operating instructions of weirs or regulated spillways in dams), preventive measures (not permitting navigation on the river transport routes) up to more interfering measures such as enforcing an alarm phase and installing an organisation that is able to take immediate actions in case of emergency situations (evacuation of inhabitants of flood prone areas).

The principle of flood forecasting is based on the analysis and interpretation of all available and relevant data at a certain moment in order to forecast at a given location in the river basin (within given probability limits) the development of a flood in the future. The time interval between the moment of collection of the data and the moment of forecasting can vary between one hour or less for river basins having a size of tens of square kilometres and a few days for river basins that are some hundreds of thousands of square kilometres in size.

Mostly, the forecast is expressed in terms of a water level or a discharge at a future point in time at a given location in the river basin and, if possible, an indication of the probability of the maximum water level that can be expected at that point of time.

Necessary data

Independent as it is of the selected method of forecasting and the collection and handling of data, the quality of the method of forecasting is very much dependent on the quality and the number of data collected about the hydrologic and hydraulic characteristics of the river basin.
The availability of data is necessary for several reasons:

- understanding of the physical phenomena in the river basin that are responsible for the development of floods;
- the choice of the method of flood forecasting;
- the choice of parameters for description of the flood;
- accuracy and reliability of the measurements of the different parameters;
- feeding of the forecasting model during the development of a flood.

The data that is most frequently used to make a forecast are derived from the following measurements:

- rainfall measurements in the catchment area basin;
- the degree of saturation of the soil at a number of locations in the basin;
- the discharges at locations in the basin upstream of the location where the forecast has to be made;
- the amount of snow and the temperature present in the basin;
- the amount of water that is retained by reservoirs present in the upstream reaches of the river basin.

It is also possible, depending on the size and shape of the basin and the adapted time of forecasting, to include weather forecasts to derive the probability of rainfall prior to and during the time of forecasting. Rainfall weather forecasts in coastal areas can also be included in storm surge models that forecast wave and wind set-up.

Most important during the period of observations and before measurements are made is to have a minimum number of distinct floods of different magnitude and type (origin and location in different sub-basins, duration and intensity of rainfall, seasonal conditions).

The number of parameters to be measured depends on the size and shape of the river basin, the nature of the hydrological characteristics, and the variety in space of the different parameters (gradients in rainfall, snow, temperature, percentage of infiltration). The nature of the measurements (rainfall every 10 min, 1h, 6h, 24h, several days) also depends on the size and shape of the river basin and the nature of the hydrological characteristics.

Rainfall-discharge models

For small river basins (10 km^2 to 100 km^2) the travelling time of the flood wave is relatively small and consequently little time is available (1/2 - 2 hrs) to monitor and measure water levels or discharges upstream in the river basin and to estimate the water level more downstream in the river basin. In this case, flood forecasting is based on a direct and simple method of translation of rainfall measurements into a discharge. By measuring the rainfall and using a standard unit hydrograph, obtained from statistical relations, it is possible to estimate the river discharge at a certain location in the river basin in a relatively short time. Several methods exist but will not be discussed here.

Propagation models

For large river basins (> 100,000 km^2) the travelling time of the flood wave from one point in the basin to another is much larger. Under these conditions it is wishful thinking to apply a model that is based on the transformation of rainfall measurements into a river discharge, as the large number of factors that will influence the flood wave characteristics will result in a large, complex and abstract model. In this case, the river discharge or part of the hydrograph upstream of the location where the forecast has to be made is taken as input and the propagation and development of the flood wave is forecast. These methods of propagation are simplified applications of the Saint Venant equations, which describe the mechanical properties of fluid flow by means of conservation of mass and momentum. Depending on the degree of simplification the application can be schematised into one-dimensional flow for straight channels or two-dimensional flow for flood plains, estuaries or to describe the development of storm surges.

Multiple regression models

These methods are based on multiple regression algorithms between output variables in the future (t+i) and input variables in the past (t-i). The variables are basically the rainfall in the river basin, the measured discharge at a station situated upstream at a time t-i and the measured discharge at a time t or t-i at the station where the forecast has to be made. These models are not based on physical principles of river flow (propagation models) but solely on the statistical correlation between the input and output variables.

Complex models

Complex models are a combination of different models as mentioned above. They are usually applied when hydraulic structures are present in the river basin that influence the development and propagation of flood waves. Such structures can be for instance weirs or dams. In these models it is necessary to describe and combine the physical behaviour of the flow under different circumstances: operation of reservoirs, flooding of land due to bank overspill or overtopping of embankments, operation of diversion channels, representation of flow in flood plains by means of 'cells'. These models are, in general, applied to large river basins or when the time of forecasting is several hours or days or when the time of calculation is not a limiting factor.

Weather forecasting

Weather forecasts have made significant progress during the last twenty years. The number of measuring stations throughout the world has increased considerably as well as the understanding of the physical phenomena involved. This understanding combined with increased computational possibilities have improved the weather forecast models. The weather forecast models can be used for the forecasting of flood waves in river basins as well for the forecasting of the development of storm surges at sea. For river basins the main interest will be the possibility of rainfall forecasting while at sea wind data will be used for models that predict wind set up along the coast.

Data collection

After selecting the most appropriate model it has to be considered how the model can be operated optimally. The logistic operation of the model consists of the following components:

- collection of basic data included in the forecast and which serve as input data of the chosen model;
- checking the quality of the data;
- transmission of the data at regular intervals to a central receiving station;
- data handling and storage in the proper format (depending on the required format used by the chosen model);
- interpretation of the data and translation into action against expected floods. At this stage the flood warning phase starts which is described in the following section.

Flood warning

Despite all flood protection measures that might have been taken it is impossible to realise a fool-proof protection against flooding. Flood warning is a way of mitigating the possible negative impacts of flooding by informing the exposed population of the risk of flooding of a future or coming flood. Warning the exposed population enables them to take certain minimal actions to protect their most valuable possessions and possibly their lives by evacuating themselves from the endangered area.

A flood warning system also requires the responsible authorities to undertake collective measures to respond to the approaching flood wave and to enforce precautionary programmes to limit the damages due to the flood, for example:

- voluntary opening of breaches in the embankments for a better protection of more vulnerable downstream areas;
- adapted operation of storage reservoirs;
- provisional protection of certain areas or possessions by placing sand bags;
- evacuation of inhabitants from the most vulnerable areas;
- enforcing first-aid organisations.

The principle of flood warning should be based on an appropriate flood forecasting model which enables the responsible authorities to foresee the development of a flood at a given location in the river basin.

Organisational aspects

A series of related tasks and procedures should function simultaneously in order to initiate a well functioning flood warning system:

- an appropriate flood forecasting model;
- a well organised Flood Warning Service;
- an assessment of all possible means of communication and transportation and their vulnerabilities;
- an assessment of all hydraulic structures in the river basin with which the flood wave can be controlled and to what extent;
- an assessment of the operating rules of the hydraulic structures, the organisations which are responsible for their operation and maintenance;
- identification of warning services;
- identification of first-aid organisations;
- identification of security measures and safeguards for the protection of viable infrastructural works and important economic activities.

Flood forecasting and flood warning in the Netherlands

In 1911, a storm surge which flooded a timber yard on the lower reaches of one of the rivers led to a request to the Government for the possibility of a warning and forecasting system to be studied. Ten years later, in 1921, an experiment in this direction was started leading, a further ten years later, to the establishment of a Storm Surge Warning Service, which was provided by the Royal

Netherlands Meteorological Institute. However, this system was very limited and related only to the lower reaches region. The 1953 disaster, which hit the south-western part of the country hardest, drowning about 1900 people, was the motive for setting up a completely new system in which the Rijkswaterstaat became involved.

The procedure followed by the Storm Surge Warning Service when storm surges are expected is as follows. The Meteorological Institute, which permanently monitors the weather, contacts the Rijkswaterstaat (i.e. the engineer in charge of the Storm Surge Warning Service) in case the weather situation develops in such a way that a storm surge along the Dutch coast may be expected. After consultation, this engineer may decide to bring the forecasting centre for monitoring the development of water levels into operation. This forecasting centre is equipped with telemetering water level recorders.

If the situation becomes such that certain critical levels are expected to be exceeded around high tide the forecasting centre issues telegrams warning the authorities responsible for the sea defences. It is required that such telegrams be issued at least six hours before the high tide concerned in order to give the responsible authorities the opportunity to take the necessary measures. There are two levels of warnings:
1) limited dyke watch;
2) extensive dyke watch.

In the case of the limited dyke watch only some of the offices are occupied, but in the event of an extensive dyke watch being declared continuous monitoring of the critical locations is organised. Conditions in situ are then inspected continuously and organisations like the Red Cross, Civil Defence, Army and Navy will also be warned.

3.3 Control of Salt Water Intrusion

3.3.1 Surface water

The control of the salinity of the surface waters in deltas and other low-lying coastal areas is one of the main problems of the water management of those areas. Salinity limits have been assessed for the use of water for various purposes. These so-called tolerance limits differ widely according to the purpose and the way the water is applied. When sodium chloride is the prevailing salt a generally accepted tolerance limit for the use of water as drinking water is 200 to 250 parts per million (p.p.m.) of chloride (Cl^-). Water for processing in food industries and breweries should not contain more than 100 to 150 p.p.m. of Cl^-. For spraying water in hothouses where tropical flowers and fruits are grown a limit of 50 p.p.m. of Cl^- should be observed. The tolerance limits for the use of water for irrigation in the open field depends on the type of crop, the climate and the soil type. These limits may range from a few hundred to a few thousand p.p.m. of Cl^-, the latter figures being valid for conditions of continuous leaching and drainage.

When considering the possible measures to control the salt water intrusion distinction should be made between preventative measures and alleviating measures. With respect to sea-water intrusion into open estuaries, which is the most generally occurring and often most important source of salt, preventative measures aim at pushing the saline intrusion back into the sea. An obvious way to realise this is by augmenting the upland discharge. This however is not an effective way of using the fresh water. To obtain an appreciable effect, high flows of water are required which are lost to the sea. For the river Mekong in Vietnam, for instance, with a minimum flow of 2,000 m^3/sec and an average flow of 15,000 m^3/sec, an increase of 500 to 1,000 m^3/sec is required to push the saline intrusion back to the sea over 900 to 2,200 metres. With this discharge some 0.5 to 1.0 million hectares could be supplied with irrigation water. Moreover, the seawater intrusion into open estuaries is maximum during dry periods precisely when there is a great need for fresh water for various purposes. Augmentation of the low flow under these conditions is only possible if more water can be released from upstream storage reservoirs, at the expense of the other water users in the river basin.

Since the length over which the saline intrusion is penetrating into the estuary is more than proportional to the square of the depth, a shallowing of the channel by filling up of deep reaches would considerably reduce this length. This requires a stabilisation of the new bottom of the channel because otherwise the fill would be scoured away by the river. This method has been applied in the Rotterdam Waterway in the Rhine delta where a dredged channel to the sea had caused erosion and deepening of the waterway.

Alleviating measures consist of shifting intakes and flushing of interior canals. They are often more feasible than the measures mentioned earlier. Shifting of intakes of water in the upstream direction is often applied when the salinity of the river water at existing intake points is no longer acceptable during low flow periods due to the deepening of entrance channels for navigational purposes. This method has been applied for the intake points along the Rotterdam Waterway before shallowing of the channel was applied. In this case, the points of abstraction were displaced several times, in total over a distance of some 40 kilometres in the course of 80 years.

The consequences of the admission of water from the estuaries with a salinity above the tolerance limit can be alleviated by periodic flushing of the canals with fresh water from upstream sources or by mixing the two waters.
The amount of saline water entering at shipping locks can be reduced by hampering the exchange of water between the outer harbour and the lock chamber during the time that the gates are open. This can be achieved with a pneumatic barrier consisting of a screen of air bubbles. The screen creates vortices with a horizontal axis which counteract the density currents (see Figure 3.12)

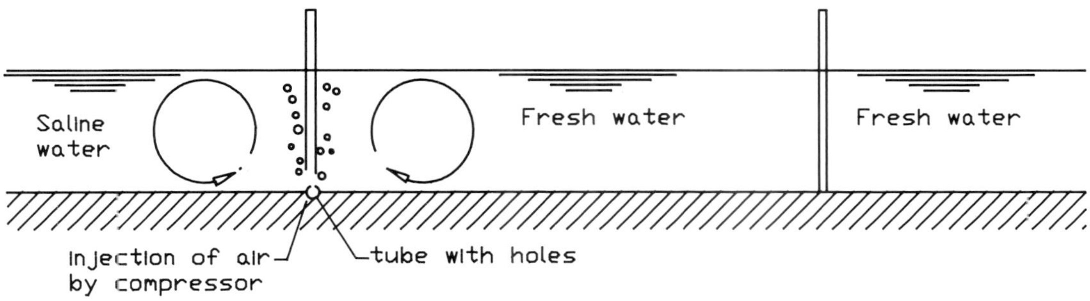

Fig. 3.12
Pneumatic barrier.

The most effective way of reducing the amount of saline water entering a lock to less than 10% of the volume of the lock chamber is to pump the saline water back to the outer harbour, replacing it by fresh water from the canal. Hendrik Stevin had this type of solution in mind in 1667 when he proposed to continuously flush the lock chamber and outer harbour with fresh water pumped by a windmill from the fresh water canal. Unfortunately, he overlooked the effect of density which keeps the heavier saline water on the floor of the lock. The only way to remove this water is to extract it from underneath and to put the fresh water from the canal on top (see Figure 3.13)

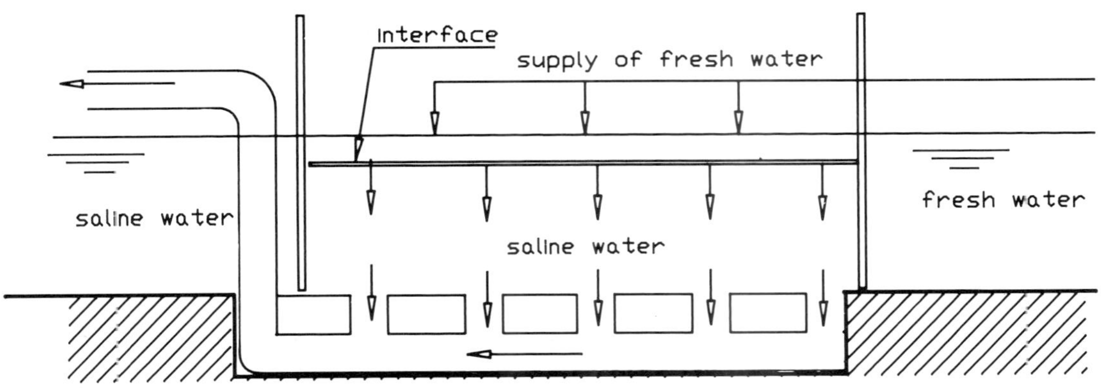

Fig. 3.13
Replacement of saline water in the lock chamber by fresh water from the canal.

This system has been applied for many years already at navigation locks at Mardiche (France) and Kreekrak (the Netherlands).

3.3.2 Groundwater

In section 2.2.6 the salinisation of groundwater has been described. Salinisation is affected by the hydraulic head and can be influenced by intervention in the groundwater flow regime. Measures against salinisation are therefore directed towards these aspects.
Salinisation problems arise most frequently at wells, where upconing of saline water may occur. In the ultimate situation, saline water is abstracted instead of fresh water. A very obvious method to avoid salinisation of wells is to limit the abstraction rate.
The solution in such cases is to apply a well-distributed systems of shallow wells. Relatively small

extractions cause little lowering of the hydraulic head and thus little upconing. Another remedy for salinisation of the wells is to increase the artificial recharge of the groundwater system (e.g. by deep well systems or by infiltration canals) whereby a more extensive fresh water body is developed (see Figure 3.14).
Artificial recharge of phreatic aquifers by infiltration canals is a simple and relatively cheap method although some purification may be necessary. Deep well infiltration is more complicated and costly, as the recharge water has to have drinking water quality in order to avoid clogging of the well system. Moreover there may be quite high pumping costs. However, environmental reasons may sometimes lead to a preference for the latter system.

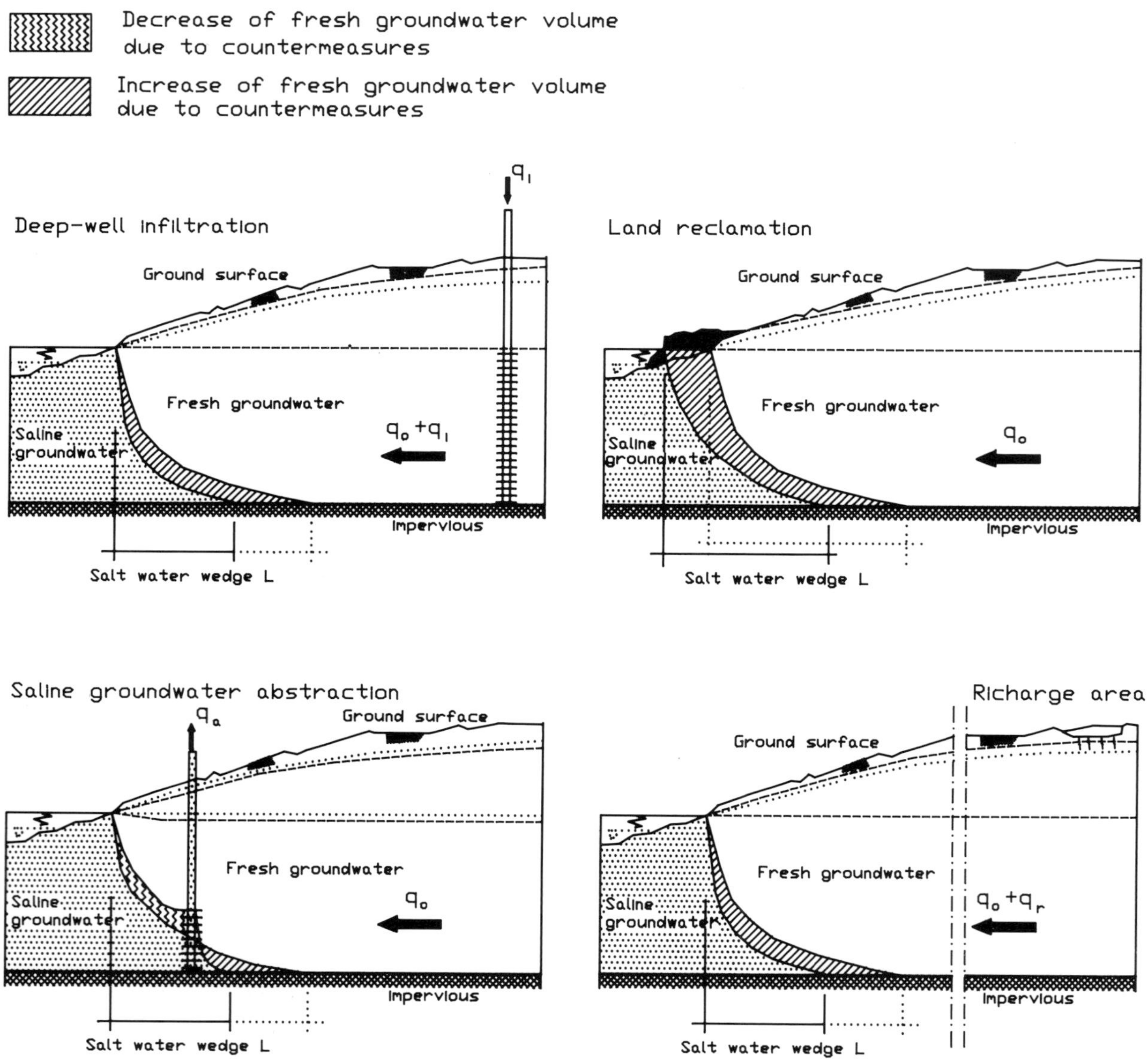

Fig. 3.14
Countermeasures against salinisation.

Salinisation of confined aquifers discharging into the sea can be controlled by the outflow rate. If a certain percentage of the fresh water flow, q_0 is extracted, the length of the salt water wedge will increase and this may endanger the well. Once more, limitation of the extraction rate or increasing the recharge of the system may avoid or diminish the danger of salinisation. The same remark can be made for semi-confined aquifers.

The reclamation of polders in deltaic areas leads very often to a lowering of the phreatic water table below mean sea level. This will promote salt water intrusion into the aquifer system. Countermeasures for these cases are difficult. Abstraction of saline groundwater will lower the hydraulic head of the saline groundwater in the aquifer. However, as well as high costs for pumping, this will also lead to a loss of fresh groundwater. Another problem that arises, is the disposal of the extracted (brackish/saline) water, which may be costly (e.g. transport costs). It is not desirable or sometimes even not allowed to discharge the saline water into the surface water. If the sea or a salinised part of the estuary is very close it may be economically feasible.

In some cases the aquifer is isolated from the sea by a fresh water body which extends to an impervious base. In this case, extra recharge will be necessary due to seepage losses.
Giving up an already reclaimed area and creating a lake with a higher water level can counteract salinisation, however this is a rather extreme solution.

An interesting possibility to oppose salinisation is reclamation of tidal forelands. In this way, the recharge area is extended, which eventually will lead to a greater fresh water volume.

Overpumping of aquifer systems should be avoided. It is very important to consider the (fresh) water balance of the groundwater system and to realise that maintaining a fresh groundwater body in a saline environment also means some losses of fresh water because there is always an outflow of fresh groundwater. The safe average extraction rate must be kept within the limits of the fresh water balance.

Figure 3.15 illustrates the effects of extraction and countermeasures such as limiting extraction rate and artificial recharge. This figure shows the occurrence of fresh groundwater in an east west cross-section of the dune area near The Hague (the Netherlands). Due to groundwater abstraction a dramatic upconing occurred between 1915 and 1954 in the dune area. The creation of infiltration ponds and the limitation of the abstraction rate cured this problem, as the 1982 curve shows. The installation of a another abstraction system further inland also affected the salinisation. To counteract this, a system of deep-well infiltration was installed recently near the wells and the situation now appears to be stabilising.

Monitoring of salt water intrusion

The occurrence of salt water intrusion should be monitored regularly, especially in case of extraction of groundwater for drinking water supply. Besides conventional methods like sampling and analyses, geophysical methods have proven to be very useful. Geo-electrical sounding is often used for regional exploration of subsurface layering and can also be applied to determine salinisation. The results obtained are vertical, electrical resistivity profiles. The specific electrical resistivity values of layers in the subsoil are strongly affected by salinity.
Another application of resistivity measurements is often used to safeguard well systems (Stark et al., 1992). For this purpose a Permanent Electrode System (PES) is installed. The system applies 4 electrodes (2 potential (P) and 2 current (C) electrodes) with one pair (C,P) located at the surface and one pair (C,P) in the subsoil. Several pairs of electrodes (C,P) may be located at fixed depth intervals in the subsoil. In this way the electrical resistivity can be measured related to depth (see Figure 3.16). Salinisation will lead to lower resistivity values. A great advantage of this system is the possibility of automation.

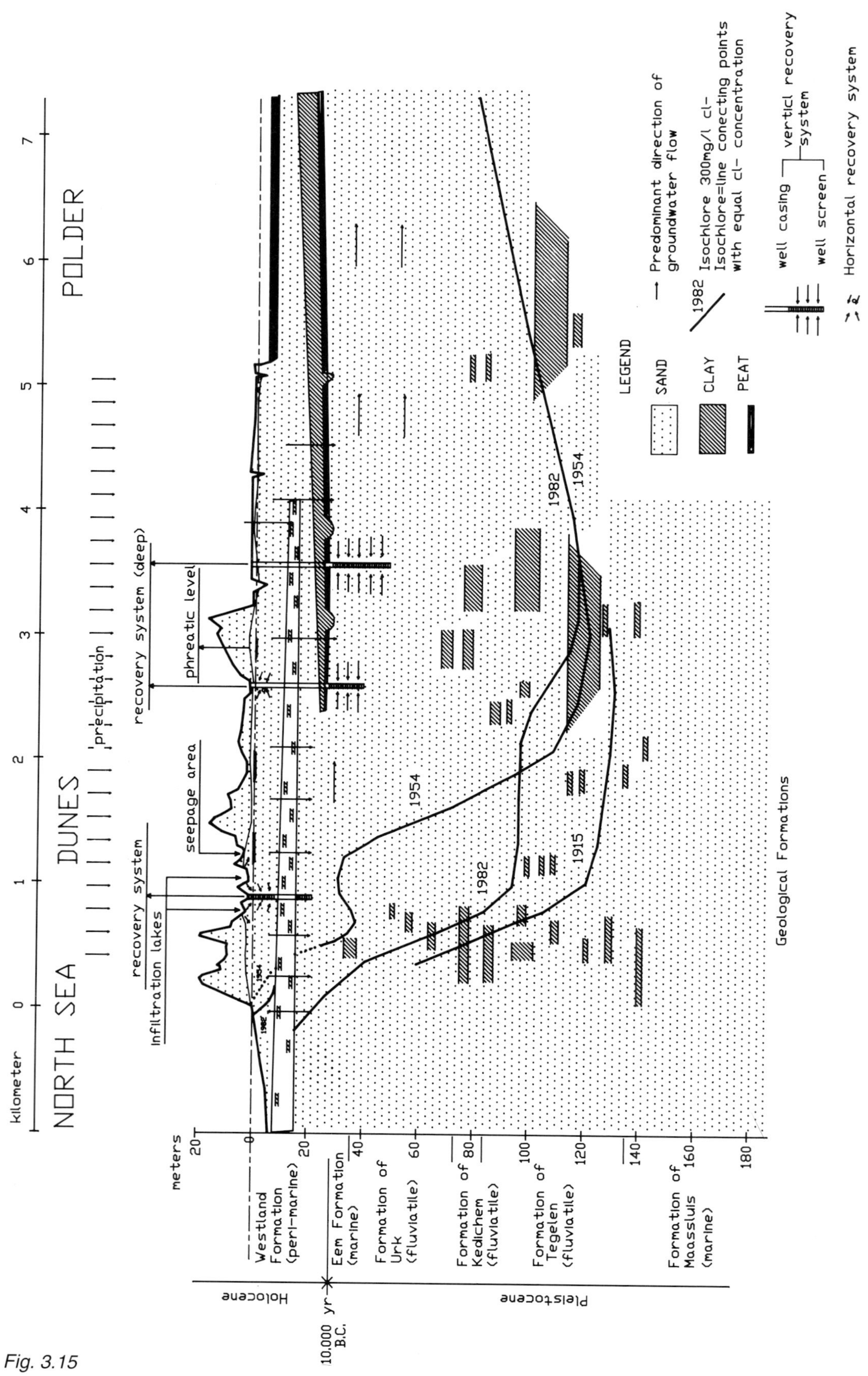

Fig. 3.15
Effects of extraction and recharge on salinisation in the sand dunes near The Hague, the Netherlands (courtesy of N.V. Dune Waterworks of South Holland (DZH).

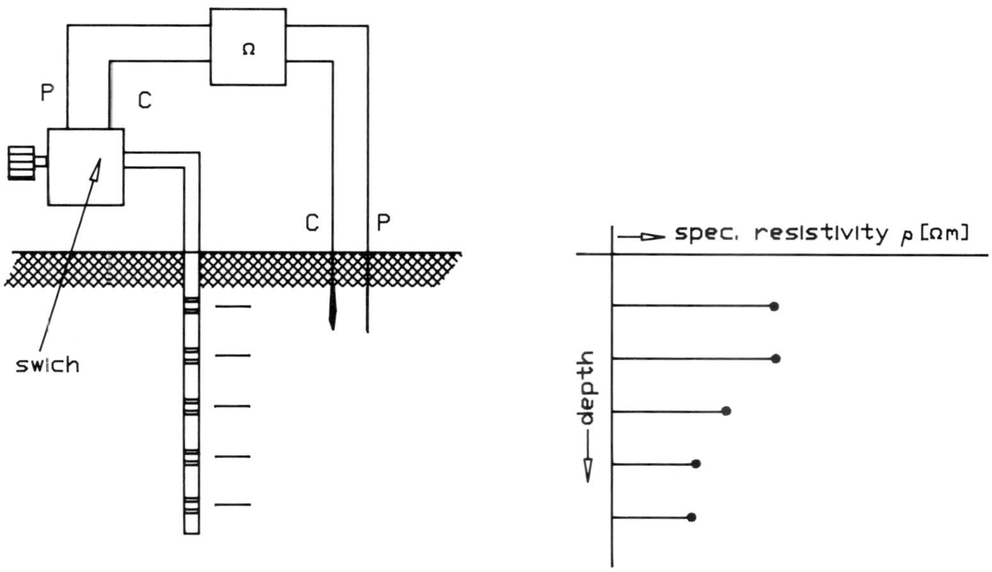

Fig. 3.16 Permanent Electrode System (PES) and measurements.

3.4 Coastal Reservoirs

3.4.1 Principles of coastal storage

The principle of storage of water in coastal areas is the same as that of storage in upstream portions of a catchment: retention of water in a reservoir during periods of water excess and release of water during periods of low flow thus augmenting the supply of water to downstream areas when there is a shortage of water. In case of coastal storage, the reservoir consists of a tidal inlet (e.g. an estuary, lagoon, or embayment) into which a river is discharging and which has been separated from the sea by a barrier or enclosing dam with an outlet equipped with discharging sluices or a pumping station or both (Figure 3.17).

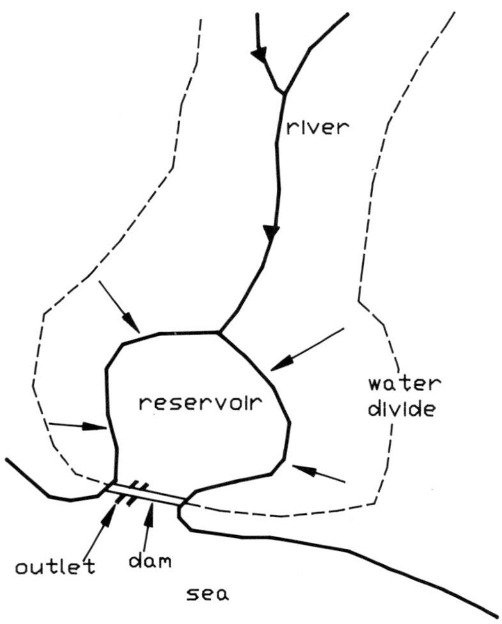

Fig. 3.17
Coastal reservoir.

Water is supplied to the reservoir by the river coming from the upstream portions of the basin and by the areas of the coastal zone draining towards the reservoir. During periods of water shortage water can be extracted from the reservoir.

Coastal storage reservoirs have been built at many places, especially in Europe and Asia, as central elements of the water management system. They are typically part of multiple purpose schemes and have the following advantages:

- fresh water is in storage in the coastal area itself and not in a distant place upstream (e.g. in another region or country with possible water users along the river before it reaches the downstream area);
- the dam near the sea halts the intrusion of sea water into the tidal inlet. The areas along these inlets which are in need of fresh water can admit water from the reservoir;
- the enclosing dam shortens the line of defence against storm surges at sea. It forms a new primary line of defence in front of the former defence system;
- under certain conditions the drainage of excess water from rainfall on the surrounding low-lying areas can be improved by enclosure of the inlet;
- the enclosing dam connects the two parts of the coastal area which are separated by the outlet to the sea and makes it possible to create a short communication by road.

Estuarine reservoirs have the same advantages as mentioned above but the area available for storage is relatively small compared with coastal reservoirs in embayments, marginal seas and lagoons. Large coastal reservoirs have been built in Bangladesh, China, India, Japan, Korea, France, Germany, the Netherlands and Vietnam. Construction of coastal reservoirs is considered in river basins where suitable sites for the construction of upstream dams have already been occupied or where objections exist against flooding of land areas.

Kojima Bay, Okayama, Japan. Discharge sluices in the enclosing dam of a coastal reservoir.

The reservoir sketched in Figure 3.17 is an on-stream reservoir. In some areas possibilities exist for off-stream reservoirs (Figure 3.18).

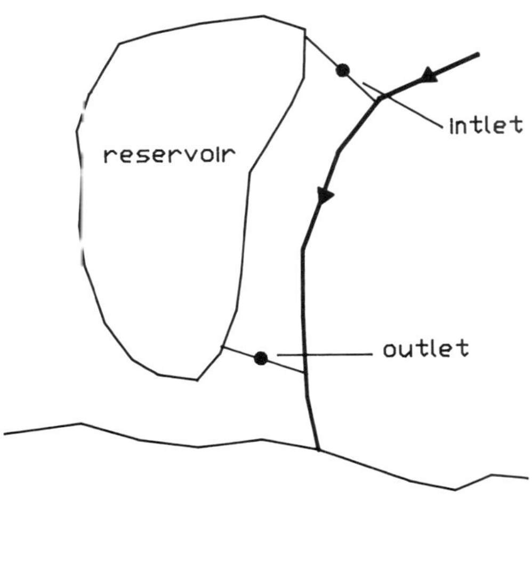

Fig. 3.18
Off-stream reservoir.

The reservoir area may consist of a lake or be created by encircling a land area with an embankment. The numerous small reservoirs traditionally known as tanks in the deltas of southeast Asia belong to this type of coastal reservoir. A modern version is the water well in the coastal area of the Netherlands where the available surface water is unfit for spraying in hot-houses with tropical fruits and flowers. Suitable water is obtained by collecting rainwater falling on the roofs of the hot-houses and storing it in small, on-land reservoirs (Figure 3.19).

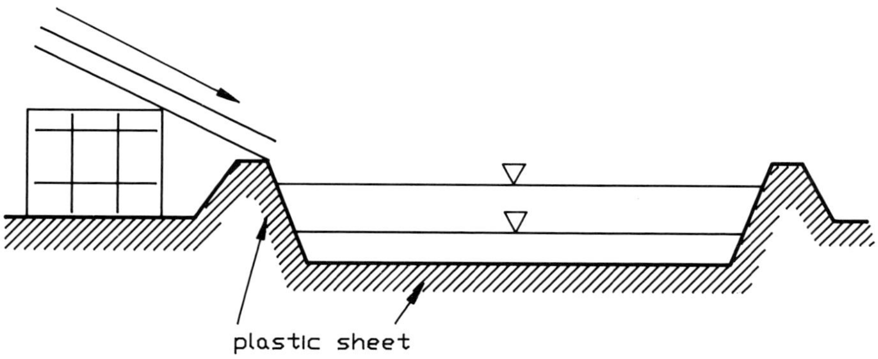

Fig. 3.19
Water collection from hot-house roofs.

3.4.2 Water and salt balance

The water balance and salt balance are the two basic concepts underlying the design and operation of coastal reservoirs. The water balance is similar to that of an upstream reservoir and contains the items shown in Table 3.3.

Water balance	
Inflow	**Outflow**
River discharge	Evaporation
Drainage to the reservoir	Drainage to the sea
Rain on the reservoir	Extraction of water

Table 3.3
Water balance in coastal reservoir.

In most cases the normal operational level of the reservoir is around Mean Sea Level (MSL) so that excess water can be drained off to the sea by gravity using the period of low water at sea (Figure 3.20).

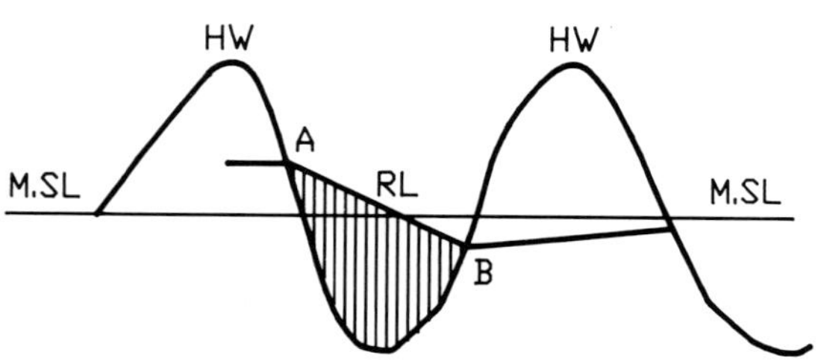

Fig. 3.20
Tidal drainage.

The sluice gates are opened at the time A and the reservoir level (RL) drops until time B when the gates have to be closed. The drainage to the sea may be hampered when the sea level is affected by landward winds and the reservoir level may rise far above the normal level (Figure 3.21).

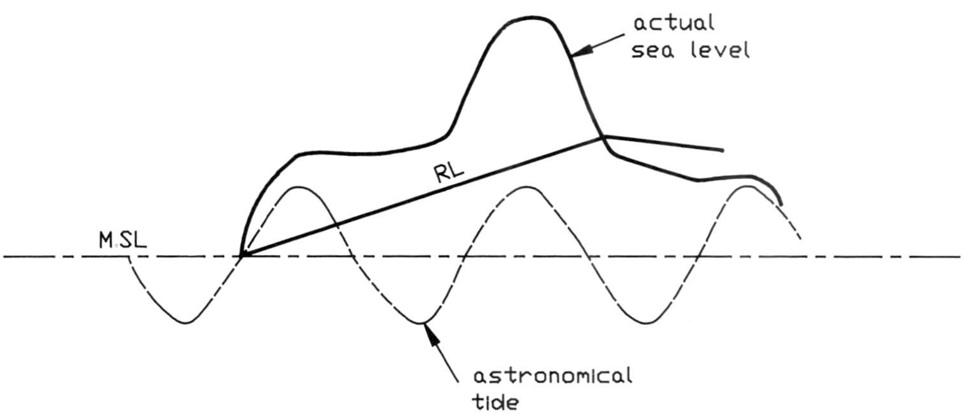

Fig. 3.21 Hampered drainage.

Similarly to upstream storage reservoirs, the hydraulic efficiency of coastal reservoirs can be increased by applying reservoir regulation and rule curves, especially when there are distinct seasonal variations in water supply. During periods with high inflow the reservoir level is kept as low as gravity drainage to the sea will allow. This facilitates the drainage of the surrounding low-lying areas to the reservoir. Towards the end of the wet season water is retained in the reservoir to have as much water as possible in storage at the beginning of the dry season. A drawback of most coastal reservoirs is that, in contrast to reservoirs in the mountains, the variation of the normal operational levels is small.

Coastal reservoirs are exposed to salt water intrusion from various sources as mentioned in Section 2.4. The salt balance contains the items given in Table 3.4.

Salt balance	
Inflow	**Outflow**
Salt load of the river	Salt drained off to the sea
Drainage of brackish water to reservoir	Salt in water extracted from reservoir
Underground inflow of salt water	
Diffusion of salt from the bottom	
Ship locks	
Leakage of sluice gates	

Table 3.4 Salt balance in coastal reservoir.

An essential point is that no salt is removed from the reservoir by evaporation. At the time of closure of the barrier dam the reservoir still contains salt water which has to be replaced by fresh water. In most cases this occurs through the inflow of fresh water from the catchment which mixes with the water in the basin and is eventually drained to the sea. In the case of the Glover Cove Scheme in Hongkong, the salt water was removed to the sea by pumping.

After the initial desalinisation of the reservoir, the salt content can be computed from the water and salt balance using the principle of continuity. In the (hypothetical) case of inflows and outflows of water and salt which are constant in time and assuming a complete mixing in the reservoir, the salinity of the water, c is given by the equation for steady conditions:

$$c = \frac{\text{rate of supply of salt to reservoir}}{\text{rate of water drained off}}$$

A representative case of a large coastal reservoir in the temperate zone is given by the IJssel Lake in the Netherlands. This fresh-water lake was formed by enclosure of the Zuiderzee, which was in open communication with the North Sea. A part of the area was reclaimed but an area of 1,200 km^2 was left as a fresh water reservoir on the IJssel, a branch of the Rhine river carrying about 10% of its discharge. Tables 3.5 and 3.6 contain the relevant data for the balances.

Water balance (in 10^9 m^3 per average year)

Inflow		**Outflow**	
River	10.1	Evaporation	0.8
Inflow from high lands	3.3	Extraction for flushing	2.0
Drainage from polders	2.6	Extraction for water supply	2.0
Rainfall on lake	0.9	Drainage to sea	12.1
Total	16.9	Total	16.9

Table 3.5
Water balance for Lake IJssel, the Netherlands.

Salt balance (in 10^6 kg Cl^- per average year)

Inflow		**Outflow**	
Salt load in river	1,097	Total evacuation to the sea and the surrounding supply areas	
Drainage from high lands	130		
	900		
Drainage from polders	9		
Salt in rain	212		
Locks and leakage			
Diffusion	100		
	2,448		2,448

Table 3.6
Salt balance for Lake IJssel, the Netherlands.

The average salinity of the water in the reservoir during an average year is then:

$$c = \frac{2448 \times 10^6 \text{ kg Cl}^-}{16.1 \times 10^9 \text{ m}^3} \equiv 150 \text{ p.p.m. Cl}^-$$

This is a very low salinity which is due to several favourable factors: the low evaporation rate, the high inflow producing a high drainage to the sea and the relatively low total salt load. As can be seen from the salt balance the inflow of salt consists of roughly 50% from external sources (the salt load of the Rhine river as mentioned in section 2.4) and 50% from internal sources (drainage of polders in the former Zuiderzee with seepage of brackish water). The figure for the salt load of the Rhine river is based on a prediction made at the time of planning of the IJssel Lake scheme (1935) with an assumed salt load of the Rhine around 200 kg/m^3 Cl$^-$. This figure was reached around 1955 but since then the salt load has almost doubled. This results in an average lake salinity of about 200 p.p.m. Cl$^-$. This is still relatively low but it should be taken into account that inflow and outflow of water and salt vary seasonally and that all figures mentioned above refer to steady conditions. Especially after winter periods with low water supply, the salinity during the ensuing summer can rise to figures exceeding 300 p.p.m. Cl$^-$. This can be computed by applying the water and salt balance to periods of one month.

The principle of continuity can also be applied for estimating the length of the period of desalinisation. Since the weather conditions during this period cannot be predicted, average conditions have to be assumed. Together with the assumption of complete mixing of inflowing water and water in the reservoir the following equation can be derived (see Figure 3.22):

$$c(t) - c = (c_0 - c)e^{-\frac{d}{V_0}t}$$

where:

$c(t)$ = salt content at the time t,
c = steady-state salt content,
c_o = initial reservoir salt content,
d = rate at which water is drained from reservoir,
V_o = reservoir volume.

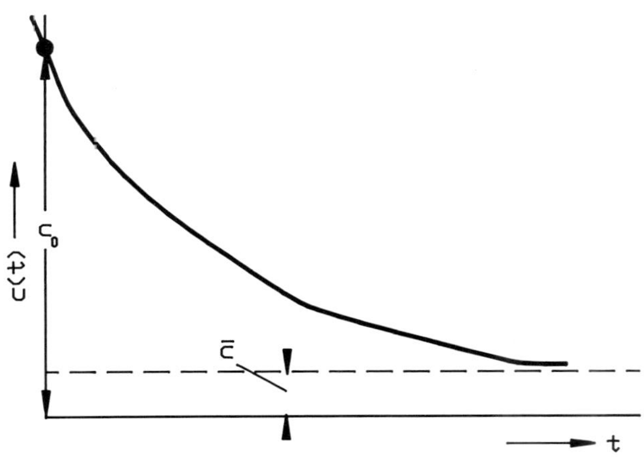

Fig. 3.22
Reservoir desalinisation by mixing.

It can be shown that for Lake IJssel, with a volume of about 13 x 10^9 m^3, the time to desalinate is about 5 years. Clearly this time period is governed by the factor d/V_0. In this example, if this factor had been approximately 0.5 instead of 1.0, then the desalinisation would have taken about 10 years. There is no underground inflow of saline water in the IJssel Lake, neither from the sea nor from surrounding higher areas with saline groundwater. This is due to the presence of semi-pervious layers at a shallow depth and the fact that the difference between M.S.L. and mean lake level is small.

The underground inflow of saline water from surrounding areas to the reservoir was the cause of the failure of a small project in the south-western part of the Netherlands. In the Braakman inlet on the Western Scheldt, it was planned to partially reclaim the area and to transform the remaining area into a fresh water reservoir (150 ha.). After enclosure a normal operational level of 0.6 m below M.S.L. was adopted so that the land could be reclaimed without embanking and pump lift drainage. This generated an inflow of saline water from the sea and especially from the surrounding areas with highly saline groundwater.

The balance equation with which the computations are carried out is based on the assumption of a complete mixing of the fresh and salt waters reaching the reservoir. Horizontal and vertical mixing is due to convective flows to the outlet, displacement of water by winds and stirring by waves. The result of non-complete mixing is a non-homogeneous distribution of the salinity in the reservoir with the highest concentrations near the points where relatively salty water is entering the reservoir. Depending on the locations of these points with respect to the outlets, and the operation of these outlets, a certain portion of the inflow of salt will leave the reservoir without completely mixing with all the waters of the reservoir. The assumption of complete mixing leads to predictions of the average salinity which are somewhat on the pessimistic side.

Another factor which is not easily accounted for is the tendency of saline waters to concentrate in local deep gullies due to density differences. It is difficult to remove this saline water by sluicing, even when these pockets are situated near an outlet. This is the reason why the desalinisation of a deep coastal reservoir at Hongkong had to be done by pumping.

3.4.3 Feasibility of coastal reservoirs

From the point of view of storage of water, coastal reservoirs are less efficient than storage reservoirs in mountainous areas. In the latter case variations of the storage levels of the reservoirs of many metres are not uncommon, whereas in the case of reservoirs in low-lying coastal areas this variation is often limited to a few decimetres. According to the present reservoir regulation of Lake IJssel, the normal operational level in summer is 0.2 m below M.S.L. and in winter 0.4 m below M.S.L., a variation of only 0.2 m. This can be increased by a few decimetres if provisions are made for the flood protection and the drainage of the adjacent areas.

Especially in (semi-)arid zones, a much greater portion of the water in storage in coastal reservoirs is lost by evaporation than in reservoirs in the mountains. There are two more reasons why the application of coastal reservoirs in (semi-)arid zones may not be feasible from a hydrological point of view. The first is that to ensure a sufficiently low salinity of the water in the reservoir a considerable portion of the inflow of fresh water must be drained off. The second reason is that the ratio between the annual drainage volume and the volume of the reservoir must not be low since this rate governs the length of the period of desalinisation of the reservoir. If this period is too long, the benefits of the project, discounted to present day values, may not justify the construction of the project. An example is the plan for a coastal reservoir in the Ensenada de la Broa in Cuba (UNESCO, 1972). This plan was rejected on the grounds that the project was neither feasible from the hydrologic nor from the economic point of view. Depending on the alignment of the enclosing dam the water area of the embayment that could be transformed into a reservoir ranges from 1,200 to 2,600 km^2. The corresponding volume ranges from 4.7 to 10.9 billion m^3. With an outflow of 1.1 to 1.3 billion m^3 per year it would take more than 25 years before the desalinisation would be complete. The ultimate salinity would not be less than 2,000 p.p.m. Cl^- because of the high salinity of the inflowing surface and sub-surface water.

Transformation of tidal inlets into freshwater reservoirs entails a drastic change of the environmental conditions. This has many consequences for the ecological systems of the inlets and estuaries and also of the surrounding areas and may be an argument against the intervention in the natural hydrological and ecological system. Enclosure of tidal inlets means an elimination of the tidal flows which generate a permanent flushing and removal of possible pollutants. The flushing that can be applied after enclosure is much smaller. Therefore much attention should be paid to the quality of the inflowing water.

3.5 Reclamation

This section reviews reclamation potentials and constraints in deltaic areas, especially focusing on tidal forelands and coastal marshes, tidal inlets, estuaries and lagoons. The reclaimed land can be used for various purposes, originally exclusively for agriculture, nowadays also for urban and industrial purposes, as well as for infrastructural works, like ports, airports and highways. In principle two reclamation methods can be followed: impoldering and land-fill. The last method will be briefly described at the end of this section, whereas the polder concept will be introduced in the first paragraph. The description of distinct reclamation techniques for the above mentioned land types is given in the following paragraphs, which also deal with flood protection, drainage and water supply after reclamation, and effects of reclamation on the surrounding environment.

A third and primitive method of reclamation which is often applied as the first phase of an impoldering process consists of installing a system of open channels. These channels, which are in open communication with each other and with the estuary, make the marshy or swampy areas accessible and provide rudimentary drainage and irrigation. This method is applied on a large scale for reclamation of lowlands in Indonesia, the Mekong River delta in Vietnam (see Appendix 2) and the Chao Phya delta in Thailand. This system can only be applied in the absence of storm surges at sea and flashy river floods.

3.5.1 The polder concept

The word polder originates from the country that is often identified with that word: The Low Lands, the Netherlands, 'Pays-Bas', 'Paises Bajos'. A polder can be defined by:
'... a level area, in its original state subject to high water levels (permanently or seasonally, originating from either ground water or surface water), but which through impoldering is separated from its surrounding hydrological regime in such a way that a certain level of independent control of its water table can be realized' (ILRI, 1983).
Based on this definition, distinction can be made between polders in areas where waterlogging (or even inundation) occurs either permanently (swamps, shallow sea and lake beds) or temporarily (tidal lowlands, seasonally flooded river plains). This leads to a distinction into three groups of polders, namely:
- impoldered low-lying lands;
- lands reclaimed from the sea;
- drained lakes.

A polder includes facilities for flood protection (dykes) and water management (drainage and irrigation, if needed). The drainage infrastructure consists of a field drainage system (surface and/or subsurface drains), a main system and one or more outlets (tidal drainage sluices and/or pumping stations; see Figure 3.23).

Field drainage system

The field system gathers the excess water from the land by means of a network of field drains. Where necessary these are supplemented by measures which promote the flow of excess water to these drains. Two principle systems may be distinguished:
- subsurface drainage systems, which are used in soils in which the excess water is able to infiltrate and percolate through the main root zone to the subsoil/substratum and then move as groundwater flow to the drains;
- surface drainage systems, which are applied where the infiltration or percolation of excess water is impeded at the surface or at shallow depth in the root zone due to the presence of poorly permeable layers.

Main drainage system

Two categories of drains can be distinguished in the main drainage system, namely collector drains and main drains. Collector drains receive water from field drains and transport it to the main drains, which in turn transport the water to the outlet. Several flow control structures, like weirs and culverts, may be installed in the main system.

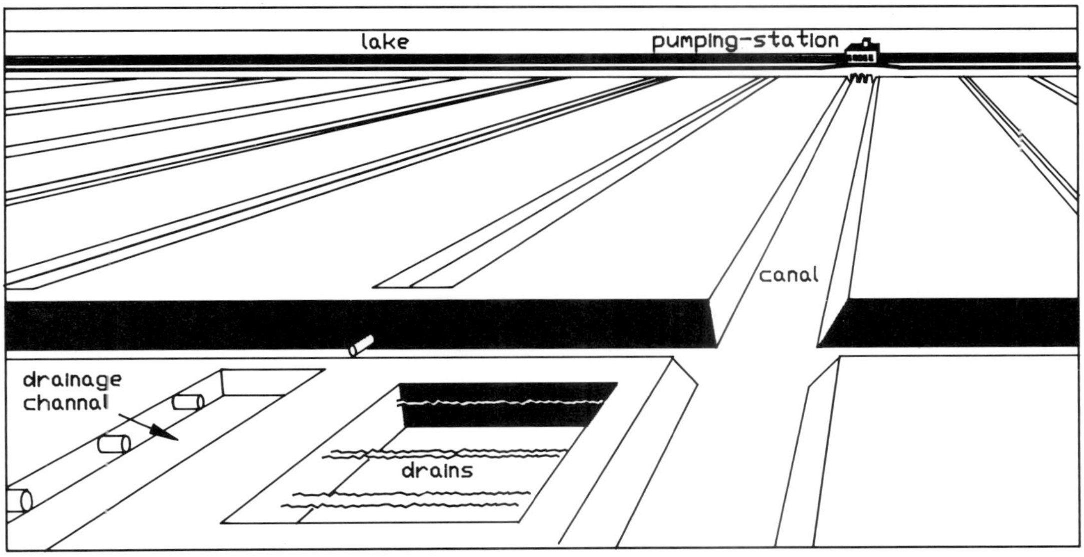

Fig. 3.23
Schematic lay-out of a drainage system in a polder (Schultz, 1982).

Outlet

The outlet is the terminal point of the whole system at which the drainage water discharges into a major element of the natural open water system of the region (river, lake or sea). The water level at the outlet constitutes the drainage base for the area concerned. This level, relative to the land level, governs the amount of hydraulic head available for the drainage flow. It determines how far the water levels may be lowered below the land surface. It also determines whether the area can be drained by gravity or requires pumping.

Assuming the presence of a reliable flood protection, the water management system forms the basis for the technical performance of a polder.

3.5.2 Reclamation of deltaic areas

Lay-out of main drainage and irrigation system

A farm turn-out from a tertiary irrigation canal in the Hachiro-Gate polder in Japan. The polder has separate drainage and irrigation canals.

Figure 3.24 indicates the possibilities for the supply of irrigation water in deltas (s is the limit of the saline reach during low flow periods). Lay-out A comprises a diversion dam in the main stem and regulators r in the branches. Upstream from these structures, main irrigation canals branch off, which are aligned on the natural levees. From these canals, secondary canals convey the water into the backswamps. In this way supply of irrigation by gravity can be realised.

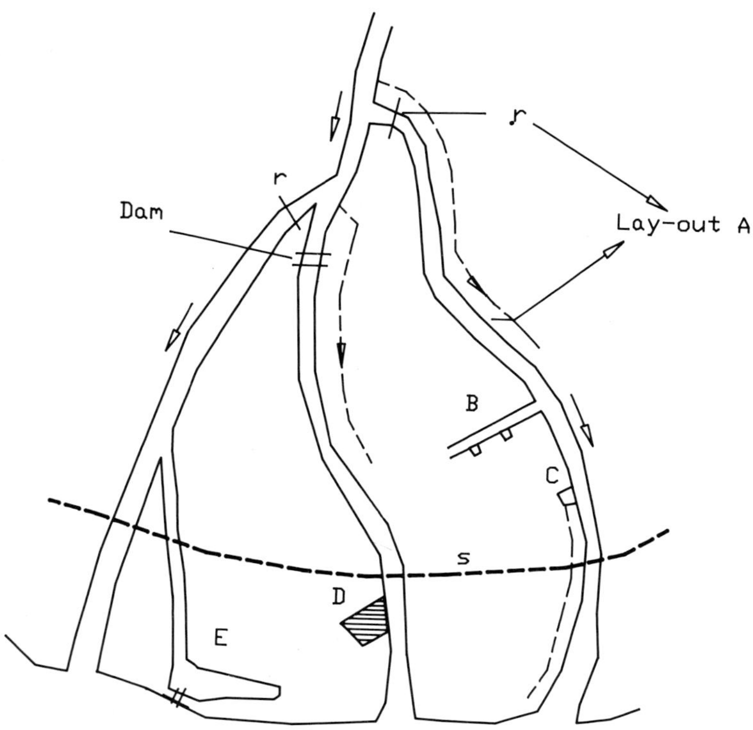

Fig. 3.24 Schematic presentation of polder development in a delta.

B is a low level irrigation canal, from where water can be pumped on to the fields using branch canals and individual or collective low lift pumps. C is a pumping station, which conveys water into a high level irrigation canal aligned on a natural levee. In the coastal zone, storage areas D can be made to store excess drainage water during wet periods. Fresh water can also be stored in the coastal zone by damming off estuaries or lagoons (E) to form coastal reservoirs (see chapter 3.4).

Possibilities for gravity drainage of excess rain water in deltas are limited, as periods of heavy rainfall generally coincide with periods of high river stages. In case the floods are of short duration, excess water can be discharged intermittently. In the coastal zone, however, the tidal variation allows for evacuation of water by gravity, even when the river is in spate. To allow for gravity drainage, the hydraulic head must be larger than the sum of:

- depth of the water level in the field drain below the land (1.0 to 1.5 m for dry foot crops and 0.2 m for paddy fields);
- required head in the drainage canals (0.2 to 0.4 m for drainage canals with a length of 5 to 10 km);
- head in the tidal sluice (this value should not be smaller than 0.3 m).

The conclusion is that for the average canal lengths, the minimum value of H is approximately 0.8 m for paddy fields and approximately 1.8 m for dry foot crops. Lands along river reaches for which the available head H is too small to allow for gravity discharge can be drained by conveying the drainage water through a canal in a downstream direction to a point where the hydraulic head is sufficiently large. In upstream reaches of a delta and in river valleys where the land gradients in the downstream direction are 10^{-4} (10 cm per km) or more, this can be effected by designing this canal with a slope smaller than that of the river and the land so as to increase the head.

Water control in the interdistributary areas

The degree of water control in the land areas depends, on the one hand, on the crops to be grown and on the yields to be obtained and, on the other hand, on the investments that are considered to be justified. A high degree of water control can be obtained by embanking the area so as to exclude the flood waters and by making a system of separate canals for irrigation and drainage (see Figure 3.25a).

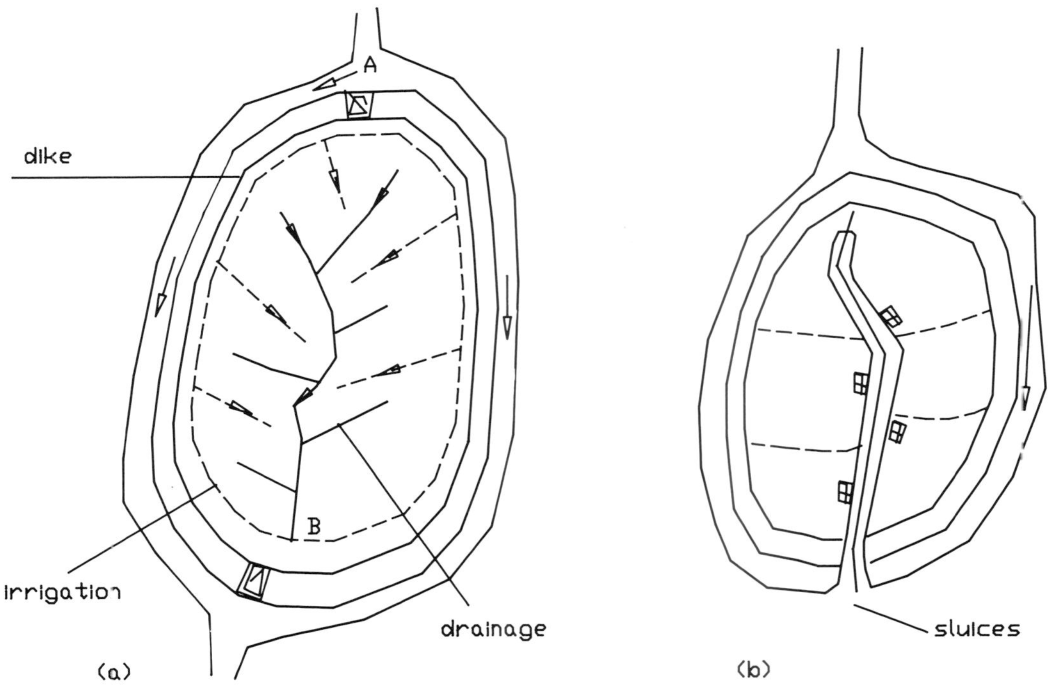

Fig. 3.25
Impoldering in
a) interdistributary areas in upstream portions of delta and
b) an alternative lay-out.

The main irrigation canal is aligned on the natural levee and supplied by pumping station A. Branch canals convey the water into the backswamps. The existing creeks are incorporated into the drainage system and act as main drainage channels, possibly after channel improvement. Branch canals are made to promote the runoff. Excess water is removed by the pumping station B. In this way the branch irrigation canals alternate with the branch drainage canals. According to this lay-out, the entire drainage system is at a low level and a single pumping station is serving the entire area.

Irrigation canal in the Sindh (delta of the Indus in Pakistan).

In the alternative lay-out of Figure 3.25b, the main drainage channel is at a high level and drains by gravity to the river. The total area is divided into subareas or sections, each provided with a pumping station to drain to the high level channel (the irrigation canals have been omitted). Under certain conditions of local topography, this alternative lay-out has advantages over the former with respect to the adaptability to topography and drainage requirements in each section, safety in operation and step-wise execution.
In the tidal reach of a delta, possibilities exist for gravity drainage. Since the natural levees are less developed in the coastal zone than in upstream areas, the main irrigation canal can be aligned outside the natural levee. Water for irrigation may be abstracted from a number of locations along the river, as long as the river water is fresh (especially during the dry season).
In the intermediate reach of a delta, the pumping facilities for irrigation and drainage may be combined in the same pumping station, which serves both the irrigation and the drainage networks.

3.5.3 Reclamation of tidal forelands and coastal marshes

Tidal forelands are under influence of coastal sedimentation or erosion. In case of sedimentation, reclamation of tidal forelands may be feasible when a sufficiently large area has reached a certain minimum elevation above mean sea level, for which the average high water level is a first indication. The smaller the difference between the land elevation and the high tide and flood levels, the easier and less costly the embanking will be and the better the possibility for gravity drainage. On the other hand, the need for new land may require earlier reclamation.

In many countries, the reclamation of tidal forelands is the traditional way of impoldering, often carried out by local people. The advantages are:
- the works are generally small and do not require any heavy or special equipment;
- the land reclaimed has a high elevation with respect to the tidal levels, so dyke construction and drainage performance are relatively easy while the damage caused by flooding is not very serious;
- the works can be carried out in stages.

The disadvantages are:
- the relatively high costs, as large lengths of dykes are needed per unit reclaimed area;
- difficulties involved in the drainage of both the adjacent areas and the polder itself;
- the delay in waiting for the necessary stage of foreshore siltation;
- the difficulty in designing a well balanced lay-out.

The problems can be alleviated by establishing a master plan, which aims at the ultimate development of the prospective tidal foreland and the existing inland areas. The development can be effected in stages, when the relevant partial schemes fit in the master plan. The latter should be flexible and subject to periodic revision in the light of the actual progress of the works and the advancement of techniques.

Several strategies to reclaim tidal forelands developed worldwide are mentioned below:

- tidal forelands in Western and Southern Europe (either along the coasts or in a delta) are reclaimed, when they have risen naturally to 0.1 to 0.3 m above mean high tide level (being the maximum possible elevation) because:
  - this elevation allows for gravity drainage;
  - there is less impact of waves on the dykes;
  - as siltation takes place on sand, a thickness of 1 to 1.5 m of clay with a sufficient retention capacity (dry periods) was considered necessary.

  Drawbacks of this policy are:
  - impoldering proceeds in small steps (1,000 to 2,000 m width);
  - intermediate periods are rather long;
  - the site is inaccessible for dredging equipment of any considerable capacity;
  - under semi-arid conditions, the presence of a thick layer of clay is no guarantee against the negative effects of drought.
- in the deltaic areas of the Ganges in India (Sunderbans in West-Bengal), islands have been reclaimed by the local farmers through constructing bunds or ring dykes. In many cases, the elevation of this land is not higher than 1.5 to 2.0 m below high tide level. As a result, the conditions for gravity drainage are poor, there are many difficulties in closing breached dykes and considerable damage may occur through flooding. Reclamation of such land is therefore considered untimely or premature; according to the present Indian regulations no land can be reclaimed, until its elevation is higher than 0.3 to 0.5 m below high water;

- in Bangladesh, the mean sea level was considered a suitable level for reclamation;

- in Japan the dykes are constructed at low water ordinary neap tide, as there is an urgent need for more arable land, which will be used for rice cultivation.

It can be concluded that there is a certain minimum land elevation which determines reclamation possibilities. This elevation is dependent on the possibilities for gravity drainage, the need for new land and the available budget. A compromise must be found between geological, pedological, hydrological, social and economic factors, to decide at which stage of sedimentation the land can be reclaimed.

Measures to promote natural siltation

Several methods have been developed to accelerate the sedimentation of the tidal foreland, which are based on the following principles:

- the suspended silt is allowed to settle during the slack water period after high tide;
- the deposit is given the opportunity to dry, so that the silt adheres to the surface and will not be removed during the next flood period. This requires some sort of drainage system.

One of the methods that can be used is the Schleswig-Holstein method, which was introduced in the Netherlands in 1935 and which was modified in a later stage (see Figure 3.26). The essential elements are open silting basins or sedimentation compartments. This system can be applied at a much earlier stage of siltation, for instance when the foreshore is still at an elevation of about mean sea level in case of a maximum tidal amplitude of 3 to 4 m. The deposition of sand and silt is promoted by various species of plants, such as *Spartina Townsendii,* which may develop at regular intervals when the fields dry out. The crests of the groynes are located at 0.2 to 0.3 m above mean high tide. The works commence with the construction of two fields, each of 400 x 400 m. The trenches in the low compartments must be larger than 0.6 m wide and 0.3 m deep. In later stages, narrower and deeper trenches can be used. Furthermore, experience shows that wide fields with wide trenches obtain better results than narrow fields with narrow trenches, and that trenches perpendicular to the dyke are more beneficial than trenches parallel to the dykes. Prevailing winds may affect siltation, like strong north-westerly winds causing low sedimentation rates in the south-eastern corner. Compartments smaller than 400 x 400 m or sub-division by earthen dams into sub-compartments of 100 x 100 m should then be considered.

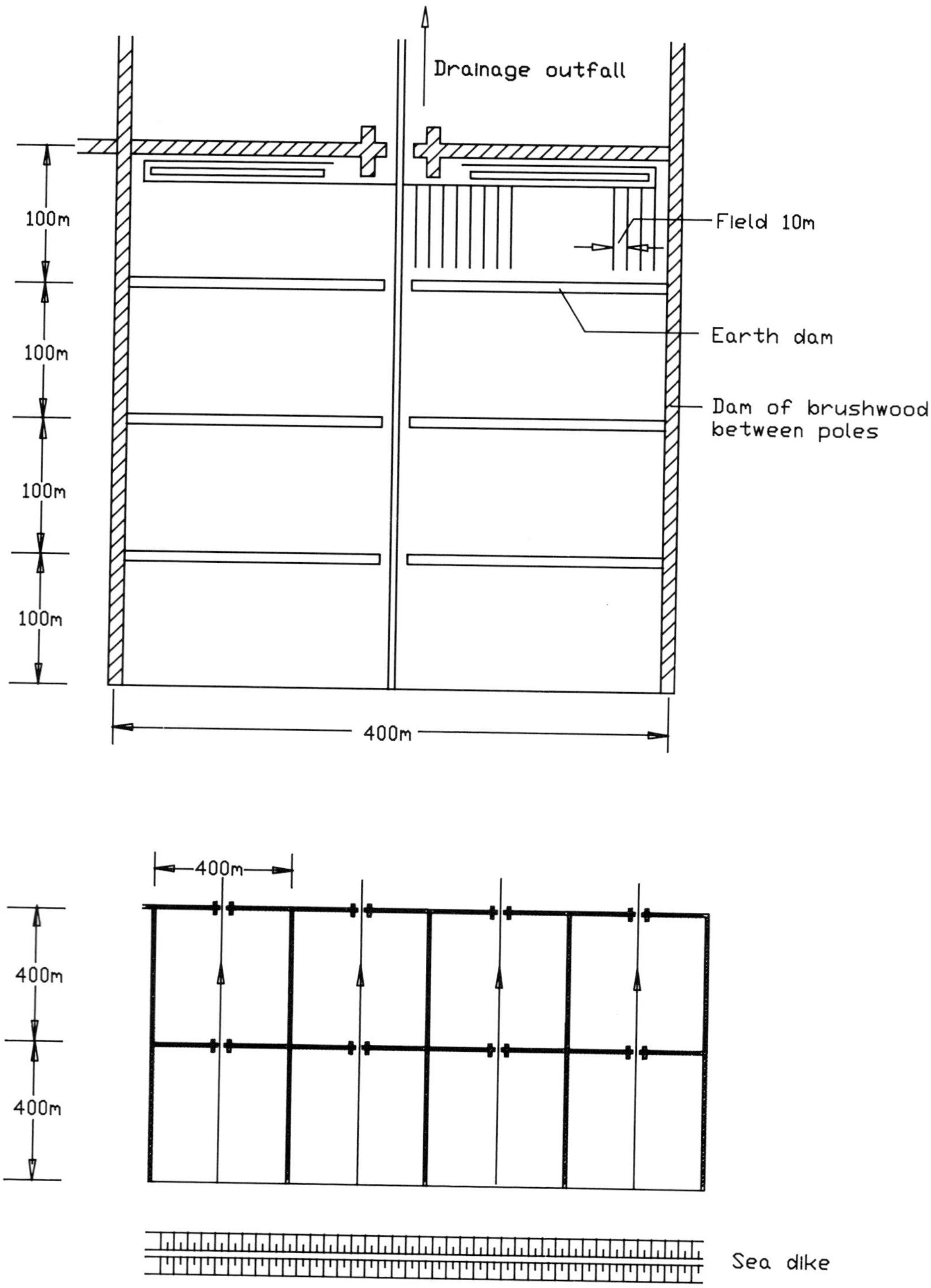

Fig. 3.26
Modified Schleswig-Holstein system.

Only tidal flats which are located well above the low water level are suitable for measures which aim at the promotion of the siltation, as construction and maintenance of the dams in other cases might be too expensive and sedimentation is not guaranteed. In countries where dykes are built near the low water level, these methods will not further increase the reclamation possibilities. The methods might then be of use for raising the foreshore and reducing the wave impact on the dykes.

Drainage and irrigation of reclaimed tidal forelands

Gravity drainage during low tide (tidal drainage) is possible for a reclaimed tidal foreland with an elevation near high tide level. Ample attention should be given to the possible siltation of the outfall channels, where the deposits may completely obstruct the outfall. In case the outfall is located on a tidal creek, sedimentation may occur, because the tidal currents which originally maintained the creek, are cut off by the dyke. This problem can be avoided by locating the outfall on a creek, which is not cut off by the embankment. If this is not possible, flushing basins can be designed for periodic flushing of the outfall channel. Such basins consist of a reservoir surrounded by a high dyke and additional gates (see Figure 3.27). To perform flushing, gate g is closed. By opening sluice s during the flood period, the reservoir is filled to high tide level. Then at the time of low tide level outside the gates s are instantaneously opened and a strong current (head z_o in the sluice) is generated, thereby removing the silt in front of the sluice. The effect is rather local, but can be increased by guiding jetties which keep the flow concentrated.

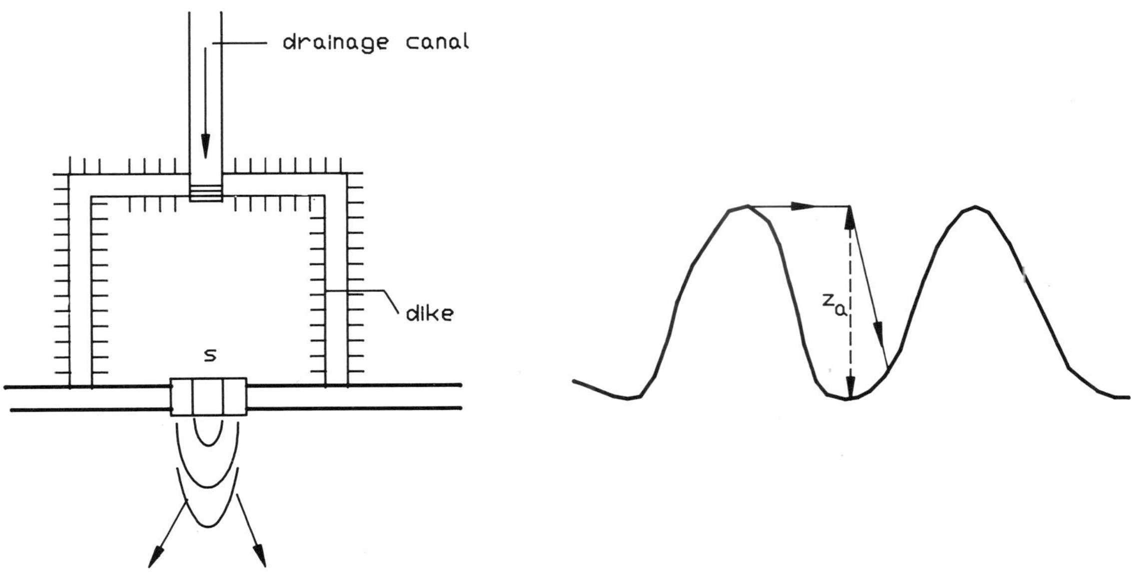

Fig. 3.27
The use of a flushing reservoir for the prevention of sedimentation in front of an outfall.

Lay-out of reclaimed tidal forelands

When a tidal foreland has attained the maximum possible elevation it is advisable to leave a strip of 100 to 300 metres width between the protection dyke and the average high water line, as this strip, especially when covered by mangrove trees, effectively reduces the impact of waves on the dykes (see Figure 3.28).

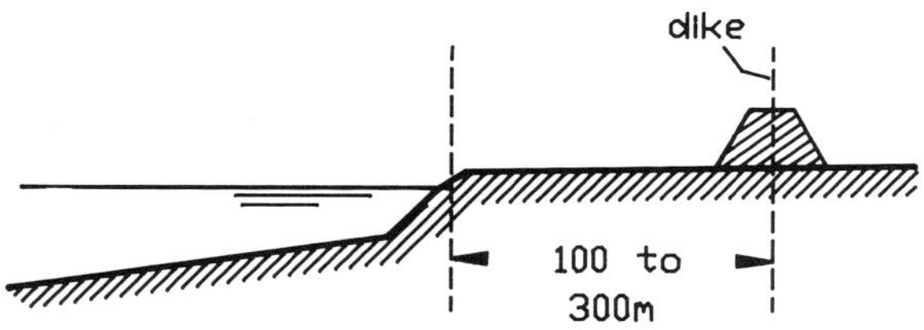

Fig. 3.28
Dyke alignment, leaving 100 to 300 m tidal foreland as protection.

In case of progressive reclamation of the forelands, the inland areas are flat. The outfall becomes longer and longer, and the loss of head increases. Moreover, because of subsidence, the older polders are at a lower elevation than the more recent ones. In such a case it may be better to apply turned back dykes to realise free outflow. The channel outside will, of course, be subject to silting up (see Figure 3.29).

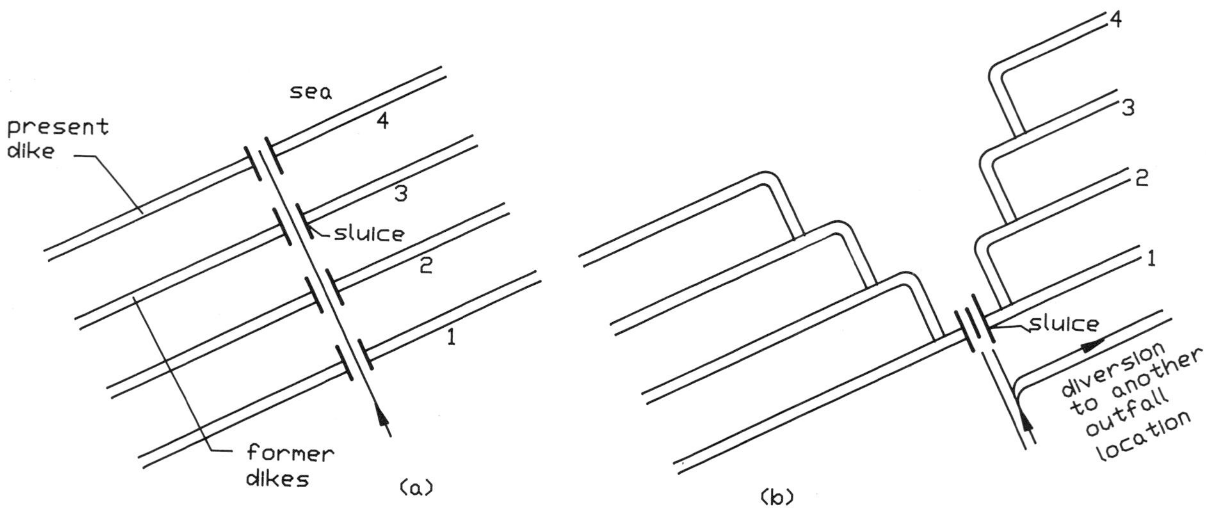

Fig. 3.29
Progressive reclamation of tidal forelands.

In case irrigation is required, a separate system of irrigation canals has to be provided. Since the ground water in the coastal areas is usually saline, the source of water will be the river where inlet of high tide level may be possible (tidal irrigation).

Dalta of the Euphrates and Tigris in Iran. Tidal irrigation of date palm gardens, which where damaged by sea water intrusion when the flow of the Karum River, which also flows into the delta, was reduced to fill a large reservoir upstream.

Reclamation of problem soils

Reclamation of tidal forelands involves a drastic change of the natural marine environment, in which a valuable ecological system is replaced by another one, which may be equally valuable but entirely different. Although the soils are generally of good quality, "problem soils" may be encountered, like acid sulphate soils, peat soils, and sandy soils, which all need their own special treatment and pose limitations as to their use.

Reclamation of acid sulphate soils (see chapter 1.4) implies special measures, often at high costs. Also after reclamation the agricultural costs are relatively high, with low to moderate yields. Reclamation of acid sulphate soils means the provision of separate canal systems for irrigation and drainage. Depending on the available funds, availability of calcium, envisaged agricultural practice, etc., it can be decided to keep the soil wet at all times, so as to prevent oxidation of the pyrite and thus acidity, or to promote pyrite oxidation, leach the acidity and neutralise the soil through application of lime. For the first alternative, the ideal agricultural practice in tropical areas is rice cultivation the whole year round, whereas in temperate zones grasslands (cattle) allow for high groundwater tables. Other possible crops in tropical areas are coconut palms, oil palms, mangos and pineapple, as their sensitiveness for acidity is relatively low. Drainage, when applied, should be limited to shallow depths (in the order of 0.45 m). Lime and potassium should be applied for de-acidification of the top layer, which means that in the first years only crops with shallow root systems can be grown. After many years, when the acid is leached from the soil or neutralised by applying calcium, the drainage depth can be increased to approximately 0.90 m, to allow cultivation of other crops. Reclamation of acid sulphate soils of a high acidity (pH 2 to 4) is very expensive and often economically unfeasible. In case reclamation is necessary, the acid soil can put on narrow ridges (2 to 6 m) and leaching can be applied with high amounts of lime and a low level in the ditches.

Because peat soils have a luxurious natural vegetation, are flat and reasonable crops grow at the perimeter of the basin peats (in deltas and lagoons), these soils are often considered to possess a high potential for agricultural production. However, peat soils can be considered to be even more problem soils than acid sulphate soils (this also applies to muck soils). In the first place, there is an economic limitation because of the high need for fertilizers, as peat is poor in basic nutrients (K, N and P). Secondly, because of the high content of organic matter, humic and organic acids are formed in an aquatic environment (like paddy fields), which damage the roots of the plants. For dry crops, these acids are removed by leaching (rain, irrigation water and drainage). The foregoing explains the bad experiences with paddy cultivation on peat soils: poor yields, even with liberal application of fertilizers or soil dressing with 10 cm clay applied on the peat (Japan Hokkaido).

Although peat soils are considered not suitable for growing rice under normal conditions prevailing in flooded paddy fields, reasonable yields could be obtained under special conditions, where favourable aeration of the soil could take place. This might occur under the effect of the tides and/or by removing the water after an initial period of flooding and keeping the soil close to saturation point. In this way a better root system could develop, as the effect of tides gradually leached out the harmful organic acids from the soils while the saturated conditions made the peaty material more compact for root holding (ESCAP, Bangkok 1977). Promising crops are tapioca, tobacco, pineapple and oil palms, provided that fertilizers and lime are applied.

Drainage and subsidence of peat areas

Planning drainage (and irrigation) in peat areas requires quite a strategy. Unless the area is kept constantly under water, subsidence of considerable magnitude will occur. Unfortunately the subsidence of peat soils is most difficult to predict. One thing is certain: as a result of oxidation, all peat will disappear ultimately. The following considerations should be addressed:

- reclamation of peat will start with the provision of shallow drainage. Because of the subsidence in time, it must be possible to easily lower the drainage level when needed. Drainage canals can be deepened, but the question is whether intakes and outlets of irrigation and drainage structures, which are usually founded on solid formations, will remain at a low enough level;
- subsidence will differ from place to place, depending on the thickness of the peat and the organic matter content. The question is whether under these conditions drainage and irrigation canals will remain functioning properly;
- ultimately all peat will disappear. What is the composition of the underlying mineral base (acid sulphate soil, sandy soil) and what is the elevation of this base. Will pump-lift drainage be necessary?

Concluding, it can be stated that peat reclamation implies that in the ultimate stage, basin peat will have disappeared and gravity drainage will not be possible any more (see Figure 3.30).

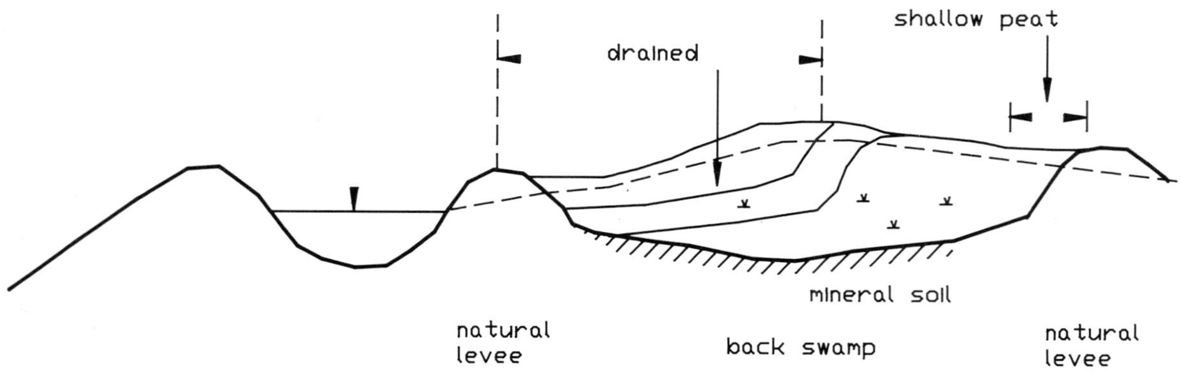

Fig. 3.30
Subsidence of peat due to reclamation.

3.5.4 Reclamation of tidal inlets, estuaries and lagoons

Deep polders in coastal areas

The areas which will be considered in this section are normally submerged and have an elevation which is generally a few metres below mean sea level and also below low tide levels. In the coastal zone, such areas are found in tidal inlets, estuaries and lagoons. Reclamation of these areas involves the creation of deep polders, whose continued existence depends on the preservation of the protecting dyke and pump-lift drainage. Such polders are created by constructing an encircling dyke and by removing the water by pumping. After the initial drying of the soil, the pumping stations serve for removal of excess water from rainfall and seepage. The new lands consist of the bottom of the former sea, lagoon or lake.

Reclamation in the Republic of Korea. Closure of last gap of the embankment using the vertical method. Following this method a horizontal stone sill is raised using a folding bridge from which stones are dumped from a narrow-gauge railway wagon.

It is a rather costly form of gaining new land and only economically justified when there is a great need for arable land and/or space to live. Owing to the low elevation of the new land with respect to high tide and storm surge levels deep polders are rather vulnerable.
The Netherlands have acquired a special reputation in this extreme field of impoldering and the largest acreage of deep polders is found in this country with minimum levels of 7.0 m below MSL (mean sea level). However, the Netherlands are not the only country where such deep polders exist; impoldering works in the Lammefjord in Denmark (6.7 m below MSL), the Vistule Delta in Poland (2.5 m below MSL), the Nile Delta in Egypt (5.0 m below MSL), Hachiro Gata polder in Japan (5.0 m below MSL), Kuttanad polder in Kerala, India (3.0 m below MSL) and polders in San Joachim Delta, Sacramento, California, USA (4.5 m below MSL) are examples of deep polders around the globe.

From the foregoing it can be concluded that the creation of such deep polders is feasible under a wide range of conditions. However, deep polders do not occur at the very coast line, facing a sea with storm surges and high waves, as such a situation would be too dangerous. Deep polders are situated in more sheltered locations like a tidal inlet, an estuary or a lagoon (see Figure 3.31).

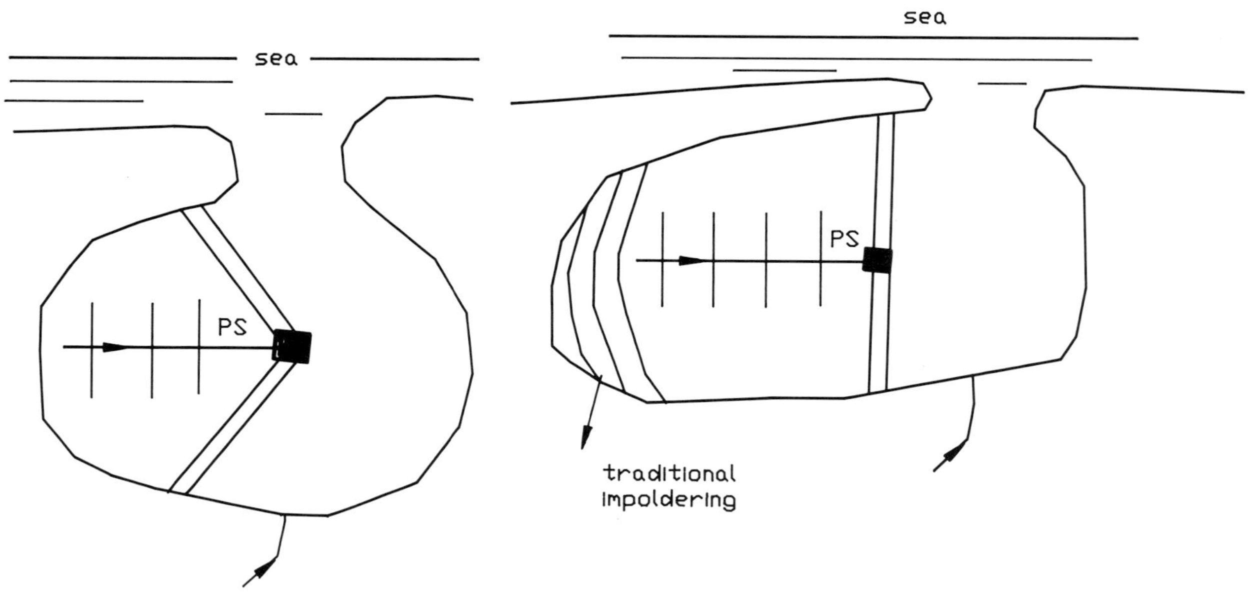

Fig. 3.31
Examples of deep polders in sheltered locations.

Bangladesh south-western region. Closure of the last opening in an enclosing dam of a tidal creek. Use of locally available materials (rolls consisting of clay, bamboo and jute).

Technical aspects of the reclamation of deep polders

Next to the technical aspects common to all types of impoldering, the reclamation of deep polders involves a number of special aspects:

- safety of the impoldered area;
  Since the existence of a deep polder depends fully on a proper functioning of the encircling dyke, special attention must be paid to its design. Depending on the number of inhabitants, whose life depend on the dyke, and on the economic value of the infrastructure, the dykes should be able to resist a storm surge with a return period in the range of 500 to 10,000 years.

Wave runup should be accounted for and moreover a liberal freeboard (1.0 to 1.5 m) should be applied to cover the subsidence and compaction that are expected to occur during the forthcoming period of say 50 years. Stability must be ensured by applying gentle slopes (1:4 to 1:8). The size of the polder should not be smaller than 5,000 to 10,000 hectares, as in case of dyke failure smaller polders are filled too quickly, giving the inhabitants no time to escape. Pumping stations and, if possible, houses should be "flood proof" (e.g. emergency top floor). During heavy storms, constant vigilance is required and emergency equipment (including army helicopters) must be available. Also a warning system is necessary;

- occurrence of saline seepage;
  In case the deep polder is bordering salt water, the original ground water will be replaced by sea water and there will be a continuous inflow of salt. If the seepage rate is high, the water in the canals may show such high salinity levels that polder construction could not feasible. If the estimated saline seepage is not too high, it can be collected in a section of the polder, which is drained to the pumping station separately from the other drainage canals;
- effect of the deep polder on the groundwater table in the adjacent areas;
  Because of the drainage of the deep polder soils, a groundwater drawdown in the adjacent higher areas might occur, with subsequent agricultural damage due to water shortages. In such cases the provision of so-called border lakes might be required;
- influence of the ripening on the functioning of the drainage system;
  A proper functioning of the pump-lift drainage is essential for the survival of a deep polder. Hence it will be wise to provide a deep polder with at least two pumping stations, operated with different fuelled engines (like electricity and diesel). One of the two can be installed on the boundary of the two sections with one or two pumps on the high section and one or two on the low. The initial soil ripening is the result of removing the water from the very top soil by pumping with the pumping stations. The required capacity is approximately the same as required for normal drainage operations (1 to 2 l/sec/ha for temperate zones and 3 to 6 l/sec/ha for the humid tropics). The initial emptying of a 4 m deep basin will then take 300 to 100 days;
- the hydrologic effect of the polder on the tidal regime of the estuary or lagoon;
  Impoldering and/or enclosure of tidal inlets and estuaries diminishes the area of the tidal basin and changes (increase or decrease) the tidal levels and currents in the channels communicating with the sea. This may have undesirable side effects, like silting up of harbour entrances, hampering of tidal drainage, or erosion by scouring. Mathematical or physical models can be helpful in the quantitative prediction of these effects.

Reclamation with enclosure and coastal reservoir

A simple solution for the safety and saline seepage problems can be the combination of reclamation of deep polders with the construction of an enclosure dam of the tidal inlet, and leaving a water area between the polder and the enclosing dam. The inflow of fresh water from the river and drainage of excess water to the sea could result in a transformation of the enclosed salt water area into a fresh water coastal reservoir (see Chapter 3.4).

Location of polders and reservoirs

In projects for enclosure and partial reclamation of tidal inlets, estuaries and lagoons, a decision has to be taken on the alignment of the polder dykes and the location of the reservoir. The following factors should be considered when deciding on the distribution of land and water areas:

- depth configuration of the tidal inlet;
  The deepest parts of the sea bottom can better be left submerged and form (a part of) the reservoir area;
- soil conditions;
  For obvious reasons, sandy soils and problem soils can better be left submerged. They partly coincide with the deep patches with gullies;
- seepage conditions;
  Areas with poorly developed semi-pervious top layers can better be avoided;
- bearing capacity of soil;

Whenever possible, the alignment of the polder dykes should avoid areas with thick soft layers of clay or peat, as they generally have low bearing capacities;

- general configuration;
  The general configuration (place where the river is debouching, desirable locations or irrigation intakes) should also be taken into account.

3.5.5 Reclamation for urban and industrial purposes

In general, impoldering provides the best solution for lands which should be used for agricultural purposes and with poor subsoil conditions. Land to be used for establishment of industries should in most cases be reclaimed by filling. The choice between reclamation by impoldering or landfill depends on:

- the envisaged land use. For non-agricultural land use in relatively small areas, landfill with sand might be more attractive than impoldering, as the expected subsidence will be smaller and realised in a shorter period, drainage of the area is a relative simple matter and sand is a good base for construction sites;
- the required safety. From a point of view of safety, reclamation by landfill to a level well above the maximum water level is preferred to reclamation by impoldering, where all depends on the strength of a relatively fragile dyke;
- the costs. Reclamation by landfill is relatively expensive, depending on the depth of fill and the acreage. If, for instance, a fill of 2 m would have to be applied, the costs would be (at a unit price 1980 of US$ 1 to 2 per m^3) $ 20,000 to $ 40,000 per ha, which is prohibitive for agricultural lands, but which may be attractive for urban or industrial development.

When the fill is applied on a soil with a low bearing capacity, more fill has to be used than corresponding with the required height, because of the subsidence of the soft layers (Figure 3.32a) or because of their replacement by sand (see Figure 3.32b).

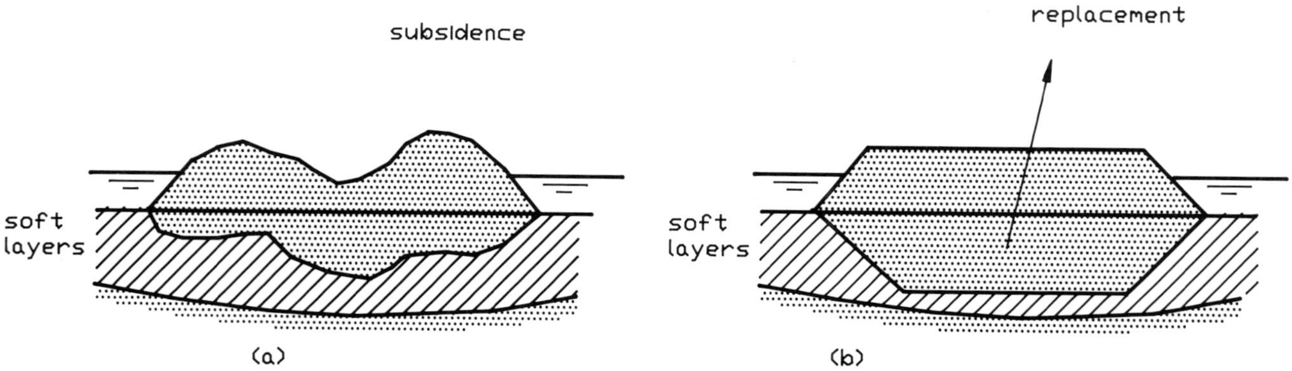

Fig. 3.32
Two methods of landfill, a) on top of the soil, causing subsidence of the sub-soil and b) landfill by replacement.

As far as urban extension and industrial sites are concerned, filling up with clean sand is preferred. However, even for this purpose the costs may be too high. Examples are:

- Amsterdam, the Netherlands;
  The ancient town centre was raised by filling to an elevation of about 1 m above mean sea level. For the new residential quarters such filling is considered too expensive and only a rather thin layer of sand is used;

- Lelystad, the Netherlands;
  Lelystad is the capital of the Zuiderzee Polders (with a ground level of 5 m below mean sea level). A part of this city was raised by landfill, but the remainder of the town is at the low level of the reclaimed former sea bottom;
- Bangkok, Thailand;
  When this city was built at the end of the 19th century in the deltaic area of the Chao Phya River, no landfill was carried out. As, due to subsidence, the ground elevation became less than 1.0 m above mean sea level, flooding of large parts of the city occurs when the river is in spate. In this case of an existing city, improvement can only be obtained by impoldering (encircling dyke and pumping stations on the canals);
- The problem of city extension by filling or by impoldering also arises in Bombay, Calcutta, Colombo and in the vicinity of New York, viz. the swamps between New Jersey and Newark.

3.6 Integrated Upstream and Downstream Development

3.6.1 An ambitious goal

Deltas owe their existence and genesis to the river basin of which they form such a characteristic part. Their morphological, hydraulic and economic development is closely related to that of the entire basin as previously mentioned. Between the deltaic area with its specific conditions and development requirements and the other parts of the basin, many common but also conflicting interests exist in the fields of flood protection, water supply and water quality.

Upstream development can be both beneficial and detrimental for the development of the delta. In the early phases of human occupancy people did not wait until appropriate measures had been taken upstream for facilitating the development of downstream areas, before utilising deltaic areas.

It must be possible to follow a policy of development which is the optimum one for the basin as a whole. Such policy should be based on hydraulic, environmental and economic objectives. This is an ambitious goal especially in the case of international rivers where the two portions of the basin are situated in different countries.

There are many examples of such rivers where serious disagreements exist about benefits and damage in the lower river reaches by human intervention upstream (Rhine, Nile, Ganges, Euphrates, Niger, Paraná, Indus, Colorado etc.). On the other hand the existence of many international river commissions and their achievements testify to the readiness of recognising the principles of integrated optimum development (see Caponera, 1992).

Special mention in this respect should be made of the Mekong Committee (at present the Interim Committee for Coordination of Investigations of the Lower Mekong Basin). This committee was established in 1957 under the aegis of the United Nations to promote joint cooperation between Cambodia, Laos, Thailand and Vietnam as riparian countries of the Mekong river. The work of the committee is based on the principle of common use of water resources and the apportionment of the benefits and costs derived from common projects. The committee deals not only with data collection, investigations, project planning including master plans for short and long term developments, but also with the funding and implementation of projects in various fields. In spite of political difficulties, and even a state of war, much can still be achieved.

3.6.2 Flood control and flood protection

As discussed in Section 3.2, flood protection of a delta can be achieved in two ways: by flood control upstream, by the construction of embankments in the delta, or by a combination of both.

Embankments are simple and relatively cheap to construct. However, they may entail a number of negative side effects as enumerated in Section 2.1. One of these effects is the almost instantaneous intervention in the environment, the water supply, and the agricultural practices and cropping pattern.

On the other hand, flood control measures upstream applying flood detention reservoirs, makes it possible through an appropriate operation, to gradually change the hydrologic conditions in the delta and to gradually build up a new infrastructure necessary to obtain the full benefits of the new environment. It gives more time to the farmers to adapt themselves.

An interesting illustration is provided by the developments during the past three decades in the Central Plain of Thailand, the delta of the Chao Phya River. The traditional staple crop was wetland rice grown during the wet monsoon. However, the amount of rainfall in a year with average meteorological conditions is insufficient for the crop and during these rainy seasons dry spells of 2 or 3 weeks may occur. The flow of the river is governed by the rainfall in the upper catchment which has a different regime from that of the lowlands. Early floods provided a welcome natural supplementary irrigation and a common saying in Thailand was: "a year with a medium flood is a good year, a year without a flood a bad one and only extreme floods are harmful". For this reason the policy has always been not to build embankments which would completely upset the water supply system and perhaps lead to a wilful breaching by the farmers as happened in Bangladesh.

In this way a good crop (Siam rice) could be obtained with simple provisions. For producing High Yielding Varieties (HYV), and for applying crop diversification, flooding cannot be tolerated and controlled irrigation has to be applied. The construction of upstream dams, the Bumiphal and the Sirikit Dam, made this possible and water control could gradually be introduced with a reduction of extreme floods as a first measure.

Another example of upstream flood control is that of the delta and the Green Valley of the Nile in Egypt. The need for more food for the growing population made it imperative to make full use of the available water, good soils and favourable climate. Crops can be grown throughout the whole year provided that perennial irrigation is applied and no flooding occurs. This was achieved through the High Aswan Dam which regulates the flow of the Nile.

3.6.3 Water supply and water management

The most delicate issue of integrated basin development, especially in semi-arid and arid regions and in regions with a prolonged dry season, is the allocation of the limited water resources. It is the question of: what is the best use that can be made of the available water, where and how much?

The problems lends itself to an optimisation analysis taking into account soil suitability, yields and costs of production. This provides in all cases, vital information for the decision process, but often the actual decision is based on considerations of a social and political nature. An example is found in Pakistan where the water supply of the Indus and its tributaries is not sufficient to adequately irrigate all areas where crops are produced. The principle of the allocation decision is the social argument not to deprive any populated areas of a supply of water, but to distribute the water in such a way that regardless of the efficiency of the use the people can survive. The result is an irrigation delivery around 0.25 litres per second per hectare, only one quarter of what is required for an optimum use.

The High Aswan Dam in Egypt mentioned earlier, has proved to be a great blessing for the delta and Upper Egypt. The dam has been much criticised because of a number of unavoidable negative side effects but it has already paid off itself during the succession of dry years from 1987-1991. Such a period would have produced a catastrophic famine in Egypt as happened so many times in its history. It was precisely for such an event that the Nasser reservoir which provides over-year storage was designed, which allowed irrigation water deliveries to continue at a 90% level, during low flow periods.

Compared with other parts of a river basin, deltas and other low-lying coastal areas need large amounts of water for a very special purpose, i.e. the repulsion of sea water intrusion. A minimum flow of fresh river water is required to keep the sea water intrusion into open estuaries within certain limits. Fresh water is also required to flush canals and other inland water courses where saline water has penetrated by seepage of saline groundwater, or by the passage of ships at navigation locks between a sea or an estuary and inland waters.

The sea water intrusion is maximum during dry periods when the river flow is low. This is precisely the period when the need for water for other purposes, primarily for irrigation, is high. Unfortunately, the water discharges required for repulsion are very high compared with other uses, and water used for that purpose has to be drained off to the sea and cannot be recycled like waste water. In the four estuaries of the Mekong River for instance, increases of 1,000 to 2,000 m^3/sec are required to push back the saline intrusion from the present position, in a seaward direction over a distance of only a few kilometres.

With a supply of this magnitude, 1 to 2 million hectares of agricultural land could be irrigated as an attractive alternative use. In the negotiations on the water allocation, claims by the water managers from the delta for water for the special purpose of repulsion of saline water would be contested by the other water users in the basin. These will deny that the sea water intrusion causes considerable damage in the delta and will exert a strong pressure to look for alternative measures for preventing saline intrusion like estuary barrages and coastal reservoirs. This is not always economically feasible as in the case of the Mekong River in Vietnam or the river branches of the Ganges-Brahmaputra in Bangladesh. The case of Bangladesh is all the more serious since during periods of low flow in the Ganges, a portion of the discharge (up to about 40%), is diverted in the direction of Calcutta by the Farahka barrage just before the river enters the territory of Bangladesh.

Deltas are situated at the downstream end of a river and receive all the waste products of the basin disposed of via the river. Water pollution in the delta from upstream sources has become a serious problem in many deltas, both in the case of national and international rivers. In particular, in industrialised river basins where the water is polluted by heavy metals and many other toxic substances alarming situations exist. On the other hand, the pollution originating in the delta does not affect the water quality upstream and water managers in the delta are often inclined to evaluate pollution from external sources differently from that which occurs from internal sources, forgetting that many deltas owe their prosperity to their position as outfalls of an industrialised hinterland.

A special form of river pollution is that caused by an upstream disposal of salts with a similar composition as sea water (Colorado, Gambia River in West Africa, Rhine in Western Europe). The salt load may be of natural origin but may also consist of waste salts of upstream industries. This is the case with the Rhine where a substantial part of the salt load comes from potassium mining in France and industries in the Ruhr area. This aggravates the salinity problem in the delta of the Rhine and Meuse in the Netherlands, where salt water intrusion is due to the very low land elevation with respect to the sea-level and can be considered as a natural phenomenon. Roughly speaking half of the salt contained in Lake IJssel (the main coastal storage reservoir of the Netherlands), is from internal sources and half comes in at the German-Dutch border. This gave rise to difficult negotiations in the International Commission for the Protection of the Rhine Against Pollution, where, from the Dutch side, much lower limits for the salt load were demanded than in case of a less special situation in the downstream country and which have been accepted as international norms.

3.7 References and Bibliography

Caponera, D.A., 1992, Principles of Water Law and Administration, National and International, A.A. Balkema, Rotterdam.

CHO-TNO, 1981, Water resources management on a regional scale. Proceedings and Information no. 27. The Hague.

CHO-TNO, 1982a, Policy analysis for the national water management of the Netherlands. Proceedings and Information no. 29a. The Hague.

CHO-TNO, 1982b, Economic instruments for the rational utilization of water resources. Proceedings and Information no. 29b. The Hague.

Colenbrander, H.J., 1986, Water in the Netherlands. CHO-TNO, The Hague.

Custodio, E., 1987, Groundwater problems in coastal areas. Studies and Reports in Hydrology 45, UNESCO, Paris.

Dam, J.C. van, 1992, Problems associated with saltwater intrusion into coastal aquifers and some solutions, Proc. ILT Seminar on Problems of Lowland Development, Saga, Japan.

Delft Hydraulics and Rijkswaterstaat, 1991, Impact of Sea Level Rise on Society - A Case Study for the Netherlands, Government Publishing Office, The Hague.

Dent D., 1986, Acid sulphate soils: a baseline for research development, ILRI-Publication 39, Wageningen, The Netherlands

Goeller, B.F. et. al., 1983, Policy analysis of water management for the Netherlands, Vol. I, Summary Report. The Rand Corporation, Santa Monica, Ca. USA.

ILRI, 1983, Proc. International Symposium Polders of the World, Wageningen, The Netherlands

Keller, W., van Duivendijk, J. and Mondada, G., 1981, Flood control on the Parana and Paraguay rivers, Water Power and Dam Construction.

Lowing, M.J., 1982, Optimal allocation of water resources. International Association of Hydrological Sciences, Publication no. 135.

Luijendijk, J., Schultz, E. and Segeren, W.A., 1988, Polders, In: Developments in Hydraulic Engineering, Vol. 5, Elseviers Applied Science, London, Great Britain

Luijendijk, J., Schultz, E., Segeren, W.A., de Vries, W.S. and Wolff, W.J., to be published, Polders Development - Worldwide Experience and Review, Elsevier Sc. Publ., Amsterdam, The Netherlands.

NEDECO, 1973, Flood Control Studies, Rharb Plain, Morocco - Mission 1. The Hague (in French).

Oude Essink, G.H.P., 1993, Effect of sea level rise on the groundwater flow system through Amsterdam Waterworks and Haarlemmermeer polder, the Netherlands. Sea Level Changes and their Consequences for Hydrology and Water Management, UNESCO, Paris, IHP-IV Project H-2-2, pp. 33-50.

Rijkswaterstaat, 1985, De waterhuishouding van Nederland. Policy report. Government Publishing Office, The Hague (in Dutch).

Rijkswaterstaat, 1986a, Omgaan met water, Government Publishing Office, The Hague (in Dutch).

Rijkswaterstaat, 1986b, Zeespiegelrijzing, Dienst Getijdewateren, The Hague (in Dutch).

Roelfzema, A. and Van Os, A.G., 1978, Effects of harbours on salt intrusion and estuaries, 16th Int. Conf. on coastal engineering. Proceedings Vol. 3, part 4, Hamburg.

Schijff, J.B. and Schönfeld, J.G., 1953, Theoretical considerations on the motion of salt and fresh water. Proc. Minnesota International Hydraulics Convention. Minnesota.

Schultz E., 1982, From natural to reclaimed land, RIJP, Lelystad, The Netherlands

Stark, J.M., Oude Essink, G.H.P. and Boekelman, R.H., 1992, Geo-hydrological modelling of the groundwater reservoir 'De Hooge Berg' on the island Texel, The Netherlands, Proc. 12th Salt Water Intrusion Meeting (SWIM), Barcelona (In print).

UN-ECAFE, 1963, Proc. Regional Symposium on Flood Control, Reclamation, Utilisation and Development of Deltaic Areas, Bangkok. Water Resources series no. 25, New York

UN-ECAFE, 1969, Proc. 2nd Regional Symposium on Deltas, Bangkok-Tokyo. Water Resources series no. 39, New York

UN-ECAFE, 1979, Proc. 3rd Regional Symposium on the Development of Deltaic Areas, Bangkok. Water Resources series no. 50, New York.

UNESCO, 1972, Investigación del balance hidrico y salinidad de la ensenada de la Broa, No. 2686, RMO-RD/SCE, Paris (in Spanish).

United Nations, 1985, Proc. Regional Seminar on Systems Analysis for Water Resources Development, Bangkok, November 1984. Water Resources Series no. 61, New York.

Veen, J. van, 1950, Dredge, Drain and Reclaim - The Art of a Nation. Martinus Nijhoff, The Hague.

Volker A., 1991, Land reclamation and land development, Lecture notes, IHE, Delft, The Netherlands.

Wind, H.G., 1987, Impact of Sea Level Rise on Society, Balkema, Rotterdam.

THE DELTA OF THE RIVER RHINE

Introduction

The location of the delta of the river Rhine is in the north-west of continental Europe. In the country of the Netherlands, the Rhine flows into the (shallow) North Sea. Geographically, the delta is situated between 51 and 53 degrees northern latitude and between 4 and 7 degrees eastern longitude. Its surface area, as calculated downstream of the apex at Lobith, measures about 20,000 km^2, which is roughly half the size of the Netherlands. About 10 million people (over 60% of the total Dutch population) live in this delta.

The river Rhine originates in the Swiss Alps and it drains (parts of) Switzerland, Germany, France, Luxembourg and the Netherlands. The total catchment area, which is shown in Figure A1.1, amounts to about 160,000 km^2. The temperate climate in this region, the gentle slopes of the river bed and the attenuating effect of precipitation storage in the Alps (by means of lakes and snow or ice) cause a gentle discharge regime of this river.

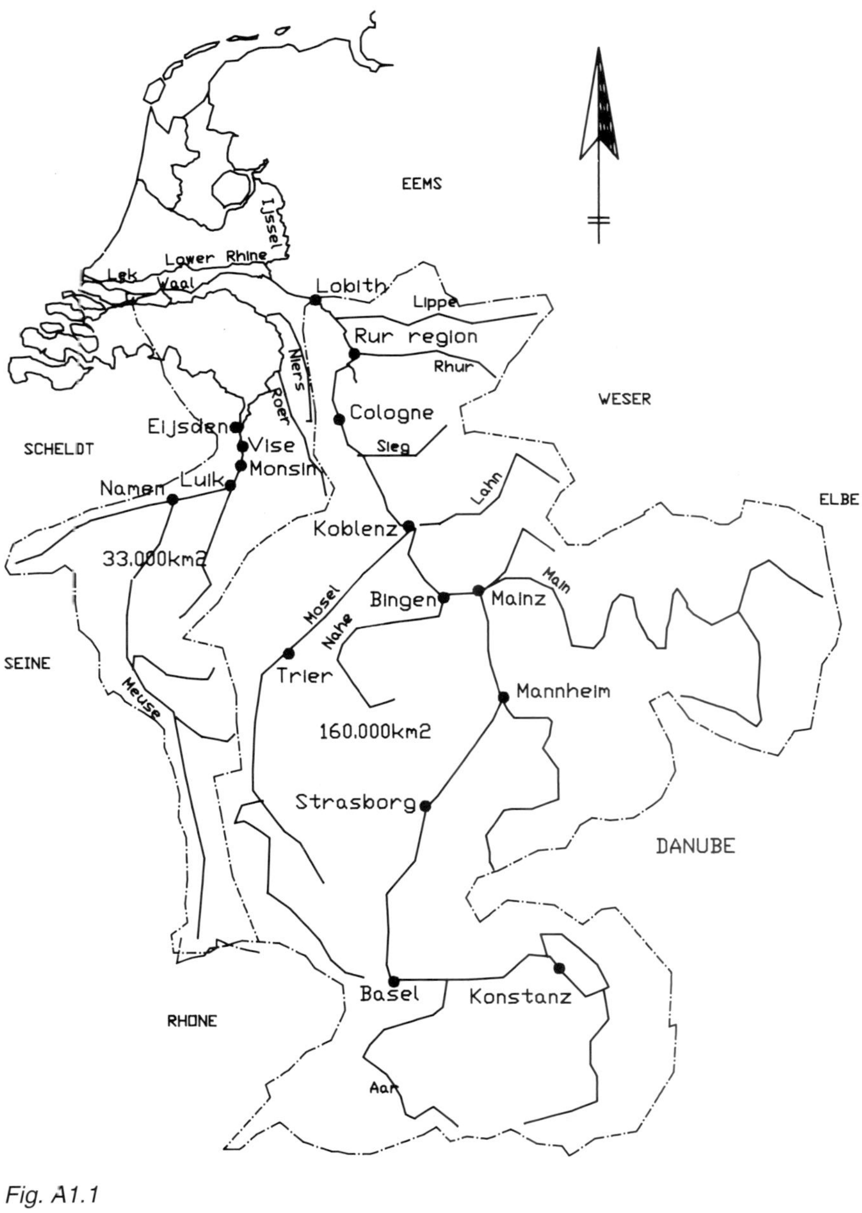

Fig. A1.1
Catchment of the Rhine and Meuse Rivers.

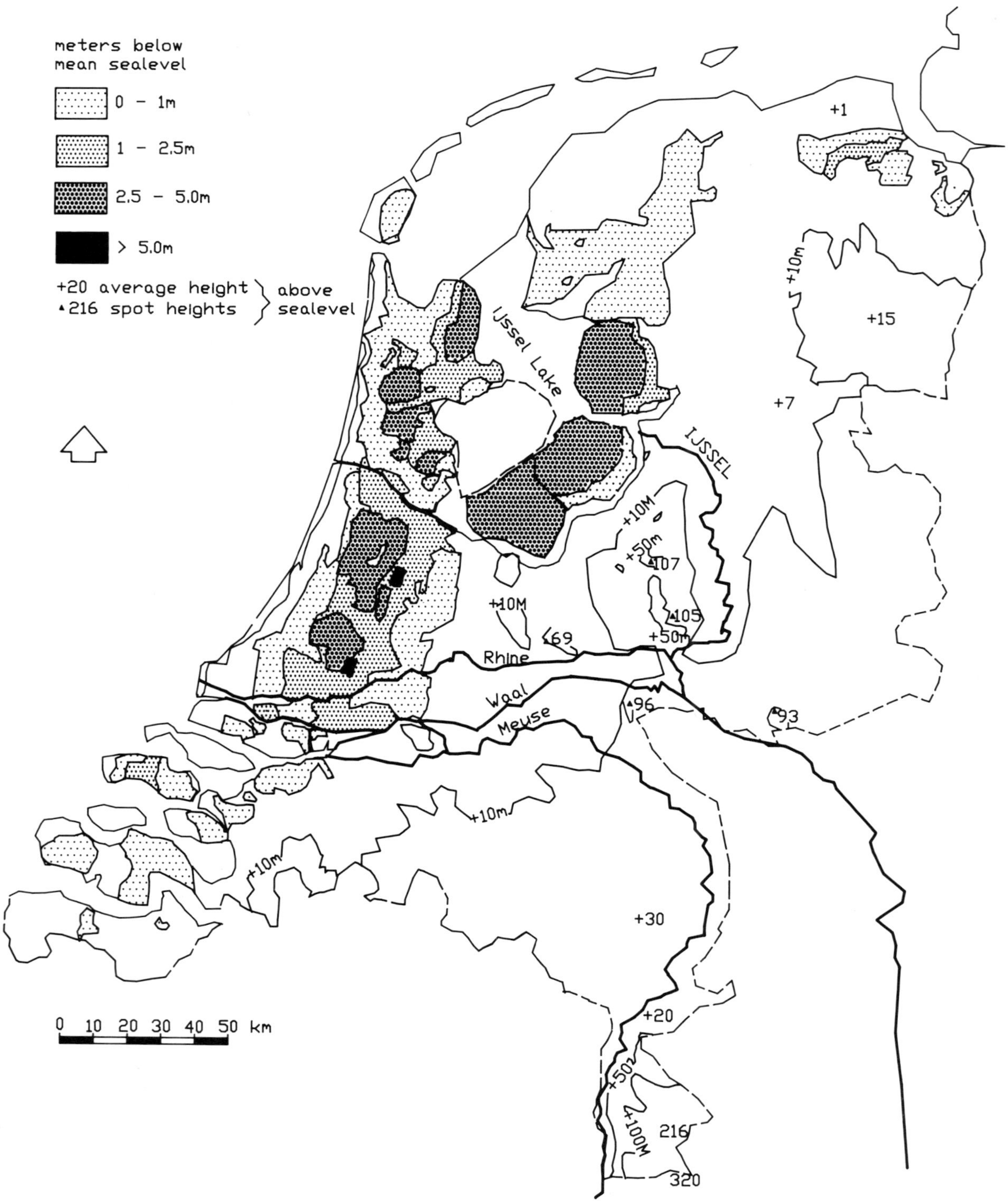

Fig. A1.2
Elevation map of the Netherlands.

Speaking of the delta of the river Rhine, one might as well call it the combination of deltas of the rivers Rhine, Meuse and Scheldt. The deltaic area in the Netherlands has been developed by these three rivers. Most clearly, the branches of the Rhine and the Meuse are connected in this delta, either in a natural way or in a man-made way. Nevertheless, the river Rhine must be regarded as the most important river in terms of delta formation and development in this area. Therefore, we will refer to this delta as the Rhine delta.

In terms of delta development this delta may be called a dead or drowning delta. Man has been interfering with the (dynamic) deltaic processes in this region since the Middle Ages. Gradually, control has been taken of these processes, protecting the delta from influences from both sides: the North Sea and the rivers themselves. As a result more than half of the Netherlands is lying under mean sea level (see Figure A1.2) and with the (relative) rising of the mean sea level a bigger part of the Netherlands will need protection.

Thanks to the very early human interference, this delta is now one of the most developed in the world. Seldom do we find in a delta such an (optimal) combination of high population density, ideal transport facilities, balanced water management and control and economic development.

Historical development

Early occupancy and passive defence (period up to AD 1000)

In the beginning of the Holocene (about 10,000 years ago), at the end of the last ice age, the sea-water level was about 25 metres below the present one. The North Sea was dry and the British Isles were connected to the continent. During the Holocene, the sea-level gradually rose and the North Sea advanced to the south. Sea intrusions occurred and large parts of the present delta consisted of peat-bogs, swamps, marshes, tidal creeks and lagoons, locally and temporarily separated from the sea by sand dunes. The delta was in its dynamic stage, literally shaped by the forces of the elements.

First settlements have been found in the north of the delta, from around 2,300 years ago, belonging to the Frisians. These people lived on natural levees or artificial mounds (terpen) which they built on the shores of the creeks. More than 1,000 of these terpen have been built, with a size of 1 to 15 hectares and an average height of 4 to 5 metres above M.S.L. (well above storm surge level). Cropping (if possible on the marshy land) was done outside the flooding season, cattle breeding and fishing were the other means of living. Gradually, the mounds developed from sheer emergency mounds for cattle and people into artificially raised mounds on which complete villages were built. The western part of the delta remained practically uninhabited until the end of this period.

Development of active defence (AD 1000 - 1250)

In this period the construction of embankments and dykes started. As such, a gradual interference of man on the natural processes started. Local river levees were interconnected and river dams were made, initiating settlements in the western part of the Netherlands (Amsterdam, Edam, Volendam etc.). Even sluice gates were constructed to discharge excess water. Inland lakes were created in the western part of the delta by removing the peat. The dried peat was used as a domestic fuel or the peat was burned until the salt was left.

Formative period and introduction of technology (AD 1250 - 1550)

Land reclamation was started by accelerating natural deposition near dykes followed by enclosure. The windmill was introduced (from the East) to supply energy for the drainage of excess water from low-lying areas. Around these areas (polders) encircling embankments were constructed necessitating the cooperation of inhabitants. This started the first institutional development and the creation of polder boards. At the end of this period this new technology was widely used. Hydraulic engineering as a science was touched upon in the first textbooks on land reclamation.

Golden Age of land reclamation (AD 1550 - 1675)

This was the period of the Netherlands as a World Power. More and more (small) inland lakes were reclaimed in the western part of the delta with the use of wind energy. Also, for the drainage of excess water (e.g. from the polders) windmills could be applied. Relative land subsidence of the reclaimed peat-soils required an expansion of pump lift. This was achieved by putting several windmills in line, creating different water levels in an expanding water control network of rivers, canals and ditches.

The birth of dredging also took place in this period. Horse-driven mud-mills were used from the beginning of the 17th century until the steam-engine era (1860).

Profits of trading and shipping in the Golden Age provided the capital for more reclamations to meet the growing demand for arable land. Dutch hydraulic techniques gained world-wide fame and application. Large and technically complex reclamation plans were discussed by Dutch pioneers.

Modern technology and industrialisation (AD 1850 - 1900)

After a period of stagnation (AD 1675 - 1850), due to the decline of the political and economic power of the Netherlands, the development of the delta gained momentum again at the end of the 19th century. This was mainly caused by the invention of the steam engine. Another contribution was the construction of the more accessible waterways to Rotterdam and Amsterdam.

Industrialisation of the hinterland (e.g. the Ruhr area in Germany) also greatly promoted the development of this delta. Steam energy was used to reclaim some bigger inland lakes (like the Haarlemmer Lake).

Challenges and responses of the 20th century

In the Netherlands, two challenges affected the development of the delta in this period. The first challenge was the storm surge of 1916. In this storm, large areas of land were inundated by a surge of the Zuiderzee, causing great economic losses and many casualties. This event triggered the decision to close off the Zuiderzee and to start reclaiming parts of the man made lake (called Lake IJssel).

The second challenge was the storm surge of 1953. During the night of 31st January, 1953, the coincidence of an astronomical high tide and the passing of a powerful storm, caused extensive flooding in parts of the south-west of the delta, resulting in the death of more than 1,800 people in the Netherlands. This disaster, taking also many lives in England and Belgium, gave political momentum to the implementation of the so-called Delta Plan or Delta Works. This plan consisted of closing off several estuaries in the south-west corner of the Netherlands with the use of dams, sluices and flood gates. The elements of this Delta Plan, which was finished with the installation of the Eastern Scheldt Storm Surge Barrier in 1986, are shown in Figure A1.3

Climate and hydrology

The climate in the delta of the Rhine is of the humid-temperate zone type. The influence of the North Sea and the Atlantic Ocean is quite apparent in the average temperatures. Mean winter temperatures are just above 10 °C and mean summer temperatures are about 16 °C.

Rainfall is evenly distributed over the year, there is no distinct dry season. Evapotranspiration, on the other hand, shows a clear seasonal pattern. Average yearly rainfall in the Netherlands (which comprises the delta of the Rhine) is estimated at about 775 mm. Average yearly (open water)

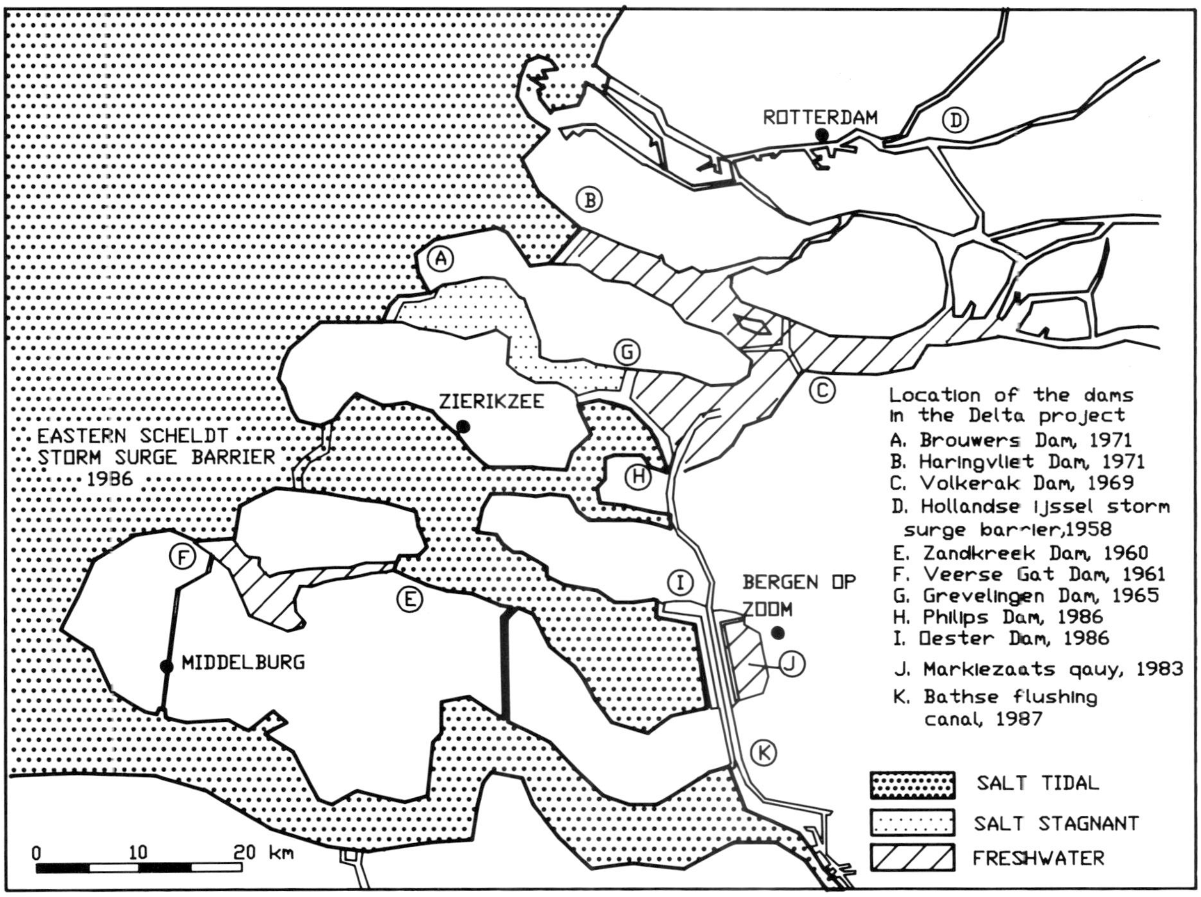

Fig. A1.3
Locations of the dams in the Delta Plan.

evaporation is about 700 mm whereas the mean actual evapotranspiration is in the order of 500 mm. Figure 3.1 shows the mean monthly rainfall and evaporation. As may also be concluded from this illustration, this variation in rainfall and evaporation usually gives rise to a water surplus in winter and a water deficit in summer.

The average yearly rainfall in the middle and upper regions of the catchment is mainly influenced by elevation. Highest mean yearly rainfall in the catchment of the Rhine is recorded in the Swiss Alps with a value of about 2,500 mm.

Wind speed and direction show the influence of the westerly wind belt. In this wind belt, although winds may blow from any direction, westerlies are definitely dominant. Storm winds (mostly westerlies) are common with the passage of low pressure ridges from the Atlantic Ocean. In general, weather in the delta may be highly changeable.

Storm-surge barrier in the delta of the Rhine and Meuse rivers (the Netherlands). The coasts of this delta are frequently hit by storm surges during which sea levels up to 3 to 4 metres above the astronomical tides may occur.

The hydrological regime of the Rhine is best illustrated with the discharge data which have been recorded at Lobith, at the apex of the delta. Figure A1.4 shows the mean monthly discharges. Average yearly flow is about 2,200 m^3/s, highest and lowest recorded flows are 13,000 and 620 m^3/s respectively.

The discharge pattern of the Rhine is quite gentle. This is caused by the relatively large proportion of melt water, as compared to rainfall water, in the actual discharge of the river. It is calculated that about 50% of the total discharge originates from the upper catchment in the Swiss Alps, upstream of Basel (with an area of about 25% of the catchment). The accumulation of precipitation in glaciers, snow fields, lakes and reservoirs gives an attenuating effect on the discharge pattern. In summer months, the base flow of the Rhine may consist of 70% melt water.

The occurrence of both high and low flows of the river Rhine have significant consequences. High flows bring the risk of flooding whereas low flows affect the quality of the water. Flooding risks in the delta have been minimised by a continuing regulation of the flow of the Rhine. Presently, in the Netherlands dyke dimensions are being adapted to statistically based flood levels of the Rhine.

Storm surges

The delta of the Rhine has not only been threatened by floods from the side of the river but also from the side of the North Sea. Over the centuries a number of floods have taken their toll. Simultaneously, these floods have been acting as starting-points or catalysts for research on and development of this delta.

In general, storm surges on the North Sea are caused by a combination of astronomical high tide and set-up by a storm wind. This latter effect accounts for the greatest influence upon flood levels threatening this delta. Gales coming from a north-western direction generate maximum set-up levels at the Dutch coast.

The highest recorded flood with respect to this delta occurred on February 1st, 1953. Other historical floods, maybe even more powerful or devastating than the flood of 1953, occurred in 1421 (Flood of St. Elizabeth) and in 1570 (Flood of All-Saints). As an example, the flood of 1953 will be described in more detail.

In 1953, on January 31st, a severe gale, producing north-western winds, developed on the North

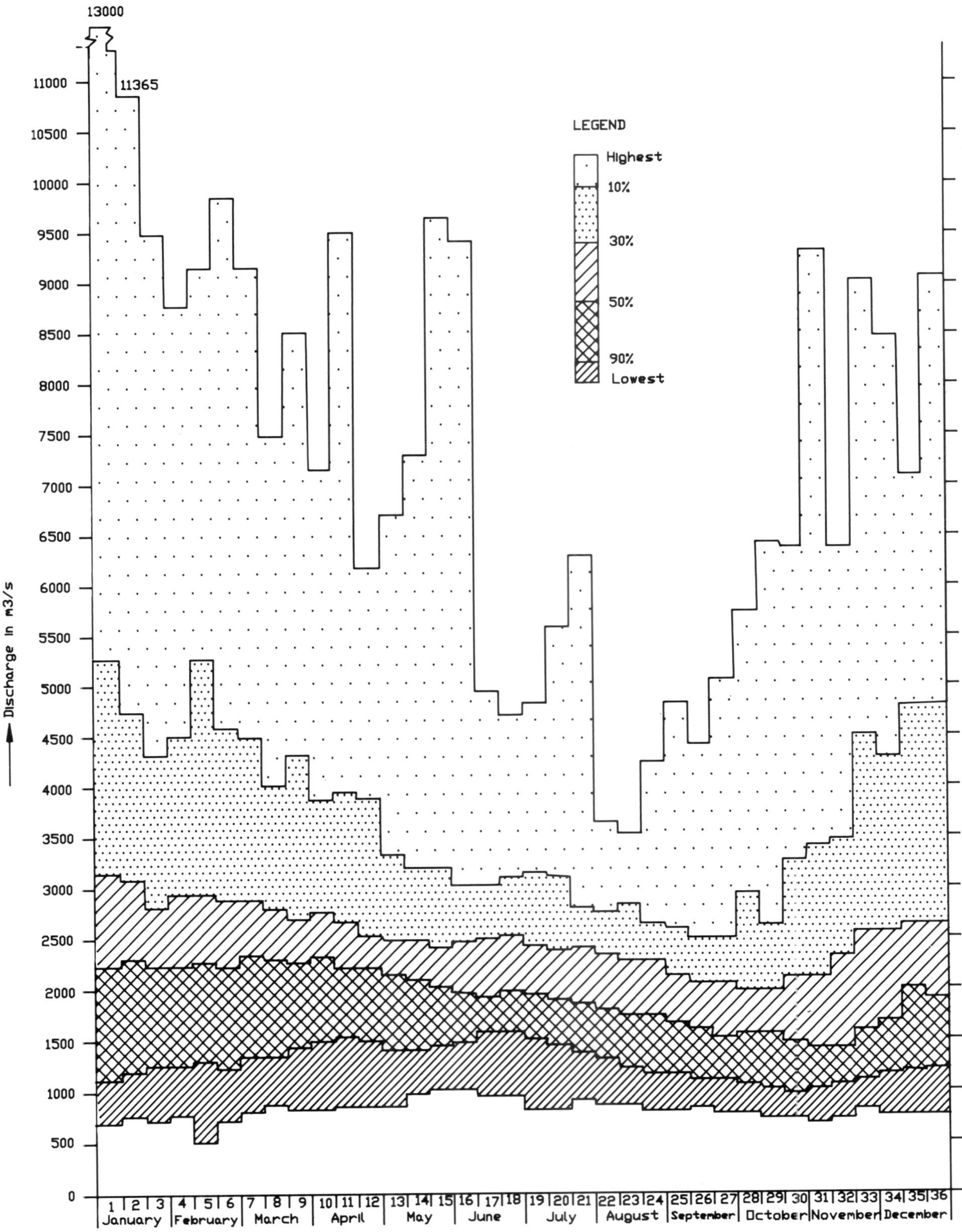

Fig. A1.4
Frequency distribution of discharges at Lobith, 1901 - 1983.

Sea. The day before, there had been full moon. During the night of the 31st, the gale reached its maximum force while passing the southern coast of the Netherlands. Wind velocities were recorded up to 35 m/s (equivalent to Beaufort 12) in Hook of Holland, near the port of Rotterdam. The combined effect of these extreme weather conditions led to a tidal level at the Eastern Scheldt estuary of more than 4.5 metres above mean sea level.

Groundwater

Geologically, the Rhine delta is situated at the edge of the North Sea basin. Due to tectonic subsidence of this basin during the Tertiary and Quaternary, huge amounts of sediments have been deposited. The base of the Quaternary lies in the north-western and the western part of the Netherlands between 300 and 500 meters below the present surface (CHO-TNO, 1979). In the eastern and southern part of the country (outside the delta), the Tertiary sediments sub-crop near the surface (see Figure A1.5).

The miocene and pliocene deposits are generally fine textured, occur rather regularly and have a nearly constant thickness. The overlying, younger strata, due to a sequence of glacial, interglacial and interstadial periods, show more variation in sedimentation. Ice-pushed ridges and deep valleys were formed (Saalien). Fluviatile sedimentation by the rivers Rhine, Meuse, Ems and Weser, as well as marine deposition during the transgression of the sea that followed the Saalien, have contributed to this rather irregular pattern. Holocene transgressions and fluviatile sedimentation during the last 10,000 years have created a very inhomogeneous top soil layer in many parts of the delta.

The groundwater table in the delta of the Rhine is man-controlled. Traditional water management in this area started many centuries ago by the creation of deep polders. These low-lying areas (mostly below average sea-level) consist of a closed embankment (dykes) and a drainage outlet for seepage or excess water. Normally, groundwater levels in these polders are controlled at about 50 to 100 cm below surface level. The excess water from the polders is transported through an extensive network of waterways (rivers, canals), eventually to the sea. At the elevated areas in the delta, like the coastal dunes and the ice-pushed ridges in the centre of the Netherlands, the groundwater table may be found at greater depths.

The geological development of the delta has also caused a rather irregular pattern of salt water intrusion in the western and northern part of the Netherlands. Salt water intrusion in this respect is regarded as both sea water intrusion through the open connections with the sea (estuaries, rivers) and saline groundwater upconing underneath the deep polders.

Water management and control

The water management of the Rhine delta has become part of the general water management of the Netherlands. Undoubtedly, the development of this delta has set out the frame of the present management system. The administrative organisation for the water management may be best described by its two main components:

- surface water;
- groundwater.

Generally speaking, the quantitative and qualitative management of surface waters is being carried out by the so-called water-boards. These water-boards should be positioned hierarchically in between provincial and national authorities. Several hundreds of them exist, mostly traditionally developed for the water management of a certain drainage area (e.g. a polder area) or catchment area. The management of the groundwater, on the other hand, is being taken care of by provincial authorities. The general legislative framework falls into separate categories for groundwater and surface water management. In this framework, general goals and objectives will be given by the national authorities while leaving enough decentralised responsibilities with the provinces and the water-boards to allow a flexible water management system in its operation.

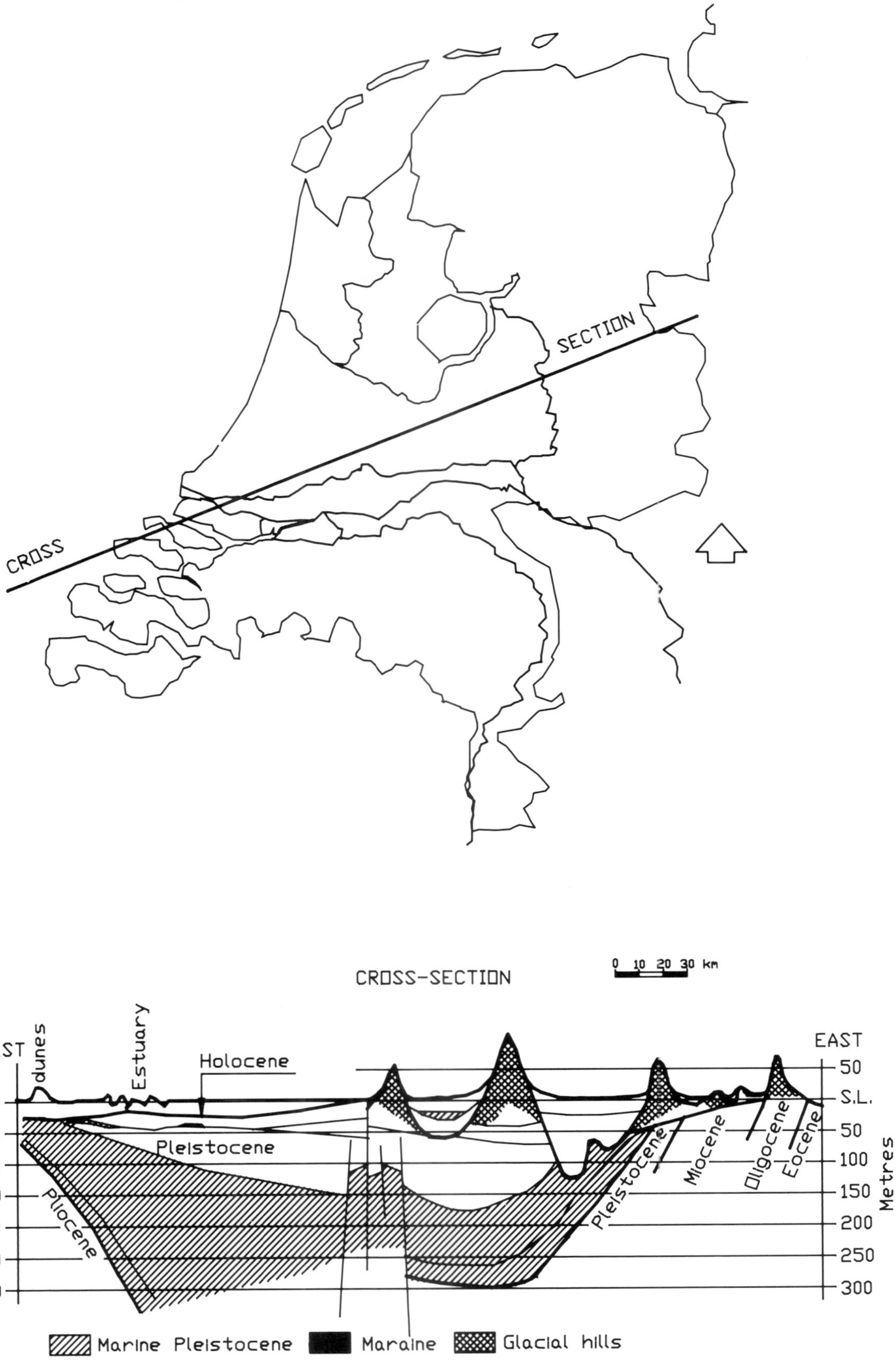

Fig. A1.5
Geological cross-section through the Netherlands.

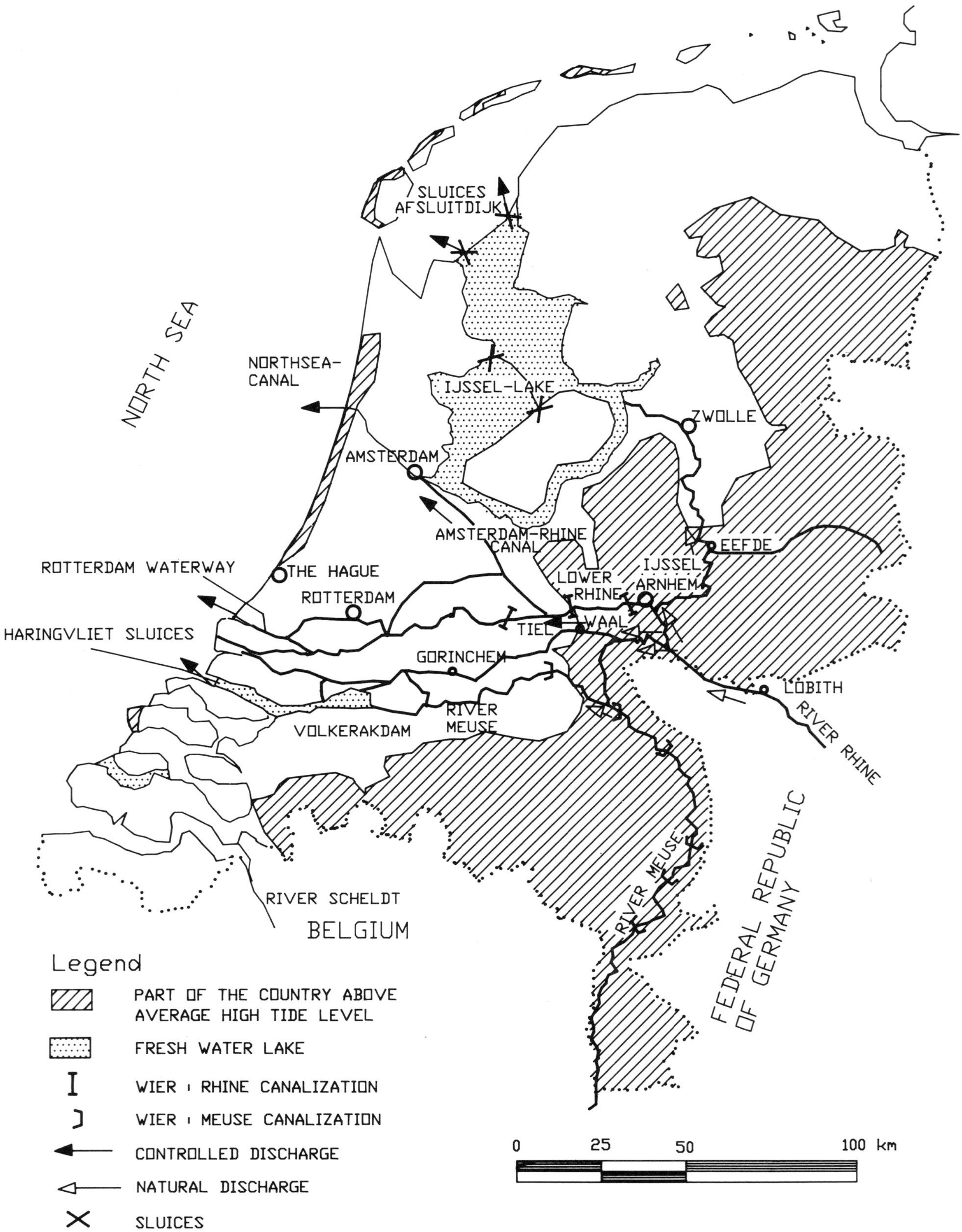

Fig. A1.6
Principal elements of the water system in the Netherlands.

The general water balance of the Netherlands for an average year may be described as follows (CHO-TNO, 1982): the total fresh water supply is about 110 x 10^9 m^3, of which 20 x 10^9 m^3 evaporates. The Rhine and Meuse account for about 75% of this supply, the rest is supplied by precipitation. The river Rhine, after passing the Dutch-German border, forks at the apex into two branches (the Waal and Lower Rhine). The Waal takes about 70% of the flow. The Lower Rhine forks again, creating the north-flowing IJssel, which accounts for about 10% of the original flow of the Rhine at the apex near Lobith.

The distribution of the flows is manipulated with the use of several weirs in the Lower Rhine (see Figure A1.6). Shipping is guaranteed by canalisation and inlets for water supply or industrial use (like energy plants) are safe-guarded against extreme variations. Another objective of this management system is the struggle against salt water intrusion through the estuaries in the south-western part of the Netherlands.

The Delta Works form part of this same control system. The biggest evacuation sluices in the world, in the Haringvliet estuary, have been constructed, mainly to withstand the salt water intrusion.

The development of the delta, after World War II, in terms of population density, industrialisation, pollution of surface and groundwater and a growing public awareness of ecological values have demanded an integrated approach for the water management in this delta. For this purpose a major policy-analysis of the water management in the Netherlands (PAWN) was conducted at the end of the seventies. This PAWN study is separately discussed in Appendix 7.

References and Bibliography

CHO-TNO, 1961, Verslag van de Technische bijeenkomst - 15, The Hague (in Dutch).

CHO-TNO, 1979, Research on possible changes in the distribution of saline seepage in the Netherlands, Technical Meeting 36, The Hague.

CHO-TNO, 1982, Policy analysis for the national water management of the Netherlands, Technical Meeting 39, The Hague.

Colenbrander, H.J., 1982, Water in the Netherlands, CHO-TNO, The Hague.

Gieske, J.M.J.,1991, De oorsprong van het brakke grondwater in het IJsselmeergebied: diffusie, dispersie of dichtheidsstroming?, H2O (24), nr. 7 (in Dutch).

ICID, 1966, Development of Deltaic Areas, 6th congress, New Delhi.

Ministry of Transport and Public Works and Water Management, 1989. The Delta Project, Preserving the environment and securing Zeeland against flooding, The Hague.

Rijkswaterstaat, 1960, Beschouwingen over stormvloeden en getijdebeweging, Rapport Deltacommissie, Deel 4, III. 1-8, The Hague (in Dutch).

Veen, J. van , 1955, Dredge, Drain, Reclaim - The Art of a Nation, 4th ed. Martinus Nijhoff, The Hague.

THE DELTA OF THE MEKONG RIVER

The delta of the Mekong river in Southeast Asia occupies a special place among the large deltas of the world. It shows the paradoxical situation of a delta which is densely populated (about 300 capita per km^2) and very productive, but still in a primitive stage of technical development. There is no flood protection and every year most of the delta is flooded to a depth between a few decimetres and a few metres.

Local transport on the main canal Phung Hiep-Camau, Mekong River Delta, Vietnam.

Another special feature of the Mekong delta is the extremely small slope of the river from the apex to the sea. This slope is in the order of 1 to 5 x 10^{-5} i.e. only a few centimetres per kilometre. There is hardly any slope in the east-west direction between the lower reaches of the river branches and the Gulf of Thailand to the west.

Different figures are mentioned with respect to the size of the delta. According to the commonly used definition of a delta as the river reaches and alluvial land areas downstream of the apex, the Mekong delta would extend downstream of Phnom Penh and occupy 16,000 km^2 in Cambodia and 39,000 km^2 in Vietnam (Figure A2.1). The river reach and flood plain downstream of Kompong Cham show typical deltaic features. If, in accordance with the definition, this is considered to be part of the delta, the area in Cambodia would amount to 20,000 km^2 and the total area to 57,000 km^2. Another definition includes the area of the Great Lake and the Toulé Sap which bears great significance for the hydrology of the delta below Phnom Penh in the deltaic portion of the river basin.

The special features of the Mekong delta explain why the present stage of technical development has taken place along lines different from those of other large deltas in the region. It shows some analogy with the present development of the so-called lowlands in Indonesia (Sumatra and Kalimantan deltaic areas).

The development, which started on a large scale at the end of the 19th century and which is still continuing, is formed by a system of canals of various sizes which are in open communication with the sea and the rivers and in which the tides and the river floods can propagate. These canals were excavated by manual labour and floating dredgers, originally gutter dredgers and, nowadays, cutter dredgers.

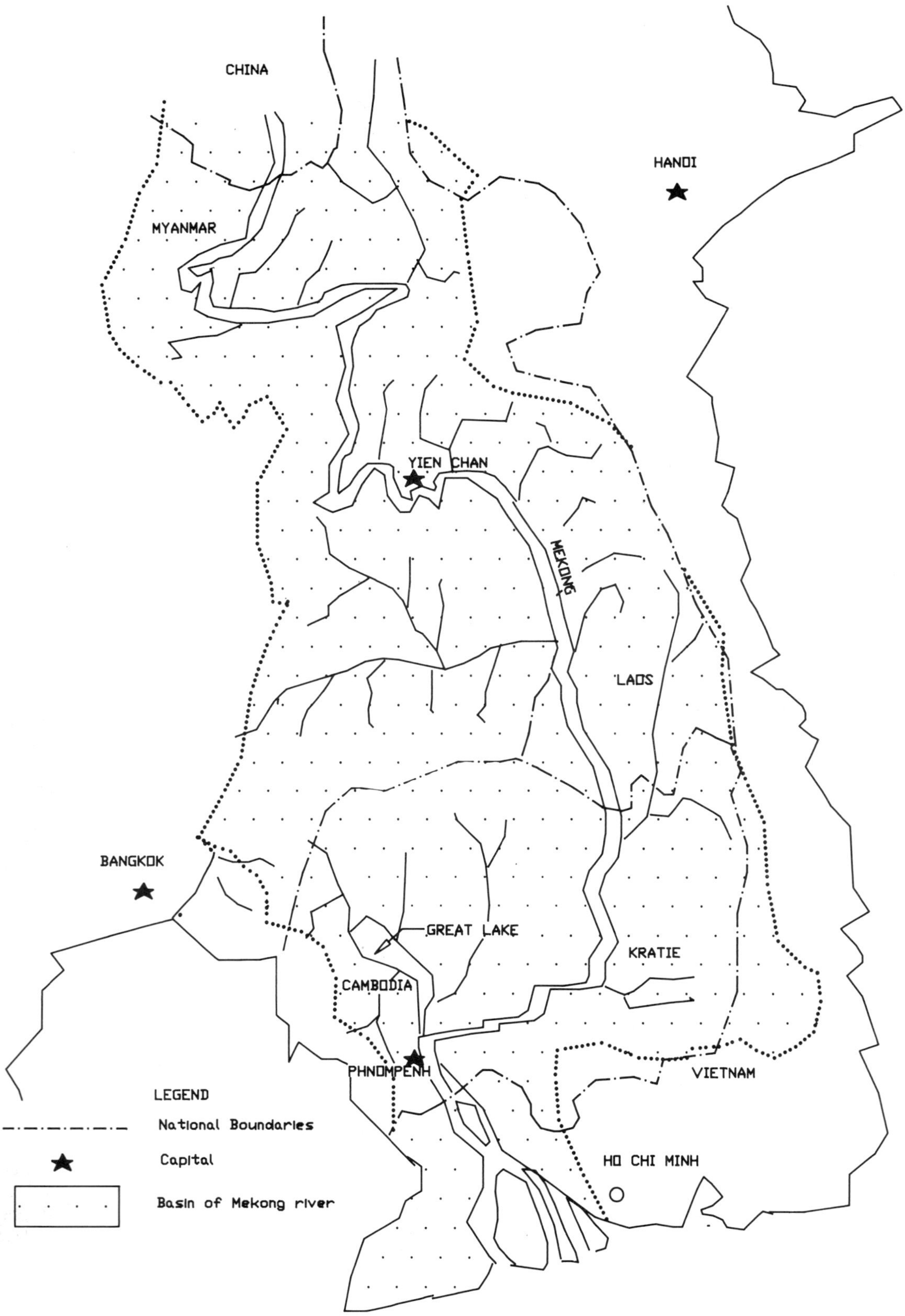

Fig. A2.1
The basin of the Mekong delta.

The total length of the main canals with a width exceeding 6 metres is more than 4,000 km. The presence of these canals has a great significance for the hydrology of the land areas. Since there are no embankments, the land areas are exposed to river floods.

In the Mekong delta a remarkable adaptation of the cropping pattern and the way of life of man to the prevailing hydrologic conditions, has taken place. There is, however, a great potential for further development through the introduction of an appropriate flood control and water management system, thus creating optimum hydrologic conditions. Such development depends partly on the development of the upstream portions of the basin of the Mekong river. As early as 1957 an international commission of the riverine countries was established under the auspices of what is now the Economic and Social Commission of Asia and the Pacific (ESCAP) with a view to coordinating the activities of the technical and economic development of the various parts of the Mekong delta. At present, this commission is known as the Interim Committee for the Coordination of Investigations in the Lower Mekong Basin. All data in this review originate from data collected or assembled by this committee.

The delta in relation to the upper catchment

The delta is a gift of the Mekong river which ranks 14th with respect to the average discharges, and 16th in terms of length, among the largest rivers of the world. The Mekong river has its origin in China on the Tibetan plateau. Before entering the delta it flows through China, Myanmar, Laos and Cambodia, draining in total 795,000 square kilometres along a course of 4,180 km. The river is a part of the border between Laos and Thailand. Below Kratié, the Mekong assumes the form and characteristics of an alluvial river with natural levees, islands and sand bars scattered along its course.

The lower Mekong basin is situated in the centre of the monsoon region of Asia and is affected by the annual regime of monsoon winds and precipitation. The north-east monsoon, lasting from November to March, is a pronounced dry period because the lower basin is sheltered by a chain of high mountain ridges on the east. The south-west or summer monsoon brings moist and warm masses of air from the Indian Ocean, which penetrates and covers the entire lower basin without interruption, but the depth of precipitation is greatly affected by orographic lifting. Thus the wettest part of the basin is in the south-west, where the monsoon strikes the mountains bordering the coast of Cambodia. The date of onset of the south-west monsoon varies from late April to June.

The annual hydrograph reflects the pattern of rainfall distribution throughout the year (Figure A2.2). The river starts to rise a few weeks after the onset of the monsoon in May or June and continues to rise until the beginning or middle of September. This is followed by two peaks of about the same magnitude or by a flattening of the hydrograph. The maximum flow usually occurs in the first half of October.

With the onset of the north-east monsoon in the upper reaches of the river basin the flow starts to decrease and in the absence of rain a very smooth depletion hydrograph can be seen, leading to a minimum flow towards the end of April and during May. During this period the river flow is derived from melting snow and ice in the upper reaches and from the outflow of groundwater. The average annual discharge of the Mekong at Kratié is 14,800 m^3/sec. The average maximum discharge is about 41,500 m^3/sec with a peak flow of 74,000 m^3/sec for a return period of 100 years. The average minimum flow in April is some 2,000 m^3/sec.

The Great Lake has a significant effect on the flow downstream of Phnom Penh (Bassac and Mekong), compared with the regime of Kratié. From June through September, with rising stages of the Mekong river, water flows from the Mekong through the Tonle Sap to the Great Lake which extends in area as it is filled. During the remainder of the year with falling stages of the Mekong there is an outflow towards the river at Phnom Penh with a maximum in December of some 6,000 m^3/sec. The Great Lake functions as a natural regulating reservoir: it reduces the peak flows downstream of Phnom Penh while it sustains the low flow during the dry period.

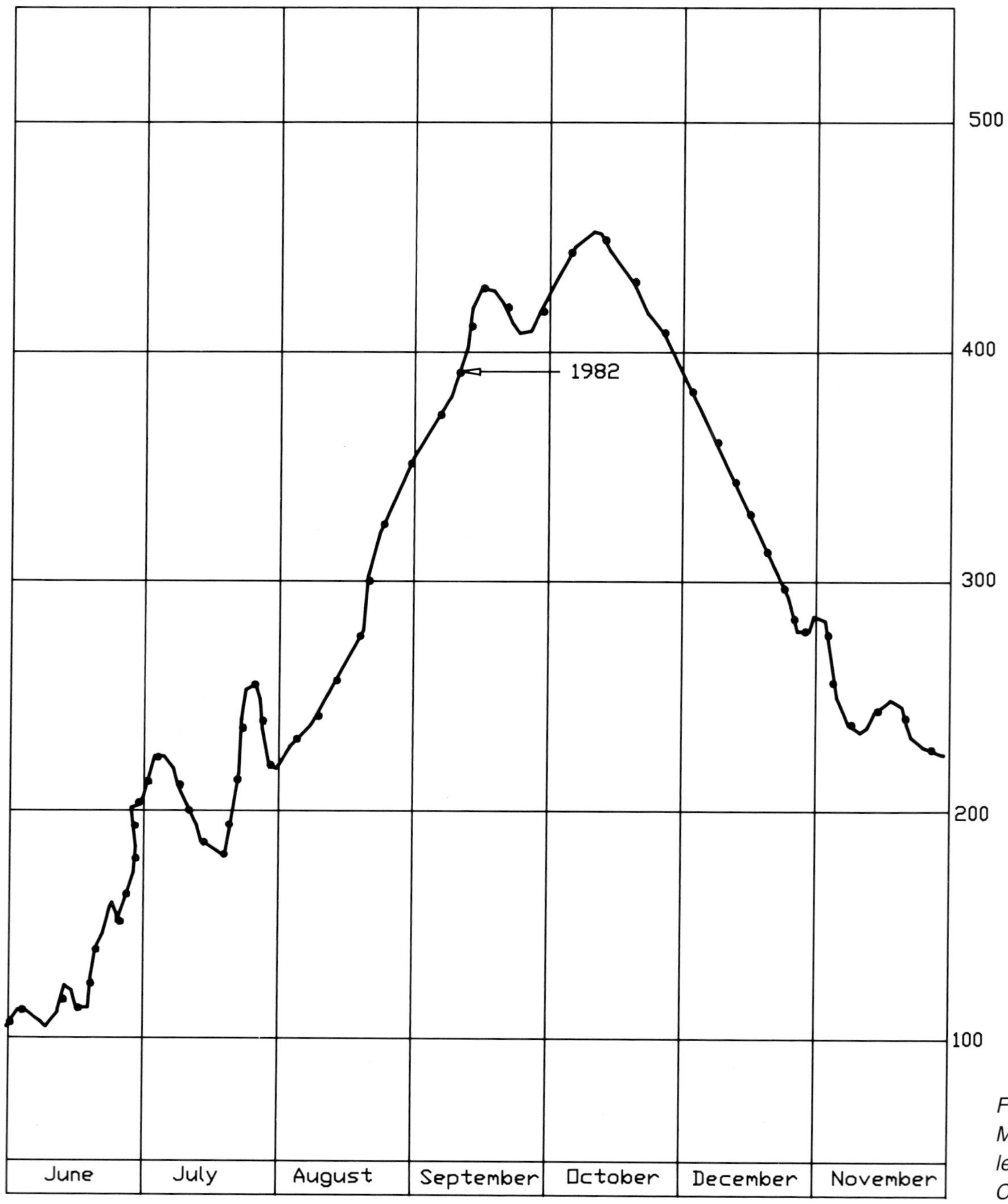

Fig. A2.2 Maximum water levels at Tan Chau in 1982.

The Mekong Committee uses a mathematical model, the SSARR-model, for flood forecasting in the Lower Mekong Basin up to Pakse. It is also used for filling in missing data and for the estimation of the design flood. The model also provides the upper boundary condition for the DELTA model which is a mathematical model simulating floods and channel flow in the delta.

Climate

The Mekong delta lies in the northern hemisphere between 8.5° and 12.3° latitude. It is situated entirely within the isotherm hot monsoon climate of the Ganges type according to Köppen's classification, where the hottest month occurs before the turn of the sun and before the summer rainy season sets in (Köppen, 1931). The wet season occurs in summer (May through October), and the dry season in winter (December through March). The transitional months are normally in April and November. The annual average temperature is 27 °C while the monthly averages do not change much throughout the year. April is the hottest month and January the coolest, the difference being less than 5 °C.

The main features regarding temperature, relative humidity, wind velocity, sunshine and rainy days are presented in Table A2.1 (averages of eight stations in the delta). Wind velocities are generally low. Strong winds occasionally occur during thunderstorms and typhoons, however, these are very rare in the delta.

The mean annual precipitation is about 1,600 mm over the entire delta but in the western part this value is higher up to 2,000-2,400 mm, in the central part it is only 1,200-1,400 mm, while in the eastern part it is 1,500-1,600 mm. During the rainy season the precipitation accounts for up to 90 to 94% of the annual rainfall. In this season, which is the main growing season, the monthly precipitation is over 100 mm.

There is no great variation in annual precipitation but there are sharp differences from one year to another, with respect to the dates of onset and cessation of the rains. In some years it begins to rain at the end of April while in others no significant rain is recorded by the end of July. Usually November marks the beginning of the dry season but in some years there were still heavy rains in November and December. Dry spells during the rainy season frequently occur around the end of July.

Season	Temp. (°C)	Relative Humidity (%)	Wind Velocity (m/s)	Sunshine hours per day	Rainy days per month
Dry season	26.3	76	2.4	8.2	2
April	28.7	76	2.3	8.2	2
Wet season	27.3	85	2.2	5.8	18
November	25.6	83	2.0	7.1	10
AVERAGE	27.0	81	2.2	6.9	Total =128

Table A2.1
The Main Features of the Climate of the Mekong delta.

The average potential evaporation in the delta, calculated according to the formula of Penman, is about 2,000 mm per year. It varies between about 1,850 mm in the Ca Mau peninsula and 2,150 mm in the northern part of the delta. The Penman evaporation compares well with the Class-A evaporation measurements and also with the pan evaporation calculated according to the formula of Christiansen.

Tides, river-flows and floods

The tidal variations around the delta and in the open canals are quite complex. This is due to the intricate network of river courses and canals, and to the difference of the tidal variations of the South China Sea (East Sea), and the Gulf of Thailand. In the former, the tides are predominantly semi-diurnal (twice daily), while in the Gulf of Thailand the diurnal components are dominant.

The water levels and flows in the river branches and in the canal system are governed by the boundary conditions of the two sides: the discharges and stages of the Mekong river at the apex and the water levels at sea.

The propagation of the astronomical tides from the seas and the lower river reaches into the systems of canals has been studied for many years by measuring water levels and discharges. Recently, a mathematical model of the flow system was developed simulating the recorded situation. Such a model will enable the prediction of the changes in tidal levels and tidal flows in case of a closure or damming off of certain canals. The model also forms the base of another model simulating the salt water intrusion into the canal system.

Flooding of the delta starts in June or July in the area near Kompong Cham. Further downstream the Mekong attains a very flat slope, in the order of 10^{-5} (1 cm per km). In Cambodia the natural levees along the river are relatively high. Behind these ridges the land slopes very gently for a few kilometres down the depressions *(bengs)*, that generally remain filled with water. The depressions are connected with the river through numerous creeks *(preks)*, which cut through the ridges. As soon as the level of the main river is above the bed level of the creeks, it starts to fill up the depressions behind the ridges. With a further rise in level, the river flow spills over the banks. In the downstream part of the delta in Vietnam, with its numerous canals, the flood waters raise the normal tide levels in the canals and the land areas become submerged.

The maximum flood level is attained in September. The extent and depth of flooding which normally occur are indicated in Figure A2.3 The total area from Kompong Cham in Cambodia to Can Tho in Vietnam which is covered by flood water amounts to 4 million hectares. Downstream of Can Tho, flooding occurs due to the slow drainage of local rainfall.

Although a remarkable adaptation to the prevailing hydrological conditions, in terms of cropping pattern and the way of life, has taken place in the Mekong delta, an optimum use of the delta for hydrological purposes requires adequate water control. Until recently, complete protection against the floods was considered an absolute necessity. For this reason, flood protection of the delta was the main item of the studies of the Mekong Committee along with the study of the hydropower potential of the upper basin. Inundations could be reduced in several ways, following one or more of three methods:

- upstream storage;
- dyking;
- floodways or improvement of drainage channels.

The effect of upstream storage in the Mekong basin on the flood conditions and water management of the delta is discussed later.

Large scale embanking entails a number of side effects in the fields of hydraulics, morphology and water management. The hydraulic effects comprise a rise of the flood levels resulting from the elimination by embanking of overland longitudinal flow and overbank storage. To predict this rise for various options of dyking and provision of floodways, a mathematical model was presented by SOGREAH under the auspices of UNESCO, as far back as 1966. The model simulates the flood conditions in the non-tidal reach of the rivers (upstream of Can Tho). Figure 2.12 in Chapter 2.1 shows the system of meshes established according to the existing system of roads, canals, ridges and topographical differences.

The magnitude of the rise at a certain location depends on the magnitude of the flood and the extent of dyking. It was found that in the case of complete dyking the 100-years' flood would increase by not less than 2.7 metres in Kompong Cham, 2.6 metres in Phnom Penh, 0.8 metres in Chau Doc, 1.5 metres in Tan Chau and in downstream stations, 0.4 metres in Can Tho and 0.6 metres in Sa Dec. Partial embanking and provision of floodways lead to much smaller effects.

Salt water intrusion

Salt water intrusion is one of the major problems of water management in the Mekong delta. The sea water penetrates into the estuaries of the river branches and from there into the local rivers draining the Trans Bassac area. It then enters the canals between the Gulf of Thailand and the Bassac, upstream of Can Tho.

The maximum extent of the sea water intrusion is attained in May when the river flow is minimum. The limit of salt water intrusion during the dry season is also shown in Figure A2.3.

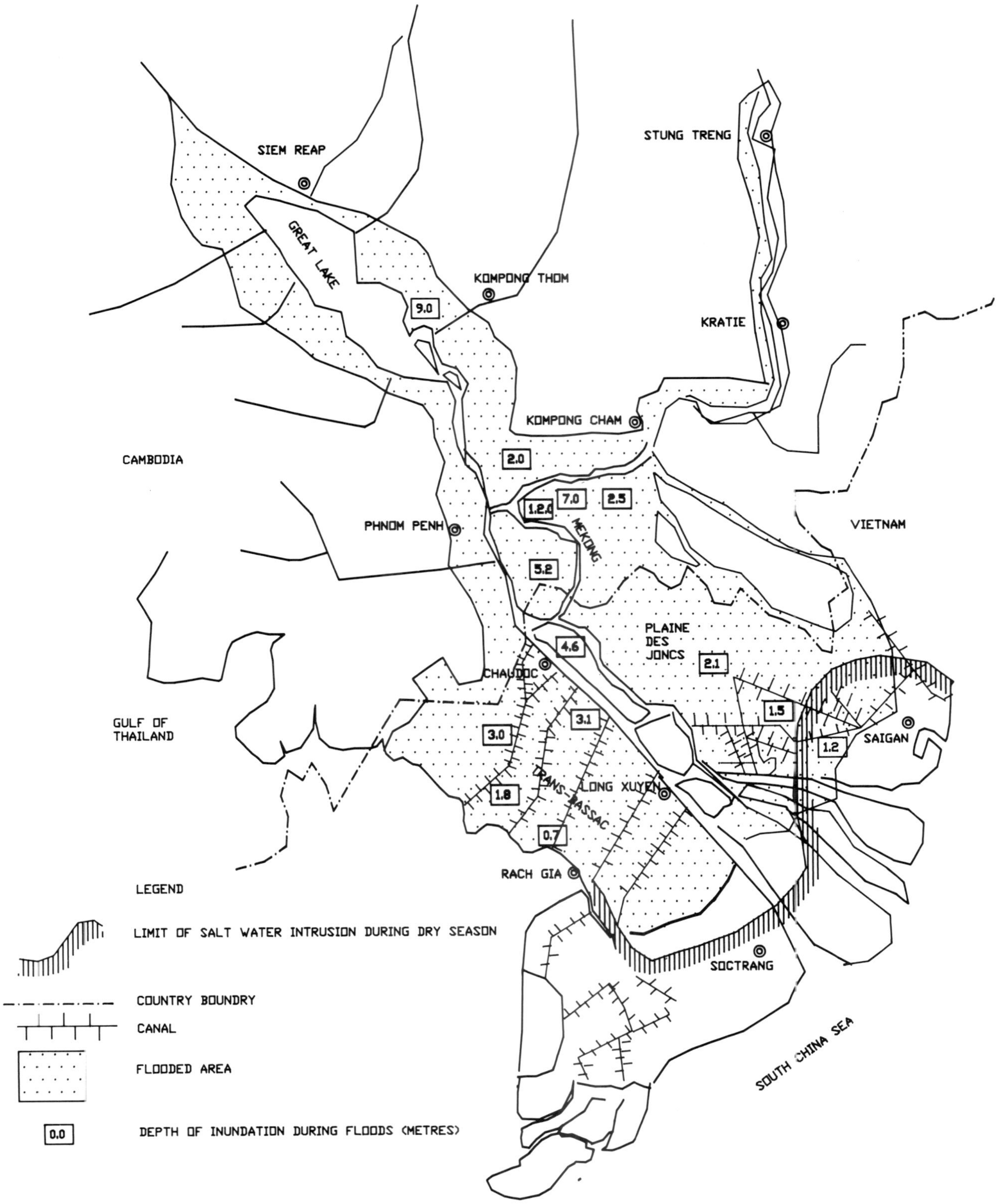

Fig. A2.3
Extent and depth of flooding.

The mechanism of the sea water intrusion is different in the river branches and the canal system. The propagation of the sea water intrusion in the canals is much more complicated because of the many connections between the canals and the unknown effect of these connections on stratification and mixing. A satisfactory simulation of the actual conditions with a mathematical model has not yet been achieved.

The salt water intrusion models and the tidal models can be used to investigate the effects of further developments in the delta, namely:
- changes in the upland discharge at Phnom Penh during the low flow season;
- abstraction of irrigation water from the river branches;
- damming off river mouths and canal entrances;
- deepening entrance channels in the river branches.

Hydrology, soils and crops

As previously mentioned a remarkable adaptation has taken place in the Mekong delta of the cropping pattern and the way of life of man to the prevailing hydrologic conditions. The staple food of the people living in the Mekong delta like in all deltas of Southeast Asia is wetland or lowland rice which can grow in water, even in water with an increasing depth. Traditionally, more than 90 percent of the cultivated area in the delta has been used for rice during the wet monsoon. This rice crop is generally planted between May and September and harvested between November and January. Between January and May (dry season), most of the land remains fallow. Many different local rice varieties are grown which differ in terms of the length of the growing period and harvesting date.

Depending on the depth of flooding, four main systems of single crop planting have been applied:
- Single transplanted main season rice.
  This variety is mainly grown in areas where flooding depths do not exceed 0.4 to 0.5 metres. Early, medium and late varieties are used for single transplanting. The greatest part of the single transplanted rice is grown in the rainy season and is not irrigated. Yields range from 2,500-3,500 kg of paddy per hectare.
- Double transplanted rice.
  Flooding depths in the area where double transplanted rice is grown generally range from about 0.4 to 1.0 metre. Late rice varieties are used for double transplanting. This is a cultivation method which allows a better control of weeds and extends the vegetative period of the rice plants, mainly because part of the roots is cut off during the second transplant. However, it is a very labour intensive method and is being replaced by broadcast.
- Broadcast main season rice.
  This group includes the floating rice and the long stem, non-floating rice varieties. Floating rice is grown in areas which are deeply flooded (more than 1.0 metre up to 3 or 4 metres). An essential condition is a slow rate of rise of the water level: not more than 5 to 10 cm per day. Yields are generally low: around 1,000 kg per hectare.
- Dry season rice *(riz de décrue)*.
  This variety is transplanted in receding flood water. It is mainly grown in Cambodia along the main rivers and around lakes *(bengs)*, which provide the water necessary for irrigation.

In the period 1980-1990, surprising changes in the traditional cropping patterns in the Vietnamese part of the delta have taken place. The introduction of modern rice varieties with a short growing season (100 day varieties) have completely changed the concepts on flood control and irrigation. It became possible to avoid cropping during the months of September and October, during the deepest flooding. It is no longer necessary to build high dykes to grow a high yielding rice crop during the wet monsoon. Partial flood protection is sufficient with submersible dykes to keep out the waters of early August floods. The flooding in September and October will continue to take place. This has many environmental advantages like no rise of the flood levels by embanking, migration of fish is not stopped, flushing of canals remains possible etc. Also the months of March, April and May, with the minimum flow and maximum sea water intrusion, are excluded from the growing period.

Depression where floating rice is grown. The hay has to be put on a stack-stand with stilts so that the hay cannot float away during the flood season.

Outside this period more water is available for irrigation and repulsion of sea water intrusion. Two and sometimes even three crops of rice or dry foot crops can be grown every year. Spectacular and unforeseen improvements have been obtained in areas with moderately and slightly acid soils that for a long time were considered as problem soils and only suitable for a crop of floating rice. With the new varieties, two crops a year are now obtained. This result is obtained by extending the canal system and connecting with the Mekong river in such a way that for most of the year flushing of canals takes place and flooding of the land occurs during the wet monsoon.

The innovations mentioned above led to a new *Green Revolution*. A more liberal economic policy after 1985 acted as an incentive to the farmer. It became attractive to increase the cropping intensity. The corresponding results are shown in Table A2.1. In the period 1980-1990 the rice production of the delta almost doubled and Vietnam is now again exporting rice. Vietnam is now the third exporter of rice of the world after the USA and Thailand.

The soils of the Mekong delta are generally suitable for growing wetland rice. Heavy clay soils occupy more than 90 percent of the delta in Vietnam. Most of these clay soils are poorly suited for crops other than rice. Problems occur in areas with acid sulphate soils such as the Plain of Reeds, the Ha Tien plain and the depression area in the southern Trans-Bassac, in great areas (U Minh forest) and in saline soils.

Percentage of area		
	1980	1990
Triple cropped	1%	4%
Double cropped	29%	54%
Single cropped	70%	42%

Table A2.2
Increase in cropping intensity in Mekong delta.

The main areas where soil and hydrological conditions permit the large scale cultivation of crops other than rice, as a cash crop, include river levees and present and former sandy beach ridges in the coastal plain. Four main groups of such crops can be distinguished:
- upland field crops (like maize, beans, peanuts, cassava etc.);
- horticultural crops (like lettuce, tomatoes, cucumber, onion etc.);
- fruit trees (mango, rose apple, citrus etc.);
- coconut palms.

Horticultural crops and fruit trees are also grown on other soils, which are often generally considered as being less suitable for such crops, and where frequent flooding occurs. Small polders are built with pump-lift drainage and irrigation by sprinkling. There is a liberal application of fertilizers, and if necessary, lime. Such areas are found in the vicinity of cities where a good market exists for such products making high investments justifiable (Ho Chi Minh City, My Tho and Can Tho).

Effects of upstream development on the hydrologic conditions of the delta

Since its establishment in 1957, the Mekong Committee has been studying and promoting the implementation of projects in the upstream Mekong Basin, in the fields of flood control, irrigation, power generation and navigation. Such projects, with storage reservoirs as a main component, may have a considerable impact on the hydrological conditions of the delta. For this reason the development of the delta cannot be considered independently from the development of the upstream portions of the basin.

The difficulties of a synchronous and harmonised development of the upstream portions of a river basin and the delta, are caused by the controversial interests in both parts of the basin, the possible occurrence of negative side effects of the upstream development on conditions in the delta, and in the correct timing of the development of both parts. In the cases of the deltas of the Nile, Chao Phya, Mississippi etc., man did not wait for the upstream development to create optimum conditions, before utilising the delta.

The upper reaches of the Mekong offer good opportunities for the creation of storage space which would make it possible to reduce the peak flow, and to augment the low flows during the dry season. Whereas storage dams in the tributaries would have only a small effect, the two possible major dams on the main stem, Pa Mong near Vientiane and Stung Treng in the north of Cambodia would have significant effects. Upstream storage reservoirs, either by tributary or main stream projects can increase the low flow by 1,000 to 2,500 m^3/sec. This would facilitate an increase in the irrigated area during the dry season and to restrict the salt water intrusion.

An accelerated rise of the sea-level will have a number of consequences with respect to morphology, flood protection, salt water intrusion and water control. In a delta like that of the Mekong river such a rise would completely upset the present delicate relation between depth and duration of flooding, and the cropping pattern. It will become necessary to embank the entire downstream portion of the delta and to apply pump-lift drainage. The embanking will cause a further rise of the flood levels although not to the same extent as embanking the upstream portions. Considerable investments will have to be made in a relatively short period of time.

References and Bibliography

Köppen, W., 1931, Grunddriss der Klimakunde. De Gruyter Publ. Co., Berlin.

UN-ECAFE, 1963, Proc. Regional Symposium on Flood Control, Reclamation, Utilisation and Development of Deltaic Areas, Bangkok. Water Resources series no. 25, New York.

UN-ECAFE, 1969, Proc. 2nd Regional Symposium on Deltas, Bangkok-Tokyo. Water Resources series no. 39, New York.

UN-ECAFE, 1979, Proc. 3rd Regional Symposium on the Development of Deltaic Areas, Bangkok. Water Resources series no. 50, New York.

THE DELTA OF THE IRRAWADDY RIVER IN MYANMAR

The Irrawaddy River in Myanmar offers a typical example of a delta in a humid tropical zone which is in an early stage of hydraulic and agricultural development. It is one of the most important regions of Myanmar from an economic point of view *(rice bowl)* and the soils in the area can be considered as the richest in the country. The density of population of the delta (around 150 inhabitants per km^2), is higher than in other parts of Myanmar but smaller than that of the other large deltas of Southeast Asia.

The delta occupies a unique position among the large reclaimed deltas of the world, as it offers the only example of a system of protection against river floods, consisting of open or horseshoe shaped embankments (Figure A3.1). This system was the result of a long controversy about the feasibility of embankments as a means of flood protection of level areas as explained in Chapter 2.1. The deltaic area downstream of the apex near Henzada covers an area of about 30,000 km^2 with a maximum north-south length of 290 km and a maximum width of 240 km. Usually, however, the deltaic area is considered to start at Myanaung at the limit of the vertical tidal effect for minimum discharge and 70 km upstream of Henzada (40,000 km^2).

Delta of the Irrawaddy River,Myanmar. Present river embankment with an inlet sluice for the controlled admission of silt laden waters.

Climate of the river basin and the delta

The rainfall on the drainage basin of the Irrawaddy is governed almost completely by the monsoon winds. Three seasons can be distinguished:

- the warm and humid rainy season from mid May to mid October, when the south-westerly monsoons convey a great deal of moisture to Myanmar across the Bay of Bengal;
- the cool, dry season when the north-east monsoon is dominant from November to February;
- the hot dry season from March to April.

The average annual rainfall shows a considerable variation within the river basin. In the upper part of the basin the areal average is about 1,750 mm with extremes of 3,750 mm in the extreme north and 840 mm at Mandalay. The central part of the basin is the driest zone of Myanmar with an average of 900 mm and a minimum of 625 mm at Pagan.

The rainfall in the delta also shows a rather great areal variation. Near the apex the average annual precipitation is 1,500 to 2,000 mm, while in the coastal areas it varies from 2,500 mm in the east to

	REFERENCES	AREA PROJECTED IN ACRES
①	KYANGIN EMBANKMENT	4,300
②	MYANAUNG EMBANKMENT	1,23,000
③	NGAWUN AND IRRAWADDY EMBANKMENT	4,14,000
④	IRRAWADDY EAST SIDE EMBANKMENT	5,00,000
⑤	YANDOON ISLAND EMBANKMENT	1,23,000
⑥	MAUBIN EMBANKMENT	1,47,000
⑦	THONGWA EMBANKMENT	41,000
⑧	SHWELAUNG EMBANKMENT	14,000

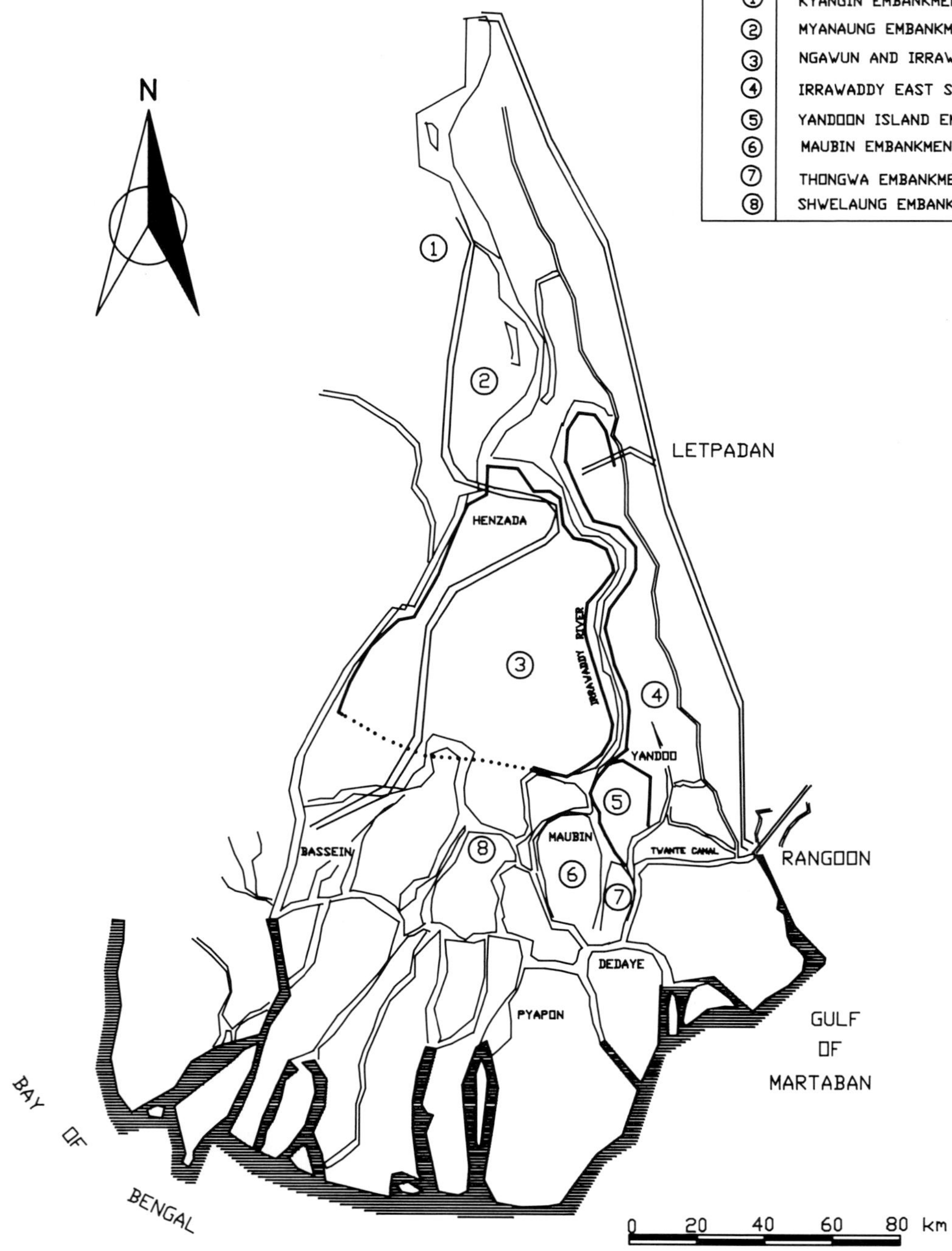

Fig. A3.1
Delta of the Irrawaddy River with protective levees.

3,500 mm in the west. This is the highest amount for deltas throughout South-East Asia. Over 90 percent of the rainfall over the delta falls within six months of the year, i.e. from May 15, to November 15. Average figures of rainfall and temperature are given in Figure A3.2.

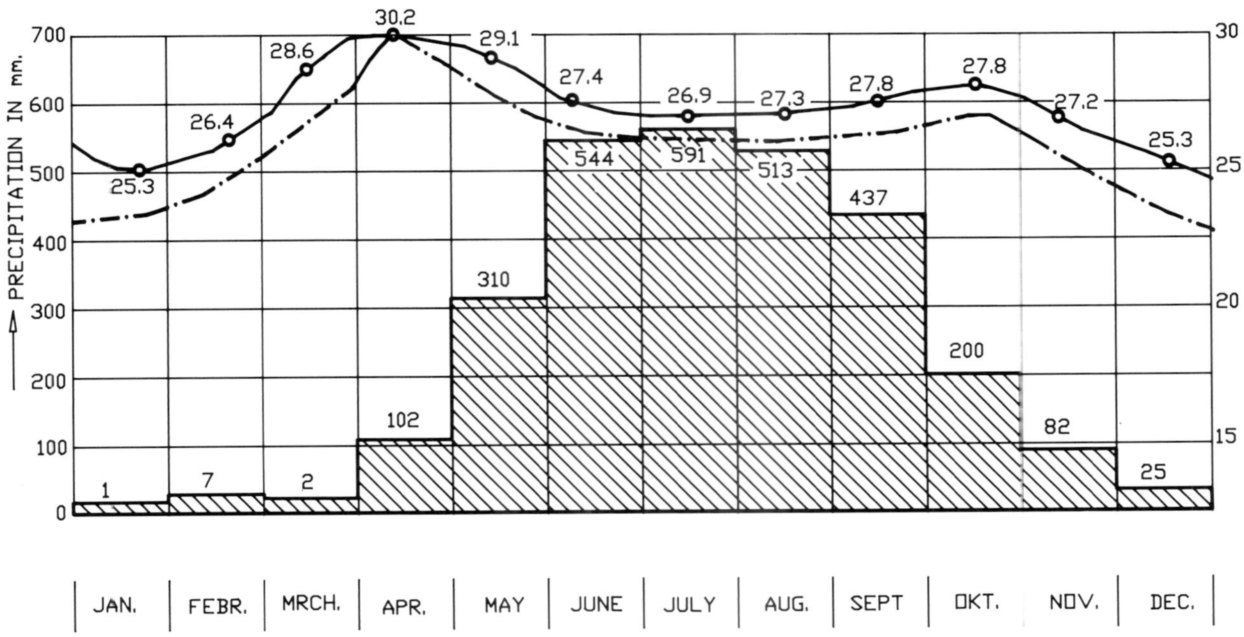

Fig. A3.2
Average rainfall and temperature at Rangoon.

The total rainfall does not vary much between years and is within some 25 percent of the normal. However, the distribution in the course of the year may be somewhat irregular. Thus the rainfall during the wet monsoon which is the main growing season is sufficient to produce a crop of rice even during rather dry years. This marks a great difference with the delta of the Chao Phya in Thailand where even during average years the rainfall is not sufficient. This explains why in the delta of the Chao Phya the flood water of the river was considered as a welcome indispensable supplemental irrigation supply, whereas in the delta of the Irrawaddy, flood water produced an excess of water supply.

Tides, river-flows and floods

The delta of the Irrawaddy is characterised by the presence of strong tides. The tidal ranges are the highest of all deltaic areas in the humid tropical zone. At Elephant Print at the entrance to the port of Rangoon and close to the sea, the mean spring tidal range is 5.8 metres and the neap tidal range 4.0 metres. In Rangoon, 72 km from the sea, these ranges are 5.1 and 3.5 metres respectively. Further inland the tides are smaller. At Bassein, 110 km from the sea, the mean spring tidal range is only 2.2 metres and the mean neap tidal range 1.8 metres.

Although there are differences in tides between various locations of the long coast (200 km), no important longshore currents are generated. An important factor in the delta-building process is the action of the waves generated by the south-west monsoon. The sediments are carried to the east where the strong ebb current of the Rangoon River, and further on of the Sittang River, carry the sediments further offshore. In this way the comparatively low rate of growth of the delta (around 50 metres annually), and the delta front configuration can be explained. The tidal currents in the estuaries and between the shoals in the coastal waters can be quite strong. At the entrance of the Bassein estuary velocities of 0.75 to 1.5 m/s occur, whereas the combined effect of the ebb current and the river flood discharge may increase velocities to up to 3.0 m/s.

The set-up of the sea level by landward winds and differences in barometric pressure is usually smaller than 1.0 metre. This is a significant difference with the sea front of the delta of the Ganges at the other end of the Gulf of Bengal.

The Irrawaddy River (drainage area 430,000 km^2), ranks 12th among the rivers of the world in terms of highest flow. The average annual discharge is around 13,000 m^3/s corresponding with an annual yield of 410 x 10^9 m^3. The average annual suspended load amounts to roughly 300 million tons and with an average salt content in the order of 700 mg/l. In spate the river carries some 1,000 to 1,300 mg/l.

The flow variation in the course of the year is similar to that of the other large rivers of South-East Asia. The lowest flow generally occurs in March to April, followed by a gentle rise in April to May and a steepest rise in May to June, as a result of melting snow and ice on the mountains in the upper catchment. The highest stage is reached in July or August, after which the flood steadily recedes, though late rains in September to October may retard the fall, as occurred in 1948.

The highest recorded flood flow is that of 1877 with a peak discharge estimated at 63,900 m^3/s. The peak stage at that time at Sailetha near the apex of the delta was 13.3 metres above mean sea level. This peak stage has been exceeded time and again since 1877, especially in 1947 and 1948, but due to the lack of gauging, these levels cannot be related to discharges.

Protection against river floods

In the delta of the Irrawaddy, three different parts can be distinguished. In the northern part starting near the apex, the extreme water levels are governed by the magnitude of the peak discharge of the river. The difference between these extreme levels and the land elevation is several metres, and deep flooding is bound to occur. In this upstream part of the delta open embankments have been built including embankments on both sides of the main river branch.

In the middle part, which is closer to the sea and in some cases very close to the sea, no embankments exist. Here the depth of flooding by the river is small. Along the coast in the south a wide mangrove belt exists. The open embankments are aligned in the shape of a horseshoe around the areas between the river tributaries which are to be protected from deep flooding. The system is open at the downstream end of the compartment. In this way, protection of the higher upstream locations of the inter-tributary area is obtained, and the lands can freely drain excess water from rainfall, in a downstream direction. In the event of extreme floods, the lower portions of the compartments are flooded, acting as a natural flood retention basin and thus slightly reducing the flood levels further downstream.

This system of open embankments was initiated as a first step towards complete flood protection while partially avoiding the negative side effects of complete embanking, like the rise of flood levels and the hampering of drainage. It also offers a compromise in the controversy between advocates and antagonists of flood protection by dyking, a problem which is encountered in many deltaic areas.

The system of embankments in the Irrawaddy Delta is of rather ancient origin following the increase of population. Until about 1850, much of the delta lands were uncultivated. Following the rush of settlers from Upper to Lower Burma, the construction of embankments to protect fertile areas from being flooded kept pace with the increase in population and of the cultivated areas. As early as 1861, a government project was initiated for the construction of a dyke on the west bank of the Irrawaddy.

Many dykes were built around 1880 and 1920, and until about 1920 dyking was considered essential and did not meet much opposition. The dyke system thus constructed, and improved on later, has successfully withstood floods exceeding 57,000 m^3/s. Breaches have occurred in times of still higher floods or for other reasons.

The views on the desirability of extending the system of embankments changed in 1926, the year when considerable flood damage occurred, mainly to the Burma Railways. 1926 was also the year of the Mississippi disaster which was attributed to the double embanking of the river. The authorities then began more and more to oppose embankments along both sides of the Irrawaddy, and further extension of the system of village embankments (built through local initiative).

In a report of 1929 by the Public Works Department, it was even recommended to demolish the village embankments on the east bank at a rate of 16 km per year. However, in view of the costs of expropriation or acquisition of land from the local landowners who would not voluntarily agree to be drowned out for the sake of future generations, this recommendation was not put into practice.

It is worthwhile to note that a similar proposal was launched around 1810 for the delta of the Rhine in the Netherlands, to cope with the problem of the rise of the river-bed and subsequent rise of flood levels. This proposal was never implemented for much the same reasons as in Myamar about a hundred years later. The matter again came to the fore after the major Irrawaddy flood of 1939. Again the idea of allowing the river to further build up the land areas was considered as being theoretically sound but impossible to be implemented. No adverse effects were noticed later on, along reaches where double embankments existed and the policy of keeping up the dykes was approved by the government.

In recent years projects were contemplated aimed at closing a number of existing embankment systems, and further increasing the areas by embankments. The difficulty was that no sufficiently accurate prediction could be made, of the effect of eliminating the storage effect of certain natural flood retention basins, on the flood levels. This only became possible when a mathematical model simulating the flooding of the delta was made, based on the simulation of actually recorded flood conditions. With such a model various measures for improving the present flood protection system can be investigated. In the meantime certain areas in the coastal zone which were subjected to flooding with saline water at high spring tide level, were provided with embankments. The effect of dykes on the flood levels in this zone is very small.

Engineering works

One of the problems in maintaining the embankments along the Irrawaddy and its tributaries, is the occurrence of bank erosion threatening the lines of defence. This is a general problem of flood protection in areas along alluvial rivers where, because of the high costs and lack of suitable material, no river training can be applied.
In the delta of the Irrawaddy the system of retired embankments is generally applied. These are embankments erected as a second line of defence at the landward side of the functioning dyke, and which come into effect should the first dike fail.

Since embankments halt the deposition of river silt and consequently the raising of the land areas, which is also considered as having an important fertilizing effect, attempts have been made to retain this benefit after their construction. This can be achieved by the construction of sluices for the controlled admission of silt laden river water. The gates can be closed to prevent flooding to an inadmissible depth. This is similar to the advantage as applied in Cambodia. A sluice was built in the western embankment of Yandoon Island facing the Irrawaddy near Mithwechaung. It effectively promoted silting in the vicinity of the inlet but the difficulty was to get the sediments further into the lowest regions of the inter-tributary areas.

THE DELTA OF THE RIO PARANÁ

The delta of the Rio Paraná is situated in the Province of Buenos Aires, Argentina. The Rio Paraná is a large river, originating from the tropical regions of Brazil, Bolivia, Paraguay and Northern Argentina. Its catchment area is 3,1 million km^2 and its length about 4,000 km. The average discharge at the delta is 79,300 m^3 /sec. It outfalls into the Rio de la Plata, a large but shallow estuary. Due to the enormous discharge of fresh water from the Paraná and Uruguay-rivers, the Rio de la Plata is fresh as far down as Montevideo, some 220 km downstream of the delta (Figure A4.1).

The delta itself has been built up and is still being enlarged by sediments from the Rio Paraná. In its lower reaches, where the river has a very flat slope, the current is unable to carry coarse sand, but large quantities of very fine sand, silt and clay are carried in suspension. The greater part of this suspended material is carried into the Rio de la Plata, where the shallow waters are filled up with these sediments. If the accumulating solids have reached a certain level, vegetation grows on the mud banks and by its calming effect on currents and waves, greatly promotes sedimentation.

In this way islands are formed, which are separated by many small creeks and a few large river branches. In the downstream regions of the delta the creeks are still open and navigable for small craft vessels. In the course of time, however, they are gradually filled with sediments and some of them are almost completely silted up, allowing the flow of water only during high floods. On the other hand, the large river branches are able to maintain their sources and remain navigable even for large sea going vessels.

In their formation the islands have been flooded by high tides entering through the river and creeks. The sediments contained in the flood waters have mainly accumulated along the water courses, thus building up natural levees, which form the highest parts of the landscape in the delta. The main natural levees bear a natural forest, the minor ones are covered with herbs. Along the principal water courses the natural levees are usually inhabited and cultivated. Here the natural forest has almost completely disappeared except for a few remnants.

The low lying backswamps, which occupy the greater part of the delta, are flooded for most of the year and consequently bear a natural vegetation of reeds and rushes.

Human activities

To the natural elements: rivers, creeks, levees and backswamps, man has added his improvements. Canals have been constructed, serving both drainage and water traffic. The spoil from these canals has been dumped alongside, forming artificial levees, which like the natural ones, are inhabited and cultivated. The lower parts of the levees have been drained by ditches, thus allowing the cultivation of otherwise unsuitable lands.

In a few places, some degree of flood protection has been achieved by surrounding the fields with dykes. These dikes are high enough to keep out the regular inundations but are unable to withstand the major floods.

The levees, both natural and artificial, are largely planted with willows and poplars, thus making the Paraná delta one of the world largest regions where these trees are found. Relatively small areas are used for other trees (mainly pines), fruit growing (mainly citrus, plums and peaches), vegetables and for grazing. These activities and their development are seriously hampered by frequent inundations to which the lands are subjected.

Living conditions

The population in the delta is scattered along the rivers, the navigable creeks and the canals. Most people live in pile dwellings due to the frequent inundations, and as a result of these precautions even the highest floods cause very few casualties.

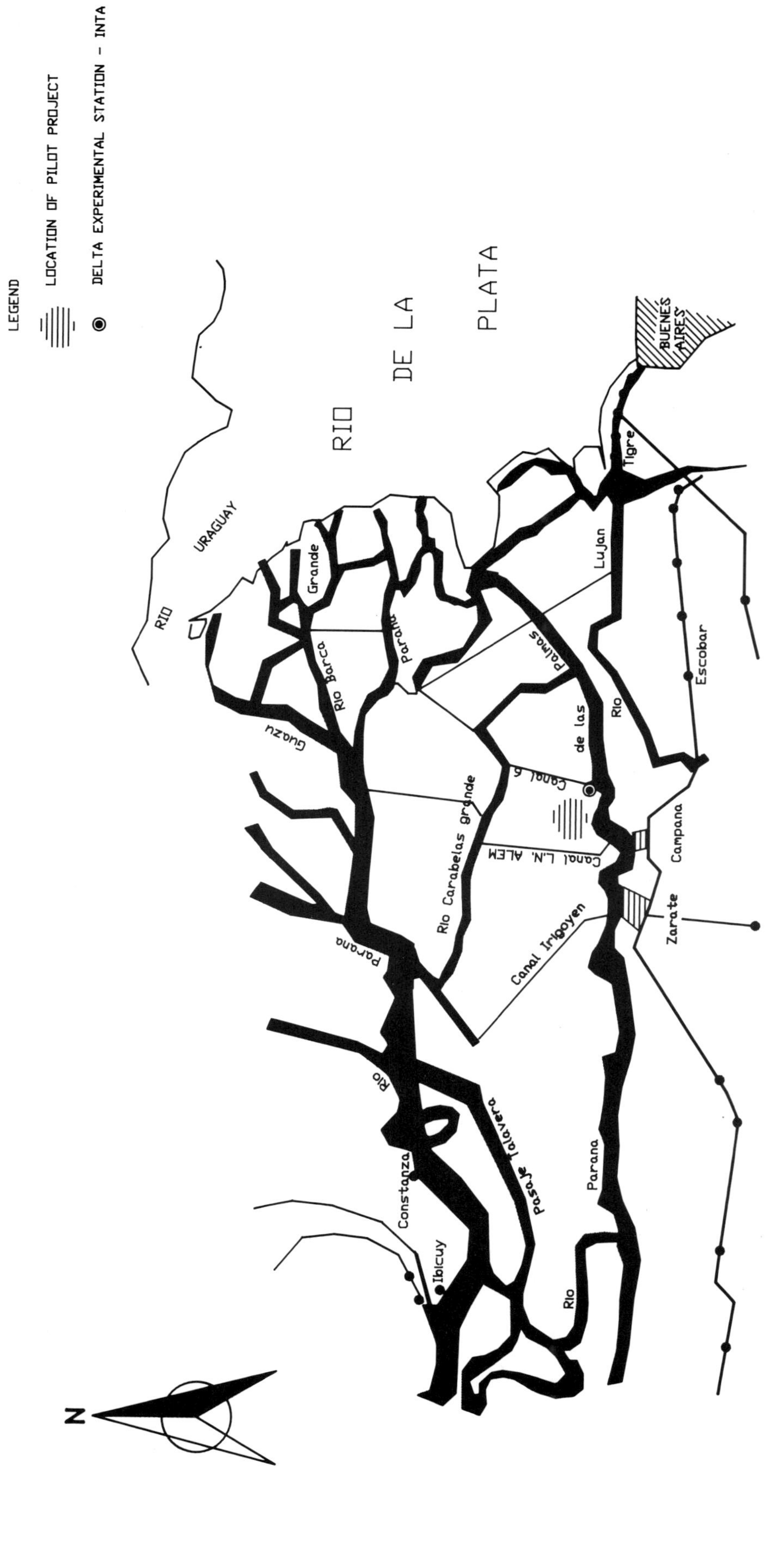

Fig. A4.1
Delta of the Rio Paraná.

Living centres like towns are absent in the delta, but public buildings are concentrated in a few places. Roads are practically non-existent, transport is exclusively by boat. For communications with the outside world each property possesses its own landing stage, which, seen from the water is often the only indication of the presence of inhabitants, as the houses are often hidden by trees.

Some parts of the delta are visited by large numbers of weekend tourists, especially those parts situated close to Buenos Aires. Boating, swimming and fishing attract thousands of people, and many residents of the capital possess weekend dwellings along the creeks.

In spite of the fertility of the soils, the favourable geographical position and tourism there is a tendency towards leaving the delta because of the flood hazard. Minor floods are expected a few times per year and impose many restrictions on cultivation. Major floods are more harmful, an example being the long lasting flood of 1959, which destroyed all fruit trees, and left many fruit plantations abandoned. Only thew willow and poplar trees are able to survive frequent or long lasting flooding without noticeable damage.

Although embankments are already present in the delta, complete protection from floods does not yet exist.

Climate

The climate of the delta may be considered as temperate. The average temperature varies from 22.4° C (January) to 9.0° C (July). The rainfall, averaging about 960 mm per year, is rather evenly distributed over the months, but the number of rainy days is low because most of the precipitation occurs during heavy showers. Cloudiness is low and sunshine is abundant. Due to the vicinity of the Atlantic Ocean and the presence of rivers and swamps, the humidity is high in all seasons, averaging about 75%.

Winds are usually slight but strong southeasterly winds can cause a considerable rise of the shallow waters of the Rio de la Plata and may lead to inundations of the delta.

Geology

The geological formations will be briefly described in order of decreasing age. The crystalline basement is assumed to be at a depth of about 145 m below sea level. The formations below a depth of about 22 metres are of Late Tertiary and Pleistocene age and consist of loam, sandy loam, silt loam and clayey layers, and some sand stone. These layers contain salty and brackish water. In some borings at depths between 22 and 40 metres natural gas was found Locally in the delta this gas is used for domestic purposes.

Above these formations, Holocene layers are found, consisting of heavy marine clay and old peat covered by non-calcareous silt to silt loam. These silt layers contain fresh ground water. Silt forms the sub-soil in the entire delta.

The formation of islands in the recent past led to a cover of sediments, partly of estuarine, partly of fluviatile origin. The rate of accretion along the coast has been estimated at 50 metres per year.

Along the coast line, especially during storms causing high water levels, the waters of the estuary deposit silt, loam or very fine sandy loam thus forming coastal ridges. Further inland the frequent inundations deposit materials, forming natural levees along the water courses. In the backswamps small amounts of very fine solids are deposited while the creeks are filled with silty materials. In the backswamps, where stagnant water conditions prevail during most of the year, the decomposition of the abundant vegetal remains is incomplete and peat is formed at the surface.

Surface water

The Rio Paraná contains only small quantities of dissolved salts (< 50 mg/1). The same holds for the Rio de la Plata, which is completely fresh as far downstream as Montevideo. Consequently even the most severe floods entering from the Rio de la Plata still bring in fresh water.

Although the salt contents of this water, after remaining stagnant in the backswamps and creek channels, becomes more concentrated due to evaporation, it remains fresh, due to its initial low salt content.

Shallow groundwater

The old sea clay forms a separating layer between the shallow and deep groundwater. The salt content of the shallow groundwater is considerably higher than that of the surface water (approximately 800 mg/l in the backswamps and 150 - 500 mg/l in the natural levees). It has been assumed that these salts originate from deeper salty formations from which they moved upwards by diffusion.

Some small patches of the natural levees and coastal ridges, however, are rather saline, especially where the soil consists of silt loam. These patches are seldom found on the highest parts of the ridges, but rather on the slopes towards the backswamps. Obviously, is such places a high evaporation of groundwater occurs during dry periods, accompanied by a considerable capillary rise. This in turn, induces a flow of groundwater from the adjacent backswamps which leads to an accumulation of salts in the soil.

Deep groundwater

The formations found underlying the old sea clay are not very permeable in general, except for a few sandy layers. The water in these layers is saline, due to the marine influences during various transgressions, and cannot be used for domestic purposes.

Soil forming processes

Environmental factors largely determine the soils of the delta. The delta can be considered young in terms of geological classification and as such the soil forming processes have not developed to any great extent. The differences in elevation caused by the morphology of the recent alluvial deposits determine the hydrologic conditions, which in turn have greatly influenced the development of the soil. Due to the great influence of water the soils are highly hydromorphic. The vegetation, forming a source of organic matter, has played a prominent part in the formation of a top layer with a considerable humus content.

Soil forming in the backswamps

The recent alluvial deposits of the backswamps consist of a thin layer of silty loam overlaying the silt loam of the fresh water sediments. They are completely reduced and have a high natural moisture content which indicates that they never or almost never become dry. The abundant vegetation, combined with the permanently wet environment, has led to the formation of a layer of recent peat, which contains considerable amounts of mineral matter, originating from the inundations.

Soil forming in the ridges and natural levees

The soils of the coastal ridges and natural levees consist mainly of silt loam to sandy loam with an accumulation of organic matter. The organic matter content, however, does not exceed 10%. The soils are slightly acid. The ridges located at the higher elevations are dry during most of the year. Therefore, the soil can dry out and air is able to enter. As mentioned before small areas can be found which are very saline.

Suitability of the soils

The soils of the delta are considered to have very favourable properties for many forms of cultivation. In their present state, however, they are far too wet, but after proper drainage and adequate flood protection they will be able to produce excellent crops. This is demonstrated in the parts of the delta which are already protected from floods. Here, yields of 80 tons of tomatoes per hectare are reported. Also peppers and cauliflowers are growing very well. The physical properties of the soils are especially favourable for growing fruits and vegetables. The soils are light, which makes them easily workable.

Flood protection

High water in the Rio Paraná can be caused by two different, though unrelated phenomena:
- high discharges from the catchment area of the river;
- strong south easterly winds, rapidly raising the water-level of the Rio de la Plata and thus heightening the waters of the Rio Paraná.

The first type of high water is usually of long duration (a couple of weeks or even a month), whereas the other type normally lasts only one or two days. Catastrophic high waters are to be expected when south easterly storms coincide with a flood in the Rio Paraná.

There is hardly any embankment in the delta to protect the land from flooding, and consequently large parts of the delta are flooded most of the year. On the right hand side of the river, however, some polders are protected.

References and Bibliography

NEDECO, 1962, Reclamation Projects in the Paraná Delta, The Hague.

THE DELTA OF THE NIGER

The delta of the Niger is situated at the southern coast of Nigeria. Nigeria has an area of 924,000 km^2 and a population of 100 million.

At the sea side the delta extends from the Benin river in the west to the Bonny River in the East. The apex is situated a few kilometres south of Aboh. Here the Niger forks into the Nunn and Forcados (see Figure A5.1). The low lying area of the delta is riddled with a large number of natural waterways, not all of which are tributaries of the Niger. Some of these creeks have their origin in the delta itself and have no open connection with the Niger or its branches. However, there are no less than 20 main outlets of the Niger river to the ocean. Each of them can be considered as a tidal basin.

The delta can be divided into three main physical regions: the fresh water zone, the mangrove swamps and the sandy ridges of the coastal zone. The northern part of the fresh water zone is a southward extension of the Lower Niger floodplain, and is affected annually by inundation from river floods. The southern part of this zone, however, has a great tidal influence but is not affected by river floods.

Most of the creeks in the fresh water zone are bordered by natural levees which are high enough to be free from river floods. Therefore they provide the site for most of the settlements in the area. Some of the lower levees are used for agricultural purposes. Crops such as cassava, yam, cocoyam and vegetables are cultivated, and are usually harvested before the river floods occur.

The land between the levees is lower and is affected by the yearly floods of the river. Large swamps with silted water courses are present in this area. Drainage is poor and therefore pools with standing water can be present for almost the entire year.

South of the fresh-water zone mangrove swamps are present, covering an area of about 10,000 square kilometres. Most of this area is very muddy and therefore uninhabitable, but there are also a number of rather high land areas which are dry throughout the year. Some of these high lands provide sites for settlements. However, these islands are affected by creek erosion, especially when the sea level drops.

Close to the open sea is a narrow strip of sandy ridges and beaches. The width of the strip varies between a few metres and 15 kilometres. The soils are rather poor, but because of the high rainfall, crops such as mangoes, coconuts, cashew nuts and palm oil trees are cultivated.

Human activities

Compared with the large deltas if the humid tropical zone in South East Asia, the delta of the Niger is sparsely populated and not very productive. The main reason may be that rice is not the staple food of West Africa.

Human activities are still largely controlled by natural conditions. In some areas these activities are limited to fishing. The swampy nature of a vast area of the delta and the high rainfall pose considerable problems to settlement and land use. Only 3% of the land is used for farming and even that is exposed to flooding during the high water period. Traditionally, the people of the delta are divided into the landsmen in the north and the watermen in the south. The landsmen are committed to farming, fishing, collecting palm fruits and hunting, while the watermen are basically fishermen and traders.

The farming activities of the landsmen in the north cannot start earlier than mid December, when the river floods have receded. Canoes are the only means of transport possible in vast areas of the delta. As such the building of canoes is an important local economic activity.

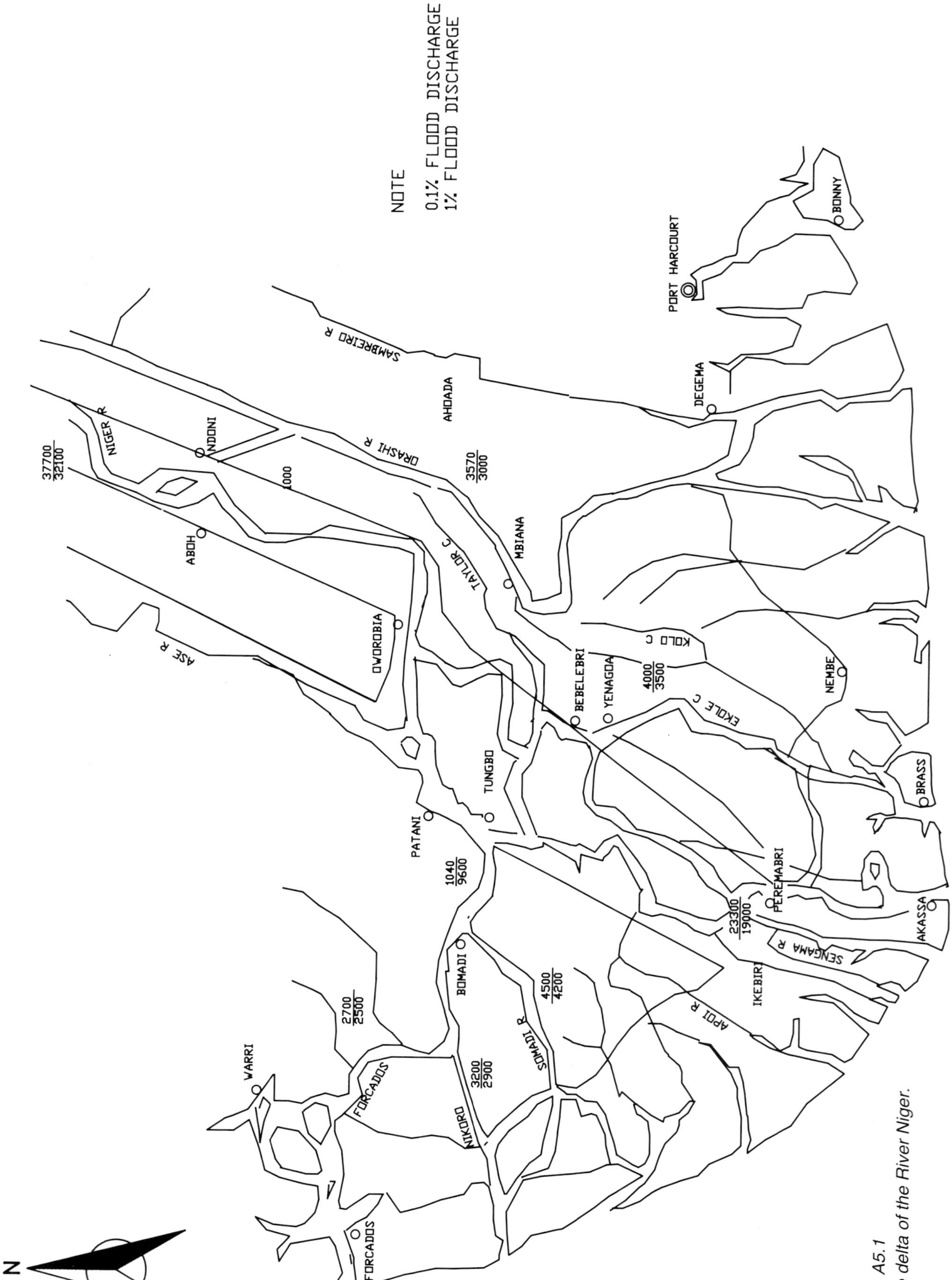

Fig. A5.1
The delta of the River Niger.

Except for some rare places in the southern part of the delta, farming is hardly possible and fishing is the main activity. Fishing rights therefore are as important as land rights are to the farming people. A non-traditional but important economic activity in the delta is the exploration for oil. The swampy nature of the delta create a lot of difficulties for transporting heavy equipment from one place to another. Some sites can only reached by helicopter. Difficulties are also encountered in the laying of pipelines through swamps and the many creeks.

Climate

The climate of Nigeria is influenced by two wind systems. One is the rain bearing south-west monsoon, which blows from the ocean, while the other is the dry north east wind which comes from the Sahara Desert. There are two seasons, the lengths of which vary from the north to the south of the country. In the southern part of the country the rainy season lasts for nine months (from March to November).
In the coastal area however, four seasons can be distinguished. There is a long rainy season (March to August), a short dry season (August), a short rainy season (September to November), and a long dry season (November to February). An important effect of this break in the rains is that it makes it possible to produce two crops a year.

The beginning of the rainy season is often marked by great heat and destructive storms. The average temperature during the year is 30.6 °C. Rainfall of as much as 40 mm/hour in this period is no exception. The relative humidity, which exceeds 80% along the coast for most of the year, makes the climate very tiring. The yearly average rainfall in Port Harcourt, on the eastern edge of the delta is 2,500 millimetres. It should be noted that the rainfall in the coastal strip of the delta can exceed 3,500 mm/year.

Geology

Like most African countries Nigeria is situated on the Precambrian African Shield. In vast areas of Nigeria this formation can be found at the surface. In some areas however, especially along the river Niger, its main tributary the river Benue and of course in the delta, this formation is covered by fluviatile sediments. Along the rivers, upstream of the apex, the width of the area covered by these sediments is 100 to 150 km.

The sediment load of the Niger amounts to some 40 millions tons per year, but in previous geological periods it may have been twice or three times as much. All these sediments have been dumped in the sea, covering the original base rock with hundreds of metres of clay, silt and sand.

Soils

The soils (and vegetation) are products of the hydrological regime of the delta. The coarsest and most pervious soils are found on the natural levees of the upper river (sand and silt soils). In the backswamps solid clays are found. Along the lower courses of the river silt and clays are deposited and the finest clay is found in the tidal swamps.

The agricultural value of these soils decreases with its grain size, due to decreasing drainage properties. Owing to the poor drainage system parts of the backswamps remain swampy throughout the year.

Hydrology

The rainfall in the delta has been discussed before. Very little is known about the evapo-transpiration but it is assumed that this varies from about 750 - 1,000 mm/year along the coast, to 1,300 mm/year in the northern part of the delta. To get an impression of the run-off, the evapo-transpiration may be subtracted from rainfall figures. The average yearly runoff amounts to approximately 2,000 millimetres. As can be seen from these figures, the amounts of water transported in the hydrological cycle are very high compared to a moderate climate. The mean fresh water discharge of the Niger river at the apex is about 7,000 m^3/sec.

Through the river outlets and creeks, saline water from the ocean penetrates into the delta. On its way this saline water meets fresh water from the river and rainfall. The saltwater zone shrinks and expands under tidal influences and under the seasonal variation of the river discharge.

References and Bibliography

ILACO-NEDECO, 1966, Report on the Soil Investigation in the Niger Delta Special Area, The Hague.

KIT, 1982, Nigeria, Koninklijk Instituut voor de Tropen, Amsterdam.

NEDECO, 1961, The Waters of the Niger Delta, The Hague.

NEDECO, 1980, Navigability of the River Niger - Investigations, Studies and Dredging Operations Downstream of Lokoja, The Hague.

THE DELTA OF THE NILE

Physiography

The Nile, a few kilometres north of Cairo bifurcates into the Rosetta and Damietta branches. These branches have their mouths on the southern coast of the Mediterranean Sea. The coastline from Alexandria in the west to Port-Said in the east, is an undulating line in the shape of an advancing delta.

The general lay-out of the Nile delta is shown in Figure A6.1. The bulge of the delta is the product of Nile sediment discharges over long geological periods, during which the Nile branches have changed their courses several times. The gradual diversion into the present Nile branches took place within the last 1,500 - 2,000 years, during which time the promontories at the estuaries were also built up (Nielsen, 1973). The directions and shapes of the promontories at Rosetta and Damietta estuaries are consistent with the relationship between sediment discharges of the river branches and the transport capacity of the sea.

Along the coast of the Nile delta there are four large lakes, all with some connection to the Mediterranean Sea. Their origin is probably nilotic. They lie at or near the mouths of the ancient outlets of the Nile, at an altitude of 0 - 3 metres above mean sea-level. From west to east the four lakes are: Maryut, Edku, Burullus and Manzalah.

The Nile discharge used to carry silt and other fine sediments to these lakes during the flood season in the period before the Aswan High Dam. The new hydrological regime brought about by the Aswan High Dam since 1965, has completely stopped the flow of the silt-laden water to the lakes. Some time is still needed before a new state of equilibrium is established.

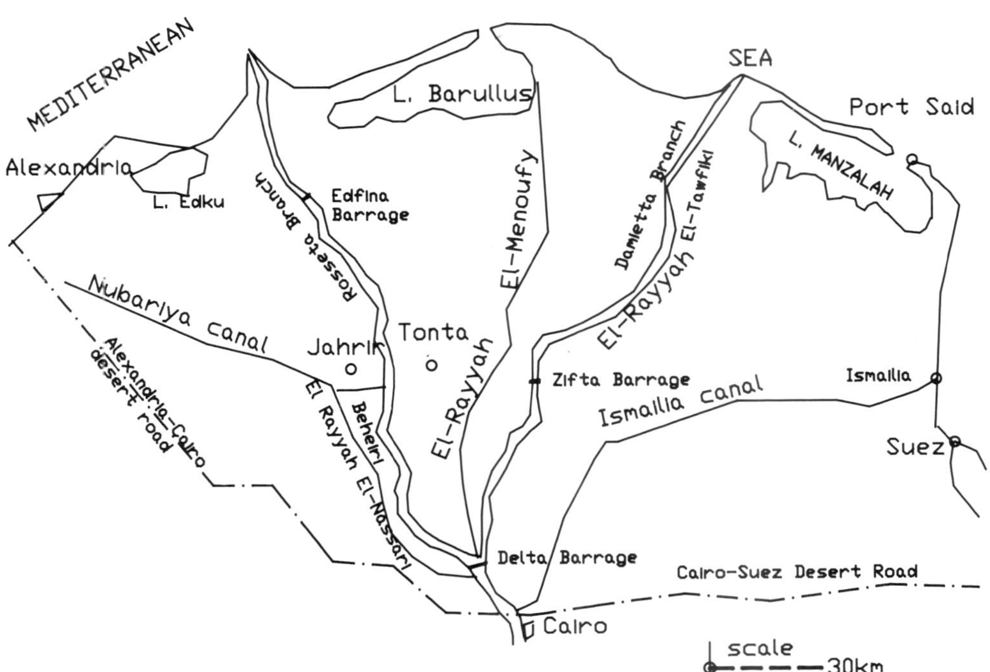

Fig. A6.1
The delta of the Nile in Egypt.

The coastline of the Nile delta has been retreating for some time, mainly due to the trapping of silt in the storage reservoirs built at Aswan. The High Dam has brought the once slow but continuous process of delta building, to an end. This means that the coastline is left open to the erosive action of the shore currents. Beach erosion near the river mouths at Damietta and Rosetta had been observed as early as 1902. Since that time the shoreline has been frequently mapped, both before and after the construction of the Aswan High Dam. According to Wassing (1964), the rate of retreat of the promontories at Damietta and Rosetta had an average of 29 m per year in the period from 1898 to 1960. The gradual development of the contour line of the shore at the mouths of the river branches is shown in Figure A6.2.

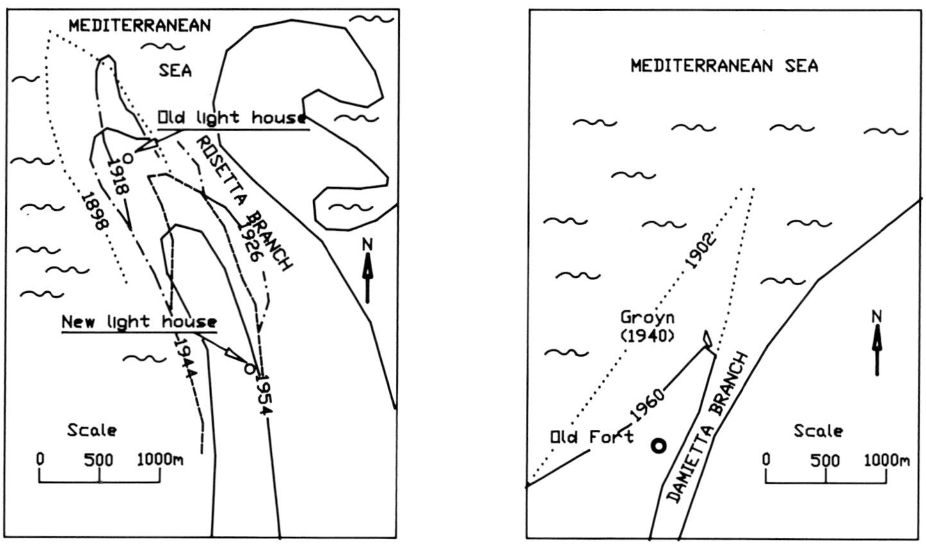

Fig. A6.2
Retreat of shoreline of Damietta and Rosetta branches.

What is usually referred to nowadays as the Nile delta is a surface of more or less triangular shape extending from just below Cairo in the south to the coast of the Mediterranean Sea. This surface is intersected by the two river branches. The land area extends east of the Damietta branch and west of the Rosetta branch. The total area of land within the village boundaries less the public utilities (including the river), and the northern lakes is approximately 20,000 km^2. More than three-quarters of this area is under perennial irrigation at the present time and the cultivated area is estimated at 16,000 km^2.

Other than the Nile branches there are the Ismailia canal extending in an easterly direction from just below Cairo to the township of Ismailia, on the southern end of the Suez canal, and the Rayyah El-Nasseri and the Nubariya canal running west of the Rosetta branch. Between the two river branches runs the Rayyah El-Menoufy and east of the Damietta branch and almost parallel to it runs the Rayyah El-Tawfiki, both large water carriers. Needless to say there is a very long and intricate network of irrigation and drainage canals of different dimensions and functions. The northern lakes receive considerable amounts of drainage water from the irrigated land in the delta.

The ground surface has a mild slope in a northerly direction. The level at the apex is nearly 14 metres above mean sea-level and the general slope of the land varies between 8 and 15 centimetres per kilometre.

Climate

The climate of the delta area varies from Mediterranean along the sea coast to arid, semi-tropical near the apex at Cairo. The mean annual air temperature varies from about 20 °C along the Mediterranean coast to about 21 °C near the apex of the delta between Cairo and the delta barrages. The average monthly temperature in summer varies between 26 °C and 27.8 °C, and in winter between 11.2 °C and 15 °C. The coastal strip is a bit cooler than the inland areas in summer, and a bit warmer in winter. It is further characterised by a narrower range of daily temperature variation than the inland.

There is no distinct pattern of variation of the average monthly air humidity especially in the post-High Dam period. The average monthly and annual relative humidity falls in the range between 65 and 85%.

In summer the north-westerly winds usually prevail over the northern part of the delta area and eastern areas as far as Alexandria. The direction of the prevailing wind over the delta changes to north-east in the spring because of the passage of the centre of the Khamasin depressions (dust storms blowing from the western quadrant and lasting some fifty days), south of the Mediterranean coastline. In winter the stations situated along the coast of the Mediterranean and in the Nile delta are strongly affected by the so-called Mediterranean depressions. Places located in front of the depressions receive south-westerly winds. Late winter and early spring are the times when gales and strong winds are likely to occur in Lower Egypt. The frequency of these gales decreases considerable from north to south. In general, autumn, especially during October, is the time of lowest wind speed.

The long-term average rainfall decreases from about 190 mm/year on the coast of the Mediterranean sea to about 50 mm/year at Tanta, in the middle of the delta, to 20 mm/year, at the delta barrages.

The free water evaporation varies from almost 5 mm/day at the apex of the delta, to 4.5 mm/day in the middle, and to around 4 mm/day along the sea coast. The potential evapo-transpiration rates are slightly less than these figures (see UNESCO (1977), Mehanna (1979) and Shahin (1985)).

Geology

During the Archezoic and Proterozoic geological areas, some 500 million years ago, the surface of Egypt was covered by igneous, metamorphic and crystallised rocks. Following that time there were alternating periods of inundation and recession. When the Mediterranean inundated the country, sedimentation of sands and pebbles took place. These sediments remained whereas most of the biological sea sediments were washed out mainly by weathering action. Frequent changes took place in the Cenozoic era, 50 million years ago, resulting in a series of transgressions and regressions of the Mediterranean. The Nile in Egypt began to form its valley in the Upper Miocene age. It was then referred to as the Eonile and it cut its gorge at a much lower level than the present sea-level. This gorge was completely filled with sediments brought by the Paleonile during the Upper Pliocene, 3.2 million years ago. Sedimentation was augmented by the torrential water carried by the Wadis, flowing down the Red Sea hills. In that period the delta became more or less even and began to acquire a gentle slope to the north. As the period from 1.8 million years to 0.7 million years ago had been extremely dry, the Paleo-Proto-Nile, as it was then referred to, ceased to flow. Consequently no further sedimentation took place. However, deposition of gravels and sands was quite pronounced in the Proto-Nile period, i.e. from 0.7 to 0.5 million years ago. The tectonic movements which took place from the end of the Protonile period, up to 125,000 years ago, helped to form some hydraulic connection between the Prenile in Egypt and the rivers in the Ethiopian Plateau. The rise in the flood volume and sediment load which it carried led to the extension of the delta into the sea.

The Nile delta occupies a great tectonic depression and is bounded on both sides by gravelly plains. On the eastern side, the delta region is bounded by a major upward zone. This is followed in a northward direction by a downward zone which occupies most of the delta region. The plains on the eastern and western sides of the delta, merge into the elevated table lands (higher than 200 metres above mean sea-level).

In the wet seasons of the Pre-Neonile period, which lasted from 125,000 up to 30,000 years ago, heavy rainfalls resulted in deposits of coarse sand and gravel. The late Acheulean pluvial which ended 3,500 years ago was an important one. The next important fluvial was the Neolithic one which lasted from about 10,000 years to around 5,000 years ago. This also gave rise to further sedimentation in the Nile valley and delta. Ball (1939), estimates the rate of increase of the thickness of Nile mud in the basin land of Upper Egypt, prior to the erection of the Aswan Dam, at 10.3 cm/century. Said (1981), by comparison with the rate of sedimentation in the valley, gives 12.5 cm/century as a probable rate of sedimentation in the delta, during the time of basin irrigation.

Geo-hydrology

The aquifer underlying the Nile delta area is a huge reservoir of ground water, with a fresh water content (salinity not exceeding 1,000 p.p.m. Cl^-) of not less than 500 billion m^3 (Solait, 1964). The thickness of this aquifer increases in a northerly direction from about 100 metres in the neighbourhood of Cairo, to about 900 metres near the shoreline (Figure A6.3). Furthermore, the reservoir has a bowl-like form, in which the maximum depth occurs almost half-way between the two river branches, and decreases in both directions, east and west. The aquifer is overlain by a layer of fine sediments, silt and clay, with thicknesses varying from zero to 50 metres depending on the location.

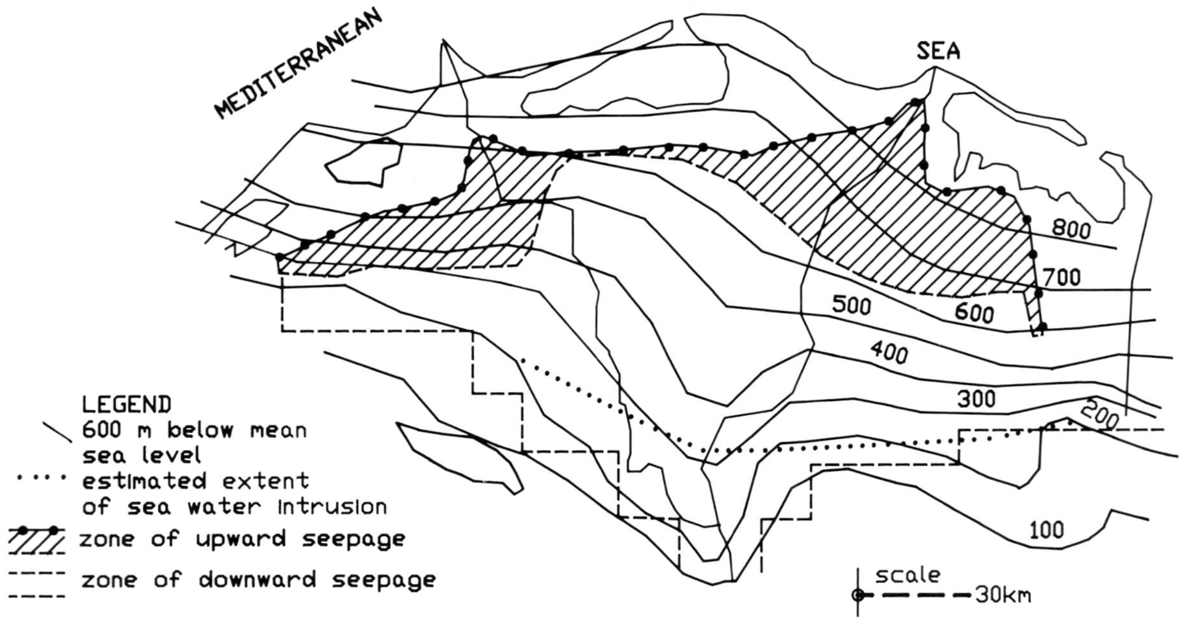

Fig. A6.3
Contour map of the main delta aquifer.

The prevailing state of affairs has led to the presence of three different types of aquifers. These are: unconfined in the southeast and southwest, confined in the area parallel to the sea coast and underneath the delta lakes with a width of not less than 40 km, and semi-confined in between. A summary of the geo-hydrological constants of the different aquifer types in the Nile delta are reported by Shahin (1983). Some of the results obtained are presented in Table A6.1. The permeability, depth, transmissivity and storage coefficient of the sand and gravel aquifer are denoted by k, D, kD and S respectively. Moreover, the permeability, depth and resistance of the confining layer (clay and silt), are denoted by k′, d′ and c respectively.

Type of aquifer	Aquifer				Semi-confining layer		
	k (m/day)	D (m)	kD (m^2/day)	S	k′ (m/day)	d′ (m)	c (day)
Unconfined	50 to 80	50 to 150	2500 to 12000	0.15 to 0.25	-	-	-
Semi-confined	25 to 100	100 to 250	2500 to 25000	0.0001 to 0.001	0.001 to 0.003	8 to 20	3000 to 20000
Confined	——— UNKNOWN ———				< 0.001	20 to 50	20000 to 50000
Greater Cairo			1000 to 10000	0.001 to 0.028	0.001 to 0.005	8 to 10	1500 to 10000

Table A6.1 Geo-hydrologic constants of the aquifers underlying the Nile delta.

The salt water intrusion from the sea is a factor which limits the extraction of groundwater from the aquifer underlying the delta area. The extent of this intrusion has been studied using different methods. Due to some inaccuracies in the data used, inconsistent results were obtained. In his paper on salt water intrusion in the delta, Kashaf (1983) indicated that the salt water wedge may intrude into the interior up to about 130 km from the sea (see Figure A6.3).

Irrigation system and water control structures

Until recently, the irrigation system that prevailed in Egypt was the basin system which produced only one crop annually. At the height of the flood season (August and September), the land was inundated by an average depth of water of about 1.5 metres. The water was retained for a period ranging from 40 to 60 days, until the level in the river fell sufficiently to allow drainage of the excess water. During that period sedimentation of silt occurred, and by the annual repetition of this process for over a century, basins with perfectly graded surfaces have been formed.

After completion of basin drainage, seeds were sown during October and November each year. The principal winter crops grown in the basins were wheat, beans, berseem (Egyptian clover), lentil, barley, fenugreek and chick. In some of the basins small-scale summer irrigation was practised using groundwater from shallow wells.

In the early nineteenth century, Mohammed Ali, the then ruler of Egypt, wanted to convert the entire delta area into a large cotton field. In order to irrigate the vast areas producing this summer crop, deep canals were excavated in the delta in the 1820's. These allowed water to be obtained from the river at its low summer levels. Additionally, a supplementary system of rather shallow canals was utilised during the flood season only. Winter irrigation was accomplished by means of the summer canals, supplemented from the lower reaches of the flood canals which still contained water.

As the summer canals are so deep, several regulators had to be built in order to raise the water to the fields. The pool formed between each pair of consecutive regulators acted as a silt trap and had to be cleared regularly to allow the water to flow easily. In order to avoid this massive task of canal clearing, the construction of the Mohammed Ali barrages began in 1833, and were finally completed in 1861. The barrages are dams of the open type, the first of its kind built in the world.

Soon after commissioning the barrages, they showed signs of weakness and instability. Between 1887 and 1890, extensive repairs had to be implemented and the barrages had to be put anew into operation by the middle of 1890. A few years later two submerged weirs were constructed downstream of the barrages. Since then the maximum head possible at the Rosetta barrage was about 4.0 metres, and at the Damietta barrage it was about 3.75 metres. It should be remembered that under conditions of natural flow in the Nile, the summer level of the former, could be 3 to 4 metres below the adjacent land level.

In 1935 it became clear that the old Mohammed Ali barrages had reached a poor state and could not take any further remodelling. Construction of new barrages called the new Mohammed Ali barrages, and later the Delta Barrages, began in 1937 and was completed in 1939. This provided a maximum head difference of 3.8 metres during the summer time.

Long ago, before the construction of any barrage, the maximum discharge of the two branches at the head of the delta were almost the same. A little lower down, however, the Rosetta branch had considerably more water than the Damietta. This was mainly due to the many channels that take water from the Damietta branch and drain to the Rosetta branch.

For the improvement and expansion of perennial irrigation in the northern part of the Nile delta, a barrage was constructed in 1902 on the Damietta branch at Zifta, 78 kilometres downstream of the Damietta barrage. Fifty years later, the old barrage was remodelled and became ready for operation in 1954.

Barrage in the Nile. Upstream of this headwork irrigation canals branch off for water supply.

Not far from the mouth of each branch at the Mediterranean Sea, is the site where an earthen dam used to be constructed each year. This was completed when the flow into the river was shut off at the Delta Barrage and all the water was diverted to the irrigation canals as it was needed for cultivation. Those banks prevented the sea water from penetrating inland, but also enabled seepage and drainage of water back into the river, from areas extending from the Delta Barrage northwards. This was collected and used for irrigating small areas to the north along the river. The earthen bank on the branch was later replaced (in 1951) by a permanent structure, the Edfina barrage. The maximum summer head used to be at 2.8 metres. One of the main objectives of the Edfina barrage was to conserve about 1.3 x 10^6 m^3 of water, which before the construction of the earthen dam, had been used to restrict the entry of sea water into the Rosetta branch (NCIDLD, 1963).

The introduction of perennial irrigation in Egypt followed by the expansion of the area under such a system, necessitated more than just the construction of a number of barrages. Many embankments were constructed and/or strengthened along both sides of the river and its branches from Aswan to the Mediterranean Sea. The largest canals (Rayyahs), were dug and equipped with a head barrage as explained earlier. A complete network of irrigation canals, from main to tributary to secondary, were excavated and equipped with the necessary structures. To have an impression of the dimensions of such a canal consider the case of a main canal serving an area of 42,000 hectares. Given a water duty of say 92 m^3/ha/day this canal should carry a discharge at the head equal to about 25.4 m^3/s. The cross-section at the head may be such that the top surface width is 24 metres, the bottom width is 12 metres, the side slopes are 2 horizontal to 1 vertical, and the depth is about 3.0 metres. The corresponding longitudinal slope of the water surface is about 7 cm/km and the mean velocity of flow is 0.45 m/s.

Summer water needed for perennial irrigation was made available mainly through storage works. These are the Aswan dam built around 1899-1900 (1 x 10^9 m^3), first heightened between 1907 and 1912 (2.3 x 10^9 m^3), then between 1929 and 1933 (5.1 x 10^9 m^3), and the construction of the Jebel Awlia dam around 1937, with a net benefit to Egypt of 2.7 x 10^9 m^3/year. This state of affairs lasted until 1965 after which the flow in the Nile in Egypt was put under a different regulation scheme, as a consequence of the construction of the Aswan High Dam commissioned in 1968.

This dam was designed as an over-year storage dam with a capacity of 164 x 10^9 m^3 with a view to:

- guarantee the necessary water requirements of the agricultural lands;
- make available additional water for the cultivation of 550,000 hectares of new land;
- enable the conversion of some 400,000 hectares from basin to perennial irrigation and the annual irrigation of 290,000 hectares of rice.

The project, which has proved to be a great success, has already paid itself off after the consecutive dry years 1987-1991, when a reduction of no greater than 10% had to be applied to the irrigation deliveries. Without the dam Egypt would have faced a disaster like those that have occurred so often in the past. It was precisely in view of such a prolonged low flow period that the dam was designed. The Aswan High Dam makes available for Egypt an annual volume of 55.5 x 10^9 m^3, of which 32 x 10^9 m^3 reaches the delta.

Irrigation water requirements

By the end of the nineteenth century the area under perennial irrigation in the Nile delta was about 1 million hectares. The total requirements for irrigation can be estimated at about 24 x 10^9 m^3/year.

As a result of the construction and operation of the various storage, control, conveyance and distribution works already explained, the area under perennial irrigation grew to 1.47 million hectares by 1940, and to 1.51 million hectares by 1945. Hurst et al. (1951) give the 10-day irrigation requirements for the 1940 area during the low flow season (February-July) as 13.94 x 10^9 m^3. In the period 1945 to 1947, the average amount of water used for basin irrigation in Upper Egypt was 6.3 x 10^9 m^3/year. The requirements for perennial irrigation in both Upper and Lower Egypt were 38.5 x 10^9 m^3/year, of which 25.3 x 10^9 m^3/year were for the delta only. Apart from that, 2 x 10^9 m^3/year were released for the purpose of navigation and 1.3 x 10^9 m^3/year for the combat of sea water intrusion, while 38 x 10^9 m^3/year were discharged into the sea.

Hurst et al. (1951) also describe the plan of the Egyptian government for agricultural expansion. This plan was thought to require about 32.4 x 10^9 m^3 for the whole year (January was not included because of the winter closure of the canals). The publication of these data have encouraged individuals, universities, research organisations and some governmental departments to investigate the question of irrigation water requirements using different approaches. El Shal (1966), based his figures of actual crop water use on soil moisture depletion studies carried out in experimental plots. The cultivated surface areas and the cropped area between 1962 and 1964 were 1.51 and 2.73 million hectares respectively. The total crop water use was then 15 to 13 x 10^9 m^3/year or 2.75 mm/day. The potential evapo-transpiration, obviously higher than this rate, was estimated by Mehanna (1979) at an average of about 4.4 mm/day. This figure is very close to the ones given by UNESCO (1977).

Kamal and Hashem (1967) describe the control and distribution of irrigation water in the United Arab Republic (Egypt). In this reference the basis of estimating the field water requirements are given. The average annual figure is 4.22 mm/day. Abul-Atta (1975), the then Minister of Irrigation of Egypt, published new figures for the field water requirements, in the post-High Dam period. These figures indicated a total field requirement of 25.5 x 10^9 m^3/year.

Abul-Atta (1975) added that the conveyance losses between Aswan and the delta are 20%, bringing the total requirements for irrigation of the Nile delta to 31.84 x 10^9 m^3/yr, measured at Aswan. These requirements are for a cropped area of about 2.94 million hectares. Finally, Abul-Atta admits that there is some wastage in the use of water supplied for irrigation.

As time goes by, the supply of water to the delta area is increasing, without a proportionate increase in the area under cultivation. The average supply during the last ten years has reached over 36 x 10^9 m^3/year.

Drainage systems

The sustainability of the Egyptian agriculture in the delta, as well as in Middle and Upper Egypt depends on the good functioning of one of the most elaborate and sophisticated drainage systems in the world. Apart from a small area near Alexandria, rainfall in Egypt is of no significance for the supply of water to the crops, leaving the Nile as main water supplier. Like in other arid regions, land drainage in Egypt is a matter of salinity control of the water in the root zone. Even good irrigation water, like the Nile water, always contains some salt and since no salt is removed by evaporation, irrigation would cause soil salinisation if no salt was removed by drainage.

Drainage machine for subsurface drainage the delta of the Nile in Egypt. To prevent soil salinisation leaching of the soil is required with fresh water from the irrigation water. This leachate has to be removed by a drainage system (field drains, canals, pumping stations to the Mediterranean Sea).

During the 19th century, crops like cotton and sugar cane were introduced which require irrigation during periods of low flow in the Nile. The climate and soil would allow two or three crops per year provided that irrigation was available almost the whole year round. This would lead to a rise of the groundwater table, waterlogging and soil salinisation if no drainage was applied. To overcome this problem, open drains were excavated and pumping stations were constructed as early as 1898. With the extension of the perennial irrigation it was found that open drains alone were not sufficient. Experiments on subsurface drainage started in 1934. Subsurface drainage was installed in six pilot areas totalling some 4,200 hectares. It was the completion of the Aswan High Dam which gave the impetus to the implementation of a programme for draining more than 50,000 hectares per year in the delta alone.

The principle of the present system is given in Figure A6.4. The salinity control is obtained by keeping the groundwater table well below land surface thus preventing capillary rise of the brackish groundwater to the root zone and use of a part of the water supplied by the irrigation system for downward percolation.

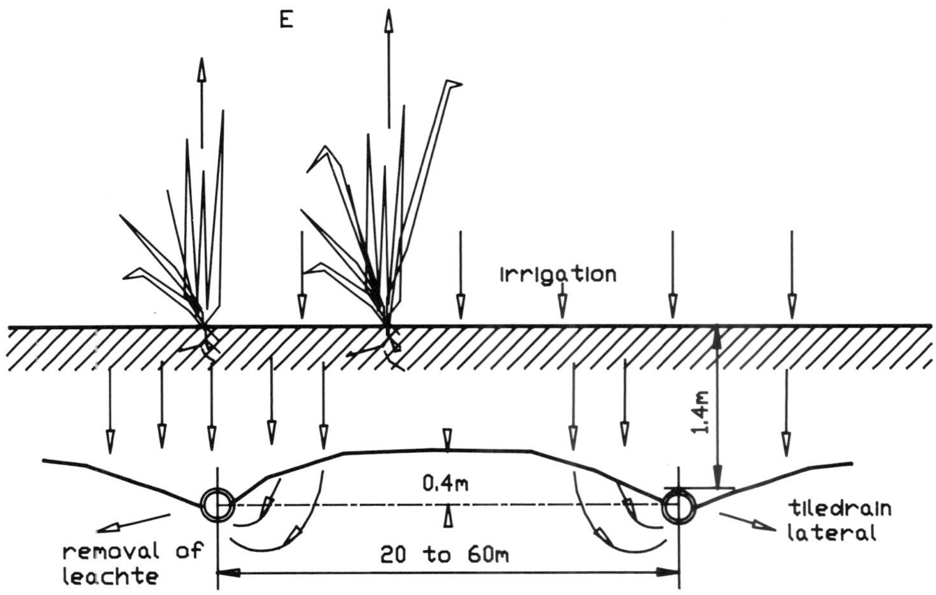

Fig A6.4
Subsurface drainage for salinity control.

This produces a leaching of the root zone. For the purpose of salinity control a drainage rate of 1 to 1.5 mm/day would be sufficient. This corresponds with about 15% of the rate of consumptive use. The lateral field drains consist of corrugated perforated PVC tubes with an outside diameter of 80 mm. The water from the lateral flows into collectors and from there to the open drains (see Chapter 3.5). Until recently, collector drains were constructed of plain concrete pipes (inside diameters, varying between 15 and 40 cm). Recently, large size corrugated plastic pipes have been introduced. The total net irrigated area of 1,610,000 hectares (total gross area of the delta is 1,980,000 hectares) has been subdivided into a few hundred drainage catchments. The drainage water has to be lifted between 1 to 3 metres from the open drains to an outfall drain. The outfall drain conveys the water to one of the four lakes or directly to the sea either as a freely flowing canal or via a pumping station. The outfall drains also receive water from upstream higher areas in the south where the open drains can discharge by gravity into the outfall drains.

In 1988, the pump-lift system consisted of 70 pumping stations with a total capacity of 1,300 m^3/sec corresponding to an average drainage capacity of 8 mm/day.

Water and salt balance

Thanks to an elaborate network of stations for measuring water levels, water flows and water quality it is possible to draw up water and salt balances for the delta as a whole and for individual catchments. Measurements for the water balance for an average year, averaged over the delta show a supply from the Nile of 5.2 mm/day and from seepage of 0.2 mm/day, while water is removed due to evaporation by 3.2 mm/day and drainage to the sea of 2.2 mm/day.

The actual evapo-transpiration of 3.2 mm/day is close to the estimates mentioned earlier. However, this figure is well below the potential evapo-transpiration because of the closure of the canals in January for maintenance, the harvesting period and the fallow lands. The evapo-transpiration rates do not vary much between the seven zones of the delta (from 2.9 to 3.4 mm/day). The average total drainage rate of 2.2 mm/day is surprisingly high considering that the leaching requirements amount to only 1 to 1.5 mm/day. The explanation lies in the fact that irrigation water which is not used during the nightly hours is discharged into the drainage canals. The drainage rates show a considerable areal variation and vary from 0.7 to 4.2 mm/day. This is due to the effect of seepage which may be upward or downward as indicated on Figure A6.3. There is downward percolation in the southern zones of the delta (0.8 mm/day) and upward seepage of brackish water in low-lying northern zones where some catchments are even situated below mean sea-level.

In the salt balance, the inflow of salt into the delta consists mainly of the salt in the Nile water. The volume of water passing the Delta Barrage is 35 x10^9 m^3 with a salinity of about 300 p.p.m. This represents an annual quantity of 10.5 x 10^9 kg of salt. The salinity of the drainage water varies from 630 p.p.m. in the south to 6,400 p.p.m. in the north. With a drainage to the sea of 12.7 x 10^9 m^3/year this represents a load of 31.5 x 10^9 kg per year. The difference from the inflow is mainly due to the seepage of brackish groundwater in the northern part of the delta and partly to the desalinisation of reclaimed lands and drainage of land with subsurface drains.

Re-use of drainage water

The present re-use of drainage water amounts to about 3 x 10^9 m^3 per year for an area of some 500,000 hectares. The average salinity is about 930 p.p.m. This water is mixed with fresh water before use and a salinity of 500 to 600 p.p.m. is obtained. Considering the enormous volume of water (12.7 x 10^9 m^3 per year) which is drained off to the sea and the pressing need for more water for food production it has been decided to increase the re-use of irrigation water. Not all outflow water can be re-used for growing crops and the statement by Napoleon that "if he were to rule a country like Egypt not even one drop of water would be allowed to flow to the Mediterranean Sea" conflicts with the concept of acceptable salinity conditions. Theoretically, a volume of 9 x 10^9 m^3 per year could be made available with a salinity of 2,000 p.p.m. or less but the consequences of using this volume should be studied further.

Groundwater

As mentioned before, the aquifer underlying the Nile delta area is a huge reservoir containing at least 500 x 10^9 m^3 of fresh water, (salinity not exceeding 1,000 p.p.m.). In the pre-High Dam era the seepage from the Nile branches during the flood season used to be a principal factor leading to the rise of the piezometric pressure in the aquifer, and to a lesser extent the water-table in the semi-confining cap. The river branches used to act as drains in the low-flow season, and their recharge from the aquifer led to the decline in the piezometric head. Furthermore, part of the excess of irrigation water was not intercepted by the drains, instead it acted as a recharge to the aquifer. Similar to the fluctuation of the Nile water from one season to another, the irrigation canals used to run at high, moderate and low levels according to the turn of the rotation. Since the duration of the summer rotation used to be 18 days, each turn lasted only 6 days. The water-table adjacent to canals used to fluctuate in a rhythmic order at close intervals. Whenever the resistance of the top layer was reasonable, the rhythmic motion used to be transmitted, though with some attenuation, to the piezometric level of the groundwater.

In the post-High Dam condition the fluctuation of the water level in the Nile and its branches is much less compared to the pre-Dam condition. The flood level is very much reduced and the summer level is much higher. The excessive application of water for irrigation is the principal source of recharge to the aquifer at present. The summer levels of 1977 compared to those 1969, shortly after the completion of the High Dam, show an average rate of rise of not less than 5 cm/year.

The aquifer recharge is estimated by Shahin (1983) at 3.8 x 10^9 m^3/year, whereas the estimate given by El-Kasher (1983) is 6.4 x 10^9 m^3/year. Furthermore, these two references show that the amounts left after deducting the groundwater withdrawal at present, are 2.2. and 3.9 x 10^9 m^3/year, respectively.

Farid (1985) developed in her thesis a management model for groundwater utilisation in the Nile delta area. An operation policy was presented based on the safe yield concept. It would be possible to extract groundwater for municipal water supply and to irrigate an additional area of 250,000 hectares using a total of 4 x 10^9 m^3 per year.

The same reference points out the plan of the present Ministry of Public Works and Water Resources to undertake more withdrawal of groundwater for irrigation and land reclamation purposes. In other words, groundwater will be considered as a new resource in the future water policy of the country. While doing so, care should be exercised to avoid any further intrusion of salt water into the aquifer underlying the Nile delta.

References and Bibliography

Abul-Atta, A.A., 1975, The Water Policy of Egypt. Ministry of Irrigation, Cairo, Arab Republic of Egypt.

Amer, M.H. and de Ridder, N.A. (eds), 1989, Land Drainage in Egypt, Drainage Research Institute, Cairo.

El-Shal, M.I., 1966, Consumptive Use and Water Requirements for some Major Crops in Egypt. Ph.D. Thesis, Cairo University, Giza, Egypt.

Farid, M.S., Hefny, K. and Amer, A.A., 1979, Hydrological Aspects of the Nile Delta Aquifer, Proc. Int. Conf. on Water resources in Egypt, Cairo, pp. 299-320.

Farid, M.S., 1985, Management of Groundwater System in the Nile Delta. Ph.D. Thesis, Cairo University, Giza, Egypt.

Hurst, H.E., Black, R.P. and Semaika, Y.M., 1951, The Nile Basin, Vol. VII, The future Conservation of the Nile. Physical Department Paper No. 51, Eastern Press (reprinted), Cairo, Egypt.

Kamal, A.A. and Hashem, A.S., 1967, Control and Distribution of Irrigation Water in the United Arab Republic, Egypt. Survey Department Press, Cairo (in Arabic).

Kashef, A.A., 1983, Salt Water Intrusion in the Nile Delta. Groundwater, Vol. 21, No. 2, pp. 160-167.

Kassas, M., 1972, Impact of river control schemes on the shoreline of the Nile delta, Proc. Conf. on Ecological Aspects of International Development - The Careless Technology, Ecology and International Development, Farvar, M. and Milton, P. (eds), Natural History Press, New York.

Mehanna, A.M., 1979, Estimation of potential evapotranspiration over Egypt. Meteorological Research Bulletin, Vol. 8, No. 1, pp. 69-93, Government Printing Offices, Cairo.

NCIDLD, 1963, Irrigation and Drainage in Egypt. Survey Department Press, Cairo, Egypt.

Nielsen, E., 1973, Coastal erosion in the Nile delta. Nature and Resources, Vol. IX, no. 1, pp. 14-18, Unesco, Paris.

Rzóska, J. (ed), 1976, The Nile, Biology of an Ancient River. Junk B.V. Publishers, The Hague, The Netherlands.

Said, N.F. and Abu-Zeid, M., 1967, Drainage of Agricultural Lands. Ministry of Irrigation, United Arab Republic, Egypt (in Arabic).

Said, R., 1981, The Geological Evolution of the River Nile. Springer-Verlag, New York.

Shahin, M., 1983, Report on a TOKTEN-UNDP Assignment to the Groundwater Research Institute, Academy of Science and Technology and UNDP Office, Cairo, Egypt.

Shahin, M., 1985, Hydrology of the Nile Basin. Developments in Water Science No. 21, Elsevier Publ.Co., Amsterdam.

Solait, M.L., 1964, Groundwater at different locations in the Nile delta and valley, Proc. 7th Arab Engineering Conference, Baghdad, Iraq (in Arabic).

UNESCO, 1977, Atlas of World Water Balance, Korzoun, V.I. (ed), Unesco Press, Paris (translated from Russian).

Wassing, F., 1964, Coastal Engineering Problems in the Delta Region of UAR. Memoranda Wl-W6, Department of Ports and Lighthouses, Alexandria, Egypt.

POLICY ANALYSIS FOR THE WATER MANAGEMENT OF THE NETHERLANDS (PAWN)

In the Netherlands the Minister of Transport and Public Works is responsible for water management policy. This task implies the care for the control of water, taking into consideration the related interests. In 1984, the Ministry of Transport and Public Works published a national policy note, in which their view on water management was described. It was necessary to make clear to all parties interested in water management how the principal policy guidelines were to be followed in the Netherlands.

In order to design the policy guidelines, the PAWN-study was carried out by RAND-Corporation (USA), Delft Hydraulics Laboratory and the Ministry of Transport and Public Works. Many others institutes contributed in their respective fields of expertise.

The PAWN-study had the following general objectives:

- to develop a methodology for assessing the multiple consequences of national water management policies (i.e. a coherent mix of measures to improve water management);
- to apply this methodology in generating and analysing alternative water management policies. Such policies consisted basically of combinations of measures on three different levels:
  - plans made by various waterboards to improve water supply and drainage in their territory (the tertiary system);
  - plans on the regional level to improve the (secondary) regional system, such as new regional canals, increase of regional intakes (e.g. for flushing), freshwater pipelines to shortage regions, etc.;
  - technical and managerial tactics on a national level, such as the canalisation of the river IJssel, level management of Lake IJssel, allocation of water to different users and regions, price regulations for groundwater extraction and water quality measures (limitation sources of pollution, etc.).

The results of the PAWN-study have been used as the basis for the water management of the Netherlands for the forthcoming period. A report has been published which can be considered as a master plan for the national water management. The environmental aspects, (aquatic and terrestrial), should play a more pronounced role in the decision making process, which should also be reflected in the PAWN-study. One of the more important new elements of the study is a much more detailed, nationwide, water quality simulation model, with which the regulation and intensity of the flow of elements and pollutants through the main water system is to be investigated.

Demand characteristics

Water management in the Netherlands has to deal with a number of different user categories, such as agriculture, navigation, households, industry, power production, environment, etc. Agriculture is by far the most important user of water, both from a quantitative and economic point of view. Most of the water for agriculture comes from natural sources (rain and soil moisture). In dry periods, water shortages can to some extent be overcome by artificial supply (sprinkling), from either surface water or groundwater sources. In the very dry year of 1976 about 14% of the area of cultivated land could have been sprinkled.
Inland navigation as a means of transportation is extensively practised in the Netherlands and neighbouring countries like Germany, Belgium and France. Considerable amounts of water are needed to maintain sufficient depth along the most important shipping routes. Sedimentation patterns can also be influenced by changing the water regime.

Households and industries use relatively small amounts of water. The main sources are groundwater and the discharges of the Rhine and Meuse, either directly or after infiltration in the coastal dune areas. Deterioration of surface and groundwater quality is of great concern.

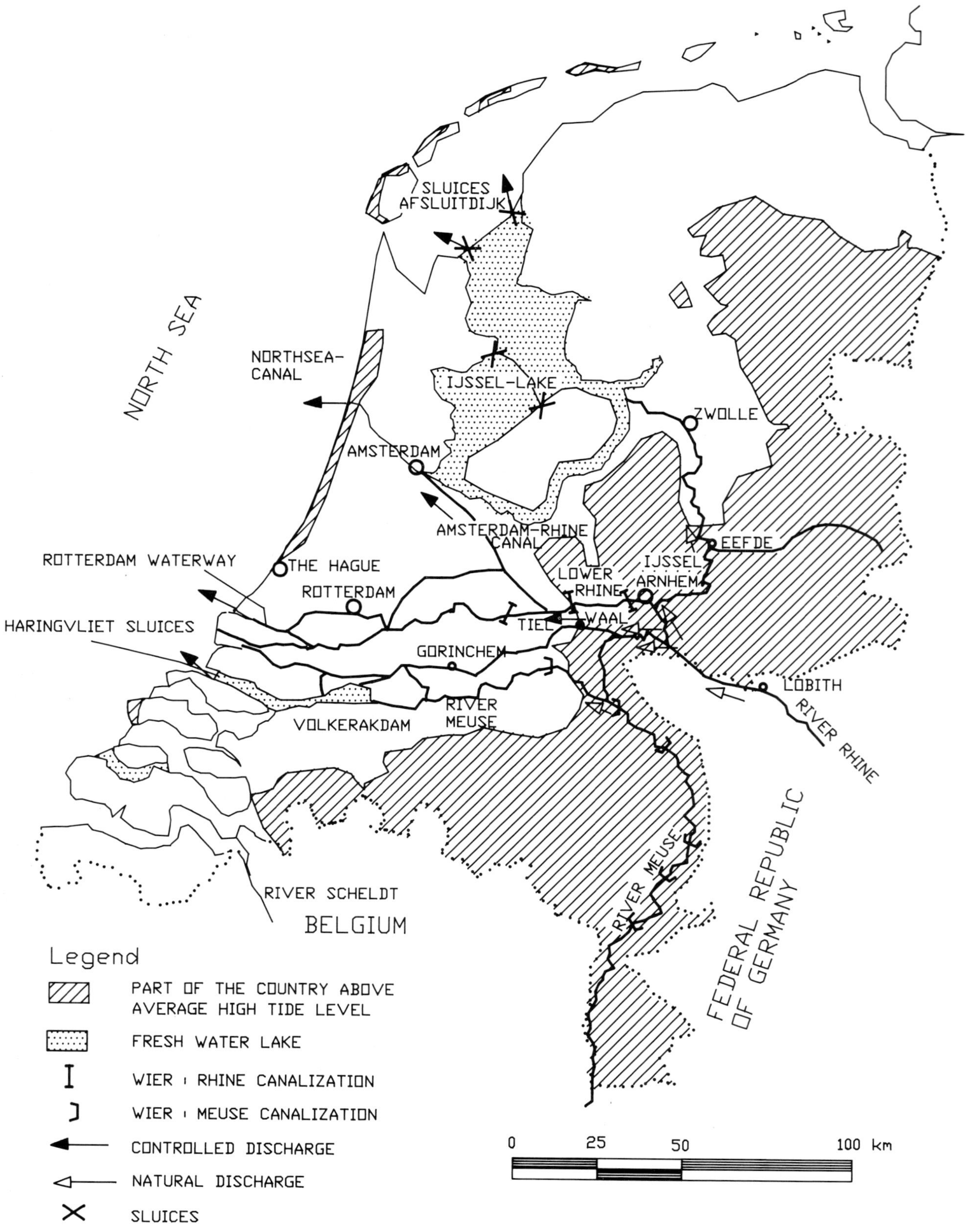

Fig. A7.1
Principal elements of the water system in the Netherlands.

Competition exists among drinking water companies, industries and agriculture for the scarce groundwater resources. Investigations have shown that the total amount of groundwater that can safely be extracted in the entire country is about 2 x 10^9 m^3 per year. The present overall extractions for domestic and industrial purposes amount to about 1.2 x 10^9 m^3/ year.

An increasing problem is the contamination of the groundwater reservoirs by industry, waste disposal and intensive agriculture (manuring), thereby reducing the available amounts of good quality water. Many power plants in the Netherlands are located along main canals, rivers, lakes or estuaries, to ensure sufficient cooling water. In order to avoid the exceedance of thermal standards the power plants have to cut back their production capacity during periods of low discharges.

The aquatic environment is directly threatened by the pollution of surface water bodies. The most important source of pollution on a national scale is the Rhine which carries a vast amount of pollutants from the industrialised areas of other European countries. Urban, industrial and agricultural waste water in the Netherlands, add to the pollution problem.

The terrestrial environment requirements relate to groundwater levels and groundwater quality. As such, it is another competitor for the scarce groundwater resources. Drying out of areas, of important environmental values can be seen especially in the high-lying areas of the Netherlands.

The water system

In Figure A7.1 a lay-out of the principal elements of the water system in the Netherlands is shown. The shaded part represents areas above average high tide sea-level while the rest is lower, up to 6 metres below mean sea-level. In the past, people have constructed dykes and dams and reclaimed many polders, which were kept dry first by windmills and later by pumping stations.

The river system shown in Figure A7.1 consists of the Rhine, Meuse and Scheldt. They are partly canalised and provided with weirs and sluices for navigation. The control of the sources of fresh water is achieved by the weirs and sluices in the rivers branches, and by the dykes and dams in the western and northern parts of the country. In an average year, the total fresh water supply to the Netherlands is 110,000 million m^3.

Groundwater is generally of a good quality, especially in the higher eastern part of the country. The total amount of fresh groundwater is, however, limited to 2,000 million m^3/year. Quality problems caused by salt water intrusion exist in the western part. In the high-lying part of the Netherlands groundwater quality is lowered by intensive agriculture (manuring). For drinking purposes, intensive use is made of artificial infiltration in the coastal dune area, so that the salinisation of groundwater aquifers can be kept to an acceptable level.

Problems related to water management

In very general terms, the main water management problem in the Netherlands is getting rid of excess rainfall in winter, and distributing the available river water and fresh groundwater to the users in summer, taking care that a good quality of both surface and groundwater is maintained.

Under average conditions enough water is available in the Netherlands to meet the various demands. Since both supply and demand vary, problems do occur in certain dry periods. These problems may be related to the limited availability of surface water or groundwater, or to the insufficient quality of the available water. Insufficient capacity or deficient infrastructure to convey the water where it is needed, may lead to local problems. The lowering of groundwater tables when excess demands for groundwater are not restricted, has implications for flora which rely on this water.

A long standing water quality problem is that of salinity. Seawater intrudes with the tides through the open estuaries and rivers. In addition, there is a constant over-pressure of saline groundwater, causing seepage flow towards ditches and canals.

These problems may have serious consequences for agriculture (by dryness or by salinity), navigation (by insufficient depth), drinking water supply and industries (both by restricted availability of groundwater). Power production will become more expensive, while water pollution will be more difficult to combat. In some cases measures can be adopted to avoid or minimise the repercussions, such as expanding infrastructure, changing management rules, etc. Most of these measures may cause damage to the environment and nature, because of the alteration in flows, levels, quality, etc.

Obviously, various user categories compete for the same water resources. All have more or less pronounced wishes concerning the minimum acceptable quality of the commodity.

The general water system approach

In recent years the way of thinking about water and water management has changed. Increasingly more account is taken of the negative repercussions of interventions in the water regime and the way the system is managed. Apart from this, there has been a growth of insight into the relationship between the use and functioning of surface water, groundwater, underwater soils and technical infrastructure. In several studies the physical, chemical and biological features and processes became better known, although a lot still remains for future studies.

This insight enables the possibility of following a water system approach, giving priority to the inter-relationships of the water systems. It also makes it easier to achieve the potential of the water systems, for in this approach account is taken of the inter-relations between:

- the different interests involved, such as sectoral interests, shipping, agriculture, industry and facets such as safety, the environment, physical planning and economics;
- the many authorities which have competence and a say in policy and management;
- the system itself which has it constraints and often has manageable potential.

The approach aims at coordinating by an integral assessment, the wishes of society with regard to the functions and the functioning of the water systems. It requires a policy aimed at the optimal use of the potentials of the systems, with the aid of a set of technical (infrastructure) and legal instruments. An inter-related management of this kind within the potential of the system itself offers possibilities for solving the most important problems at the moment, such as eutrophication, the pollution of water soils and the pollution from diverse sources.

Computational framework

Extensive use was made of mathematical models in the PAWN study, because of the many aspects to be incorporated, and the often complex relations between the various components of the water management system. A review of the general modelling structure is given in Figure A7.2, showing the most important components and their inter-relations. Each box in the diagram represents one or more models which are all connected with the central water distribution model.

In the models the relationship between the different sectors and water management are described. This can be the way the sector can be influenced by water management, but there is also a possibility that the sector influences the water systems, e.g. by pollution or extraction and discharge. Manuring in intensive agriculture influences the quality of ground and surface water.

In the distribution model (DM), the main rivers, canals and lakes are represented as a network of links and nodes. The water distribution over the country for the various water use categories is simulated with this model for time steps of 10 days. Basic inputs to the DM are external supply, demand and allocation rules. Connected to the DM are a series of models, of which the major ones are:

- The District Hydrologic and Agriculture Model (DISTAG), which computes the surface water and salt balances of 77 regions called districts, the basic hydrologic entities of PAWN. The districts are connected to the nodes on the distribution network in the DM.

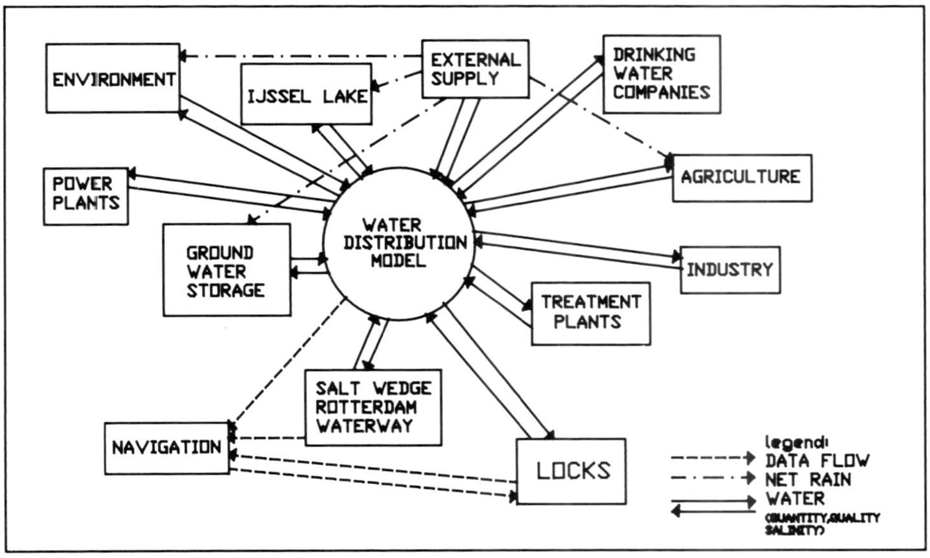

Fig. A7.2 PAWN system diagram.

- The response Design Model (DRISIM) which, given groundwater extraction quotas and groundwater charges (if any), determines the most economically efficient allocation from a national point of view of the available groundwater among public water supply companies.
- To investigate the effect of price and regulation tactics on the water demands and production costs of industrial firms, PAWN used the Industrial Response Simulation Model (IRSM).
- The Electric Power Reallocation and Cost Model (EPRAC), which computes the least costly way of using the inventory of power plants to satisfy the demand for electric power, without discharging so much heat at any node so as to violate the thermal standards.

In addition, separate studies were carried out and models constructed for:
- Navigation and shipping locks: various models were developed to determine the economic consequences for the navigation sector as a result of changes in water depth and lock cycles (cost of delay and extra storage, the long-run required fleet capacity), to calculate the costs for dredging, and to determine the salt intrusion at salt/fresh locks.
- The salt wedge in the Rotterdam Waterway: the progression of the salt wedge in the river was calculated as a function of changes in river flows. The place of the salt wedge itself was necessary for determining possible locations for intake of surface water into the polder area.
- External supply: a hydro-meteorologic analysis was carried out to select a number of external supply scenarios containing regional rainfall and evaporation, discharges of the various rivers and salt concentrations of the river Rhine.
- Analysis of agricultural benefits: a benefits computation program (BENCOMP) was developed to calculate the change in crop prices resulting from a water management policy.
- Water quality analysis: the transport of pollutants in the PAWN network was modelled within the DM. The quantities calculated are the concentration of each pollutant at each node, in each time period, and the amount of each pollutant transported to and from each node.

All models described above were linked with the DM either as subroutines in the main model or as pre or post-processing modules. With the set of models developed, the impact of any infrastructural measures and/or particular water distribution policy, could be determined.

The analysis

The PAWN study considered all water user categories interested in and influencing both quantity and quality of surface water and groundwater. PAWN determined the consequences for these categories of possible alternative water management policies. In this context a policy or strategy was defined as a combination of tactics, where a tactic referred to a single measure aimed at improving water management.

Three different kinds of tactics were considered:
- technical tactics: changes in the infrastructure, e.g. digging new canals, installing weirs or sluices for the diversion of water, or increasing pumping station capacities;
- managerial tactics: changing the operation of the water management infrastructure, e.g. operation rules of weirs, sluices or locks;
- price and regulation tactics: imposing taxes, levies or quotas to change water demand or the quality of water discharged.

The general approach with respect to the generation and evaluation of water management policies took place in various stages of analysis as depicted in Fig. A7.3. The major stages were screening of alternatives, policy design and impact assessment. In the screening stage, a great number of possible tactics were roughly evaluated to identify promising tactics (mainly based on benefit-cost and implementation criteria). In the policy design stage, promising tactics were combined in different ways to obtain a number of promising policies. In the impact assessment stage, the selected promising policies underwent a more detailed analysis to determine their consequences for all relevant user categories.

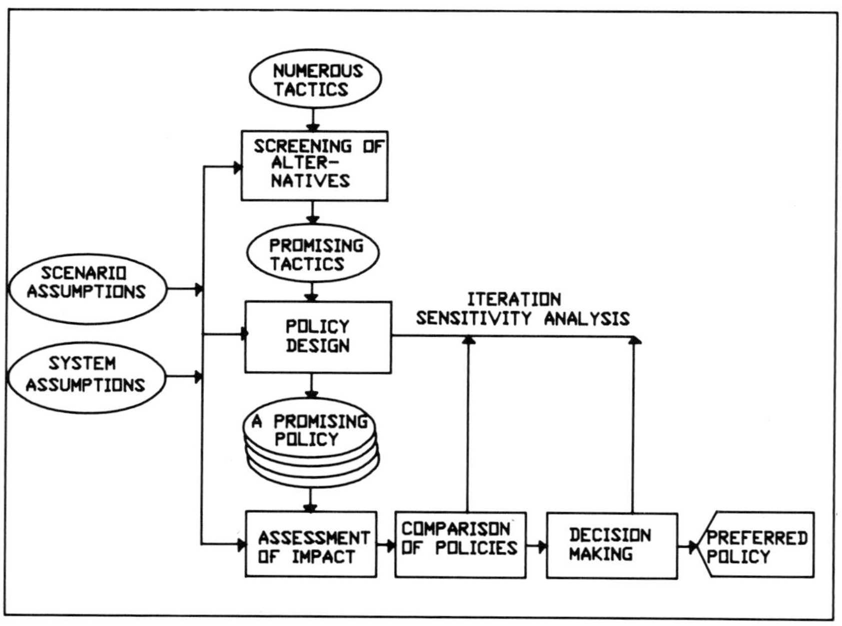

Fig. A7.3
Stages of policy analysis in PAWN.

To find the effects of tactics/policies both in the screening and in the impact assessment stage, the performance of the water management system was assessed with and without the tactic/policy under consideration. The major criteria in the evaluation of different strategies included:
- national net benefits;
- levels of Lake IJssel and the variations of it;

- frequency, magnitude and geographic extent with basic standards for several parameters (nutrients, micro-pollutants, salt, temperature, etc.);
- amount of groundwater to be extracted;
- the percentage of groundwater in the total public water supply;
- the number of surface water projects necessary for public water supply.

Impacts on different water use categories were evaluated with criteria like allocated costs and benefits as well as regional distribution of benefits and costs. Scenarios are necessary to describe the future state of external input variables that do not depend on the behaviour and decisions within the water management system. The most important scenarios in the PAWN-study were the following:

The external supply

The external supply scenario contains assumptions about rainfall and evaporation, discharges of the river Rhine and other smaller rivers, and the salt concentration of the river Rhine. The following four characteristic years were chosen for modelling:

- 1967 (average dry year: likelihood for drought losses 63%);
- 1943 (moderately dry year: likelihood for drought losses 21%);
- 1959 (very dry year: likelihood for drought losses 7%);
- an artificial year called DEX (extremely dry; likelihood for drought losses 2%), obtained by combining the climatologic conditions of 1976 and the minimum Rhine discharges per decade in the three worst years 1934, 1949 and 1976.

The losses in the year 1943 were found to be roughly comparable to the average losses in 44 years. Therefore, in many computations the 1943 scenario was used to approximate the average impacts.

Sprinkling behaviour

About 20% of the 20,000 km^2 of farm land in the Netherlands was sprinkled in 1985 as a form of supplementary irrigation. The amount and type of sprinkler installations had to be estimated for the future context (with and without the so-called water board plans). Based upon a model analysis of the investment behaviour of farmers, three sprinkler scenarios, each denoting a different area eligible for surface water and groundwater sprinkling, were composed: high, medium and low.

Dumping of salt in the Rhine, upstream of the Netherlands

As mentioned previously, the most important source of water in the Netherlands is the river Rhine. Unfortunately, the quality of this water is poor, especially with regard to the high contents of chloride caused by the dumping of large amounts of salt from potassium mines in France, and the discharge of salt water from mines in Germany which creates problems for the downstream users. In the PAWN-study two scenarios were used: a reference scenario of 311 kg/sec for the present and future context, and a high scenario of 365 kg/sec (future context without a treaty), for sensitivity analysis in the impact assessment stage.

Growth rate of industrial production

The industrial activities now and in the future are important for water resources management, in two ways. Firstly, industry is a water user and/or consumer, so it has a demand with respect to quantity and quality; secondly, it may cause pollution of groundwater and surface waters. For the future groundwater demand, two scenarios (high and low) were used, which were based upon sectoral growth rates developed by the Central Planning Bureau of the government.

The PAWN-study analysed:

- the present situation with the present water management policy;
- the future situation with the present policy (problem definition);
- the future situation with a variety of possible future policies.

The present situation is defined as the year for which the most complete accurate and up-to-date set of data is available. The future is taken to be about 10-15 years ahead.

Out of the enormous amount of possible combinations of alternatives, which could be used for elaboration of the last mentioned situation, a number of primary cases have been chosen, covering different managerial strategies (rules that govern the operation of the infrastructure), different infrastructural measures and different scenarios for the use of groundwater and surface water for sprinkling. Each alternative looks at the water management from a special point of view:

- policy optimal for the aquatic environment;
- policy optimal for the terrestrial environment;
- policy optimal for agriculture;
- policy optimal for all other sectors.

For all of them, calculations of impacts on the different user categories and on regional and national economics accounts, have been made under different hydrologic conditions. Finally a summary scorecard is drawn up for the most important criteria.

The scorecard method of presentation of the results of the analysis was already introduced by the RAND-corporation in the policy analysis with respect to the Eastern Scheldt closure. The results of the calculations of the effects of the proposed strategies are organised in a set of scorecards, listing in a vertical direction the considered criteria and horizontally one or more alternatives. Eventually the scores on the criteria are marked with different colours to indicate their relative attractiveness. Different scorecards are drawn up for different scenario conditions or system assumptions.

The results of the PAWN-1 study

An important result of the PAWN study was that a general methodology became available to assess the impacts of changes in water management policy, on various user categories. In addition to this:

- a set of models became available to analyse specific water user categories, system components and the water management system in the country as a whole;
- an organised data base of data directly or indirectly related to water resources management is now available for future use;
- a number of cost-benefit data on promising tactics to improve water management were obtained;
- priority and allocation rules for surface water distribution over the country have been formulated;
- new water management policies and their impacts on various interests became clear.

PAWN not only produced specific results but also provided useful tools for subsequent studies. As good water management requires a continuous adjustment of policies, its use in this respect is certainly the most valuable result of the PAWN study.

Some important conclusions of the study with respect to the water resources management of the country were:

- Agriculture dominates all other categories when it comes to potential benefits associated with the implementation of technical and managerial tactics.
- Sprinkling of agricultural crops seems to be cost-effective in many cases from the farmers point of view, and an increase in the use of sprinkling equipment in the Netherlands can therefore be expected.
- Many local plans of water board, to enlarge the area with access to surface water seem to be cost-effective. This and the expected increase in the use of sprinkling equipment may result in substantially higher future demands for water for agricultural purposes.
- All the large and expensive infrastructural works to improve water management were found not to be cost-effective.

- Measures on a national scale to improve surface water quality are not very effective due to the fact that the bulk of the pollution load of Dutch water originates from the Rhine.
- Stiffer competition for the use of the limited groundwater resources among drinking water companies, industries and agriculture is to be expected in the near future.
- Imposing tight restrictions on industrial and agricultural activities to meet desired environmental standards, will result in large losses to the various users.

The success of the original PAWN studies has lead to a continuation of this programme. The most recent developments are summarised in the National Policy Document on Water Management (see Ministry of Transport, Public Works and Water Management, 1989).

References and Bibliogaphy

Goeller, B.F., 1983, Policy Analysis for the Water Managements of the Netherlands (PAWN), Volume 1- Summary Report, Rand Corporation, Santa Monica.

Ministry of Transport, Public Works and Water Managements, 1989, Water in the Netherlands: a time for action, Summary of the National Policy Document on Water Managements, Goverments Printing Office, The Hague.

nr.	title	price*
CONCRETE		
Research reports (B5-format)		
16A	Concrete slabs subjected to concentrated loads, directives for analysis	Dfl 25,00
16B	Concrete slabs subjected to concentrated loads, explanatory notes	Dfl 25,00
17	Vibration problems in prestressed concrete	Dfl 25,00
39	The behaviour of prestressed concrete beams	Dfl 30,00
40	Loading tests on a full-size suspended beam and a model of this beam for a Metro viaduct at Rotterdam	Dfl 30,00
46A	mathematical analysis of results of relaxation tests on drawn heat-treated prestressing wires	Dfl 25,00
49	Cases of damage due to corrosion of prestressing steel	Dfl 45,00
119	Guide to concrete dike revetments	Dfl 105,00
123	Influence of high pressure in sea-water on the stability of hardened cement paste	Dfl 57,50
124	Acoustic inspection and monitoring of prestressing tendons and bars in concrete structures	Dfl 62,50
127	Detection of reinforcement corrosion and of surface cracks in concrete under water	Dfl 97,50
144(E)	Fly ash as addition to concrete	Dfl 112,50
Research reports (A4-format)		
85-1	Demountable construction in precast concrete	Dfl 40,00
88-3	Properties of concrete at very low temperatures. A survey of the literature	Dfl 52,50
89-6	Concrete and concrete structures under impact loading. Rate effects	Dfl 47,50
RECOMMENDATIONS		
20	Determination of the bond strength of mortars in concrete	Dfl 30,00
21	Concrete repairs with polymer-modified cement mortars	Dfl 45,00
ENVIRONMENT		
Research reports (B5-form)		
144(E)	Fly ash as addition to concrete	Dfl 112,50
HYDRAULIC ENGINEERING		
Research reports (B5-format)		
119	Guide to concrete dike revetments	Dfl 105,00
130	Manual on artificial beach nourishment	Dfl 110,00
	Background information on artificial beach nourishment	Dfl 205,00
140	Guide to the assessment of the safety of dunes as a sea defence	Dfl 75,00
141	Probabilistic design of flood defences	Dfl 110,00
142	Guide for the design of river dikes. Volume 1: upper river area	Dfl 162,50
157	Sand closures	Dfl 240,00

*** all prices are including bankcharges and mailing costs.**

BROCHURES

Manual on the use of rock in coastal and shoreline engineering	Dfl	25,00

VIDEOFILMS

Epoxy coating for reinforcing steel	Dfl	30,00

RESEARCH REPORTS IN PREPARATION

155(E) Design manual for pitched slope protection
C 67 Use of rock in hydraulic engineering (II)
162(E) Ground construction on soft subsoil

After receiving your cheque we send you the ordered reports !

For more information, please write, call or fax to:

Stichting CUR,
P.O.Box 420,
2800 AK GOUDA,
HOLLAND
Telephone number (+31) 1820 39600
Fax number (+31) 1820 30046